T0292612

Smart Innovation, Systems and Technologies

Volume 49

Series editors

Robert J. Howlett, KES International, Shoreham-by-Sea, UK
e-mail: rjhowlett@kesinternational.org

Lakhmi C. Jain, Bournemouth University, Fern Barrow, Poole, UK, and
University of Canberra, Canberra, Australia
e-mail: jainlc2002@yahoo.co.uk

About this Series

The Smart Innovation, Systems and Technologies book series encompasses the topics of knowledge, intelligence, innovation and sustainability. The aim of the series is to make available a platform for the publication of books on all aspects of single and multi-disciplinary research on these themes in order to make the latest results available in a readily-accessible form. Volumes on interdisciplinary research combining two or more of these areas is particularly sought.

The series covers systems and paradigms that employ knowledge and intelligence in a broad sense. Its scope is systems having embedded knowledge and intelligence, which may be applied to the solution of world problems in industry, the environment and the community. It also focusses on the knowledge-transfer methodologies and innovation strategies employed to make this happen effectively. The combination of intelligent systems tools and a broad range of applications introduces a need for a synergy of disciplines from science, technology, business and the humanities. The series will include conference proceedings, edited collections, monographs, handbooks, reference books, and other relevant types of book in areas of science and technology where smart systems and technologies can offer innovative solutions.

High quality content is an essential feature for all book proposals accepted for the series. It is expected that editors of all accepted volumes will ensure that contributions are subjected to an appropriate level of reviewing process and adhere to KES quality principles.

More information about this series at http://www.springer.com/series/8767

V. Vijayakumar · V. Neelanarayanan
Editors

Proceedings of the 3rd International Symposium on Big Data and Cloud Computing Challenges (ISBCC – 16')

Editors
V. Vijayakumar
School of Computing Sciences
VIT University
Chennai
India

V. Neelanarayanan
School of Computing Sciences
VIT University
Chennai
India

ISSN 2190-3018   ISSN 2190-3026   (electronic)
Smart Innovation, Systems and Technologies
ISBN 978-3-319-30347-5   ISBN 978-3-319-30348-2   (eBook)
DOI 10.1007/978-3-319-30348-2

Library of Congress Control Number: 2016932320

Printed on acid-free paper

This Springer imprint is published by SpringerNature
The registered company is Springer International Publishing AG Switzerland

Preface

The 3rd International Symposium on Big Data and Cloud Computing (ISBCC'2016) was held at VIT University, Chennai on 10 and 11 March 2016. As an exciting initiative in the area of big data and cloud computing, this symposium series is becoming a leading forum for disseminating the latest advances in big data and cloud computing research and development. The aim of the conference is to bring together scientists interested in all aspects of theory and practise of cloud computing and big data technology, providing an international forum for exchanging ideas, setting questions for discussion, and sharing the experience.

The symposium attracted high-quality original research papers on various aspects of big data and cloud computing. This year, we received 154 submissions. After a rigorous peer-review process undertaken by the programme committee members, 42 papers were accepted, representing acceptance rates of 27 %. We congratulate the authors of those accepted papers and sincerely thank all the submitting authors for their interest in the symposium.

The 3rd ISBCC symposium consists of eight full session presentations. All the presentations reflect perspectives from the industrial and research community on how to address challenges in big data and cloud computing. The submissions were split between various conference areas, i.e., computer security, WSN, cloud computing, image processing, big data, IOT, and artificial intelligence. Track 1, *Computer Security,* incorporates all security domains including information security, network security, cloud security and mobile security. Track 2, *Wireless Sensor Networks (WSN),* focuses on novel approaches, quality and quantitative aspects of WSN. Track 3, cloud computing focuses on energy efficiency, scheduling and optimization. Track 4, *Image Processing,* focuses on pattern recognition, tracking, computer vision, image retrieval and object detection. Track 5, *Big Data,* incorporates data management, mining approaches, search techniques, recommendation system, data classification and social graphs. Track 6, *Internet of Things (IOT),* focuses on key fields of IOT implementation such as healthcare systems and interactive systems; and link management. Track 7, *Artificial Intelligence,* focuses on prediction and decision making algorithms using fuzzy theory. Track 8, *Software Engineering,* focuses on business tools, requirement analysis and metrics.

We are honoured to feature the following high-profile distinguished speakers for their inspiring and insightful keynotes:

- Dr. P.A. Subrahmanyam (FIXNIX, USA)
- Dr. Ron Doyle (IBM, USA)
- Dr. Shajulin Benedict (St. Xavier's Catholic College of Engineering, India)
- Dr. Wahiba Ben Abdessalem (Taif University, Saudi Arabia)
- Dr. Surya Putchala (ZettaMine Technologies, Hyderabad)
- Mr. Chidambaram Kollengode (DataXu Inc, India)
- Dr. Jemal H. Abawajy (Deakin University, Australia)
- Dr. Mohan Kumar (Trendwise Analytics, India)
- Dr. Md. Fokhray Hossain (Daffodil International University, Bangladesh)
- Dr. Nilanjan Dey (Techno India College of Technology, India)
- Dr. Dinesh Garg (IBM Research Laboratory India)
- Dr. Kapilan Radhakrishnan (University of Wales Trinity St. David, UK)
- Mr. P. Ravishankar (CTS, India)
- Dr. Neeran M. Karnik (Vuclip, Inc. Canada)
- Dr. B. Ravi Kishore (HCL, India)
- Dr. Prof. Md Abdul Hannan Mia (Agni Systems, Bangladesh)
- Dr. R. Venkateswaran (Persistent Systems, India)
- Dr. A. Clementking (King Khalid University, Saudi Arabia)
- Dr. V.N. Mani (DEIT, India)
- Prof. P.D. Jose (MET's School of Engineering, India)
- Ms. Kiran mai Yanamala (Freelance Consultant, India)
- Dr. Praveen Jayachandran (IBM Research, India)
- Dr. Ajay Deshpande (Rakya Technologies, India)
- Dr. Murali Meenakshi Sundaram, Consultant, CTS India
- Prof. Dr. Lorna Uden, Staffordshire University, UK
- Dr. Tulika Pandey, Department of Electronics and IT, India
- Mr. Narang N. Kishor, Narnix Technolabs Pvt. Ltd, India
- Mr. S. Kailash, CDAC, India

In addition, the symposium features workshops covering many emerging research directions in big data and cloud computing by both industry experts and academia.

We are deeply grateful to all who have contributed to this amazing technical programme in one way or another. Particularly, we profoundly appreciate the great contribution by the following international scientific committee members: Dr. P.A. Subrahmanyam (FIXNIX, USA), Dr. Ron Doyle (IBM, USA), Dr. Shajulin Benedict (St. Xavier's Catholic College of Engineering, India), Dr. Wahiba Ben Abdessalem (Taif University, Saudi Arabia), Dr. Dey (Nilanjan, Techno India College of Technology, India), Dr. Siti Mariyam Shamsuddin (Universiti Teknologi Malaysia, Malaysia), Dr. Robert James Howlett (KES

International, UK), Dr. Thomas Ditzinger (Executive Editor, Springer, Germany) and Dr. Valentina Emilia Balas (University of Arad, Romania).

We appreciate the contribution by programme committee members and many external reviewers. They reviewed and discussed the submissions carefully, critically and constructively. We thank the contribution from all our sponsors for the generous financial support. We thank all the members of the organizing committee. We also thank our volunteers without whom this mission would not be possible. Last but not least, we thank all attendees who joined us in ISBCC 2016. The attendants of the ISBCC'2016 shared the common opinion that the conference and workshops were extremely fruitful and well organized.

V. Vijayakumar
V. Neelanarayanan

Organization Committee

Patrons

Chief Patron
Dr. G. Viswanathan, Founder—Chancellor of VIT University

Patrons
Mr. Sankar Viswanathan, Vice President, VIT University
Ms. Kadhambari S. Viswanathan, Assistant Vice President, VIT University
Dr. Anand A. Samuel, Vice-Chancellor, VIT University
Dr. P. Gunasekaran, Pro Vice-Chancellor, VIT University, Chennai Campus, India

Organizing Committee

Dr. V. Vijayakumar, VIT, Chennai, India
Dr. V. Neelanarayanan, VIT, Chennai, India

Programme Committee

Dr. Backhouse Helen, Northampton University, UK
Dr. Bates Margaret, Northampton University, UK
Dr. Burdett Sarah, Northampton University, UK
Dr. Dravid Rashmi, Northampton University, UK
Dr. Garwood Rachel, Northampton University, UK
Dr. Hookings Lynn, Northampton University, UK
Dr. Tudor Terry, Northampton University, UK
Dr. Balaji Rajendran, CDAC, India
Dr. Balakrishnan, NUS, Singapore
Dr. Muthukumar, Meenakshi University, India
Dr. Nithya, Vinayaka Missions University, India
Dr. S. Prabhakaran, Vinayaka Missions University, India

Dr. Pramod S. Pawar, University of Kent, UK
Mr. Rupesh, Thirdware Technologies, India
Dr. Rao, IT University of Copenhagen, Denmark
Dr. Sivakumar, Madras Institute of Technology, India
Dr. N. Subramanian, CDAC, India
Dr. Subramanisami, Sastra University, Thanjavur, India
Dr. Vijayakumar, Anna University, India
Dr. J. Priyadarshini, VIT University, India
Dr. V. Viswanathan, VIT University, India
Dr. P. Nithyanandam, VIT University, India
Mr. Chidambaram Kollengode, Nokia, Bengaluru
Dr. James Xue, The University of Northampton, Northampton
Prof. S.R. Kolla, Bowling Green State University, USA
Prof. Akshay Kumar Rathore, NUS, Singapore
Prof. Vignesh Rajamani, Oklahoma State University, USA
Prof. Debasish Ghose, IISc, Bangalore
Dr. Kunihiko Sadakane, University of Tokyo, Japan
Dr. Arvind K. Bansal, Kent State University, USA
Dr. Bettina Harriehausen, Darmstadt University of Applied Sciences, Germany
Dr. Hamid Zarrabi-Zadeh, Sharif University of Technology, Iran
Dr. Maziar Goudarzi, Sharif University of Technology, Iran
Dr. Mansour Jamzad, Sharif University of Technology, Iran
Dr. Morteza Amini, Sharif University of Technology, Iran
Dr. G. Anderson, University of Botswana, Botswana
Dr. A.N. Masizana, University of Botswana, Botswana
Dr. Shareef M. Shareef, College of Engineering, Erbil
Dr. Bestoun S. Ahmed, College of Engineering, Erbil
Dr. Syed Akhter Hossain, Daffodil International University, Bangladesh
Dr. Md. Fokhray Hossain, Daffodil International University, Bangladesh
Dr. Abdullah b Gani, University of Malaya, Malaysia
Dr. Seyed Reza Shahamiri, University of Malaya, Malaysia
Dr. Adam Krzyzak, Concordia University, Canada
Dr. Mia Persson, Malmo University, Sweden
Dr. Rafiqul Islam, Charles Sturt University, Australia
Dr. Zeyad Al-Zhour, University of Dammam, Saudi Arabia
Dr. Imad Fakhri Al-Shaikhli, International Islamic University, Malaysia
Dr. Abdul Wahab, International Islamic University, Malaysia
Dr. Dinesh Garg, IBM Research Laboratory India
Dr. Abhijat Vichare, Consultant, India
Dr. Subhas C. Nandy, Indian Statistical Institute, India
Dr. Dan Wu, University of Windsor, Canada
Dr. Subir Bandyopadhyay, University of Windsor, Canada
Dr. Md. Saidur Rahman, BUET, Bangladesh
Dr. M. Sohel Rahman, BUET, Bangladesh
Dr. Guillaume Blin, University of Paris-Est, France

Dr. William F. Smyth, McMaster University, Canada
Dr. Burkhard Morgenstern, University of Göttingen, Germany
Dr. Basim Alhadidi, Al-Balqa Applied University, Jordan
Dr. N. Raghavendra Rao, India
Dr. Radhakrishnan Delhibabu, Kazan Federal University, Russia
Dr. Dong Huang, Chongqing University, China
Dr. S. Kishore Reddy, Adama Science and Technology University, Ethiopia
Dr. V. Dhanakoti, SRM Valliammai Engineering College, India
Dr. R. Nandhini, VIT Chennai, India
Dr. S.V. Nagaraj, VIT Chennai, India
Dr. R. Sakkaravarthi, VIT Chennai, India
Dr. S. Asha, VIT Chennai, India
Dr. Y. Asnath Victy Phamila, VIT Chennai, India
Dr. Renta Chintala Bhargavi, VIT Chennai, India
Dr. B. Saleena, VIT Chennai, India
Dr. R. Vishnu Priya, VIT Chennai, India
Dr. V. Viswanathan, VIT Chennai, India
Dr. P. Sandhya, VIT Chennai, India
Dr. E. Umamaheswari, VIT Chennai, India
Dr. S. Geetha, VIT Chennai, India
Dr. Subbulakshmi, VIT Chennai, India
Dr. R. Kumar, VIT Chennai, India
Dr. Sweetlin Hemalatha, VIT Chennai, India
Dr. Janaki Meena, VIT Chennai, India
Dr. G. Babu, AEC, Chennai, India
Dr. S. Hariharan, VIT Chennai, India
Dr. S. Kaliyappan, VIT Chennai, India
Dr. S Dhanasekar, VIT Chennai, India
Dr. V.N. Mani, DEIT, India
Dr. R. Venkateswaran, Persistent Systems, India
Dr. B. Ravi Kishore, HCL, India
Prof. P.D. Jose, MET's School of Engineering, India
Mr. P. Ravishankar, Director, CTS, India
Ms. Kiran mai Yanamala, Freelance Consultant, India
Dr. Praveen Jayachandran Researcher, Distributed Systems Research, IBM Research, India
Dr. Ajay Deshpande CTO, Rakya Technologies India

Advisory Committee

Dr. P.A. Subrahmanyam, FIXNIX, USA
Dr. Ron Doyle, IBM, USA
Dr. Shajulin Benedict, St. Xavier's Catholic College of Engineering, India
Dr. Wahiba Ben Abdessalem, Taif University, Saudi Arabia

Dr. Dey, Nilanjan, Techno India College of Technology, India
Dr. Siti Mariyam Shamsuddin, Universiti Teknologi Malaysia, Malaysia
Dr. Robert James Howlett, KES International, UK
Dr. Thomas Ditzinger, Executive Editor, Springer, Germany
Dr. Valentina Emilia Balas, University of Arad, Romania
Dr. Anca L. Ralescu, College of Engineering University of Cincinnati, USA
Dr. Emmanuel Udoh, Sullivan University, USA
Dr. Mohammad Khan, Sullivan University, USA
Dr. Ching-Hsien Hsu, Chung Hua University, Taiwan
Dr. Kaushal K. Srivastava, IIT(BHU), India
Dr. R.K.S. Rathore, IITK, India
Dr. Tanuja Srivastava, IITR, India
Dr. Radu-Emil Precup, Politehnica University of Timisoara, Romania
Dr. John Rumble, Codata, USA
Dr. Diane Rumble, Codata, USA
Dr. Jay Liebowitz, University of Maryland, USA
Dr. Sarah Callaghan, STFC, UK
Dr. Jemal H. Abawajy, Deakin University, Australia
Dr. Rajkumar Buyya, Melbourne University, Australia
Dr. Hamid R. Arbania, University of Georgia, USA
Dr. Mustafa Mat Deris, Malaysia
Prof. Ekow J. Otoo, University of the Witwatersrand
Dr. Weiyi Meng, Binghamton University, USA
Dr. Venu Dasigi, Bowling Green State University, USA
Dr. Hassan Rajaei, Bowling Green State University, USA
Dr. Robert C. Green, Bowling Green State University, USA
Dr. Hyun Park, Kookmin University, South Korea
Dr. R.S.D. Wahida Banu, GCES, India
Dr. L. Jegannathan, VIT, Chennai, India
Dr. N. Maheswari, VIT, Chennai, India
Dr. Nayeemulla Khan, VIT, Chennai, India
Dr. V. Pattabiraman, VIT, Chennai, India
Dr. R. Parvathi, VIT, Chennai, India
Dr. B. Rajesh Kanna, VIT, Chennai, India
Dr. Bharadwaja Kumar, VIT, Chennai, India
Dr. R. Ganesan, VIT, Chennai, India
Dr. D. Rekha, VIT, Chennai, India
Dr. S. Justus, VIT, Chennai, India
Dr. S.P. Syed Ibrahim, VIT, Chennai, India
Dr. M. Sivabalakrishnan, VIT, Chennai, India
Dr. L.M. Jenila Livingston, VIT, Chennai, India
Dr. G. Malathi, VIT, Chennai, India
Dr. A. Kannan, Anna University, India
Dr. E.A. Mary Anita, SAEC, India
Dr. Kay Chen Tan, NUS, Singapore

Dr. Abel Bliss, University of Illinois, USA
Dr. Daniel O'Leary, USC Marshall School of Business, USA
Dr. B.S. Bindu Madhava, CDAC, Bangalore
Dr. Murali Meenakshi Sundaram, CSC, Chennai
Dr. Ajay Ohri, CEO, Decisionstats, Delhi
Dr. Dharanipragada Janakiram, IITM, Chennai
Dr. Surya Putchala, CEO, ZettaMine Technologies, Hyderabad
Dr. Patrick Martinent, CTO, CloudMatters, Chennai
Dr. R.S. Mundada, BARC, Mumbai
Dr. Natarajan Vijayarangan, TCS, Chennai
Dr. Karthik Padmanabhan, IBM, India
Dr. Rajdeep Dua, VMWare, Hyderabad
Dr. Shuveb Hussain, K7 Computing Pvt. Ltd, India
Dr. Mohan Kumar, Trendwise Analytics, Bangalore
Dr. Wolfgang von Loeper, Mysmartfarm, Southafrica
Dr. Pethuru Raj, IBM Resrarch Laboratory, India
Dr. Ravichandran, CTS Research Laboratory, India
Dr. K. Ananth Krishnan, TCS, India
Dr. Zaigham Mahmood, University of Derby, UK
Dr. Madhav Mukund, CMI, Chennai
Dr. Bala Venkatesh, Ryerson University, Toronto, Canada
Dr. Zvi M. Kedem, New York University, USA
Dr. Bennecer Abdeldjalil, Northampton University, UK
Dr. Lakhmi C. Jain, University of Canberra, Australia
Dr. Mika Sato-Ilic, University of Tsukuba, Japan
Dr. Manuel Mazzara, Innopolis University, Russia
Dr. Max Talanov, Kazan Federal University and Innopolis University, Russia
Dr. Salvatore Distefano, Politecnico di Milano, Italy
Dr. Jordi Vallverdú, Universitat Autònoma de Barcelona, Spain
Dr. Vladimir Ivanov, Kazan Federal University, Russia
Dr. Meghana Hemphill, WILEY, USA
Dr. Alexandra Cury, WILEY, USA
Dr. Erica Byrd, WILEY, USA
Dr. Cassandra Strickland, WILEY, USA
Dr. Witold Pedrycz, University of Alberta, Canada
Dr. Sushmita Mitra, Indian Statistical Institute, India
Dr. ArulMurugan Ambikapathi, UTECHZONE Co. Ltd, Taiwan
Dr. Neeran M. Karnik, Vuclip, Inc. Canada
Dr. K.V. Ramanathan, University of Washington Foster Business School, USA
Dr. Prof. Md Abdul Hannan Mia, Agni Systems, Bangladesh
Dr. A. Clementking, King Khalid University, Saudi Arabia

Reviewers

Dr. G. Anderson, University of Botswana, Botswana
Dr. P. Nithyanandam, VIT, Chennai, India
Dr. E. Umamaheswari, VIT, Chennai, India
Prof. Abdul Quadir Md, VIT, Chennai, India
Dr. B. Rajesh Kanna, VIT, Chennai, India
Dr. Dong Huang, Chongqing University, China
Prof. T.S. Pradeepkumar, VIT, Chennai, India
Dr. S. Asha, VIT, Chennai, India
Prof. Tulasi Prasad Sariki, VIT, Chennai, India
Dr. M. Kaliyappan, VIT, Chennai, India
Dr. Syed Akhter Hossain, Daffodil International University, Bangladesh
Prof. P. Rukmani, VIT, Chennai, India
Prof. G. Anusooya, VIT, Chennai, India
Prof. K.V. Pradeep, VIT, Chennai, India
Prof. Alok Chauhan, VIT, Chennai, India
Dr. P. Vijayakumar, VIT, Chennai, India
Prof. V. Kanchanadevi, VIT, Chennai, India
Dr. R. Sakkaravarthi, VIT, Chennai, India
Dr. G. Singaravel, KSR College of Engineering, India
Dr. G. Malathi, VIT, Chennai, India
Dr. Bharathwaja Kumar, VIT, Chennai, India
Dr. M. Janakimeena, VIT, Chennai, India
Dr. V. Dhanakoti, SRM Valliammai Engineering College, India
Prof. S. Harini, VIT, Chennai, India
Dr. A. Clementking, King Khalid University, Saudi Arabia
Prof. Ekow J. Otoo, University of the Witwatersrand
Dr. Arvind K. Bansal, Kent State University, USA
Dr. S.P. Syed Ibrahim, VIT, Chennai, India
Dr. R. Parvathi, VIT, Chennai, India
Prof. M. Rajesh, VIT, Chennai, India
Dr. E.A. Mary Anita, SAEC, India
Dr. Y. Asnath Victy Phamila, VIT Chennai, India
Dr. M. Sivabalakrishnan, VIT, Chennai, India
Prof. M. Sivagami, VIT, Chennai, India
Dr. Subramanisami, Sastra University, Thanjavur, India
Dr. Imad Fakhri Al-Shaikhli, International Islamic University, Malaysia
Dr. Sweetlin Hemalatha, VIT, Chennai, India
Dr. James Xue, The University of Northampton, Northampton
Dr. V. Pattabirman, VIT, Chennai, India
Prof. A. Vijayalakshmi, VIT, Chennai, India
Dr. V. Viswanathan, VIT, Chennai, India
Dr. Abhijat Vichare, Consultant, India
Dr. Wahiba Ben Abdessalem, Taif University, Saudi Arabia

Dr. G. Babu, AEC, Chennai, India
Prof. S. Bharathiraja, VIT, Chennai, India
Dr. J. Priyadarshini, VIT, Chennai, India
Prof. M. Nivedita, VIT, Chennai, India
Dr. Hyun Park, Kookmin University, South Korea
Prof. B.V.A.N.S.S. Prabhakar Rao, VIT, Chennai, India
Prof. K. Khadar Nawas, VIT, Chennai, India
Prof. A. Karmel, VIT, Chennai, India
Prof. R. Maheswari, VIT, Chennai, India
Dr. Jenila Livingston, VIT, Chennai, India
Prof. S. Shridevi, VIT, Chennai, India
Prof. S. Geetha, VIT, Chennai, India
Prof. R. Gayathri, VIT, Chennai, India
Dr. S.V. Nagaraj, VIT, Chennai, India
Dr. Josephc Tsai, National Yang-Ming University, Taiwan
Prof. Rabindra Kumar Singh, VIT, Chennai, India
Dr. Pramod S. Pawar, University of Kent, UK
Prof. S. Sajidha, VIT, Chennai, India
Dr. Dinesh Garg, IBM Research Laboratory India
Prof. M. Nishav, VIT, Chennai, India
Prof. J. Prassanna, VIT, Chennai, India
Dr. Madhav Mukund, CMI, Chennai
Dr. T. Vigneswaran, VIT, Chennai, India
Prof. M. Vergin Raja Sarobin, VIT, Chennai, India
Prof. S. Rajarajeswari, VIT, Chennai, India
Mr. Ajay Ohri, CEO, Decisionstats, Delhi
Prof. S. Maheswari, VIT, Chennai, India
Dr. N. Maheswari, VIT, Chennai, India
Prof. S. Nithyadarisinip, VIT, Chennai, India
Prof. M. Priyaadharshini, VIT, Chennai, India
Prof. G. Suganya, VIT, Chennai, India
Dr. Valentina Emilia Balas, University of Arad, Romania
Prof. B. Sathiskumar, VIT, Chennai, India
Prof. S. Rajkumar, VIT, Chennai, India
Dr. Adam Krzyzak, Concordia University, Canada
Prof. S. Dhanasekar, VIT, Chennai, India
Prof. M. Sakthiganesh, VIT, Chennai, India
Dr. R. Dhanalakshmi, VIT, Chennai, India
Dr. Renta Chintala Bhargavi, VIT, Chennai, India
Dr. Seyed Reza Shahamiri, University of Malaya, Malaysia
Dr. Jordi Vallverdú, Universitat Autònoma de Barcelona, Spain
Prof. Rajeshkumar, VIT, Chennai, India
Dr. P. Sandhya, VIT, Chennai, India
Dr. Robert C. Green, Bowling Green State University, USA
Dr. Sriniavasa Rao, VIT, Chennai, India

Dr. Pethuru Raj, IBM Resrarch Laboratory, India
Prof. B. Gayathridevi, VIT, Chennai, India
Dr. Hamid Zarrabi-Zadeh, Sharif University of Technology, Iran
Prof. K. Deivanai, VIT, Chennai, India
Prof. I. Sumaiya Thaseen, VIT, Chennai, India
Ms. Jayalakshmi Sankaran, VIT, Chennai, India

Event Coordinators (VIT University)

Dr. M. Sivabalakrishnan
Dr. D. Rekha
Dr. Umamaheswari
Dr. S. Geetha
Dr. P. Nithyanandam
Dr. S.P. SyedIbrahim
Prof. R. Sridhar
Prof. Ramesh Ragala
Prof. G. Suguna
Prof. J. Prassanna
Prof. G. Anusooya
Prof. K. Punitha
Prof. Abdul Quadir Md
Prof. S. Raja Rajeswari
Prof. S. Usha Rani
Prof. K.V. Pradeep
Prof. J. Christy Jackson
Prof. Rajiv Vincent
Prof. Alok Chauhan
Prof. Rabindra Kumar Singh
Prof. R. Gayathri
Prof. Aparna
Prof. Sakthi Ganesh
Dr. R. Nandhini
Prof. V. Kanchana Devi
Prof. A. Vijayalakshmi
Prof. T.S. PradeepKumar
Prof. R. Prabakaran
Ms. Jayalakshmi Sankaran
Mr. Arunkumar Goge
Mr. Sivakumar
Mr. S. Arunkumar
Mr. Jeeva
Ms. Bharathi
Ms. Durgalakshmi

Contents

Part I
Computer Security: Information Security, Network Security, Cloud Security, Mobile Security

A Preventive Method for Host Level Security in Cloud Infrastructure

S. Ramamoorthy and S. Rajalakshmi

Abstract Cloud Virtual Infrastructure management is the emerging technology in the IT industry today. Most of the IT business process demanding on the virtual infrastructure as a service from the cloud service providers including compute, network and storage etc., The dynamic adoption of computing resource and elimination of major investment cost to setup the physical infrastructure attracted the IT industry towards cloud infrastructure. Live VM Migration techniques allows more frequently move the virtual machines from one physical location to another to avoid the situations like load balancing, Fault tolerance, edge computing, virtual migration etc., The major challenge which found on this process security issues on the VM live migration. The proposed model trying to eliminate the security challenges in the pre-copy migration strategy by introducing the network addressing level hashing technique to avoid the critical part of migration process. The dirty VM memory pages are released to the destination server only after the authenticated network process.

Keywords Cloud computing · VM live migration · Hashing · Pre-copy · Memory pages · Network addressing

1 Introduction

Cloud Computing Data Center model will provide the facility to operate the IT business services by offering its infrastructure like Compute, Storage, Network etc., Number of physical Servers are massively utilized to perform many number of applications with the hypervisor support on the physical Environment. Every

S. Ramamoorthy (✉) · S. Rajalakshmi
Department of Computer Science and Engineering, SCSVMV University,
Kancheepuram, India
e-mail: sriraram2@gmail.com

S. Rajalakshmi
e-mail: raji.scsvmv@gmail.com

© Springer International Publishing Switzerland 2016   3
V. Vijayakumar and V. Neelanarayanan (eds.), *Proceedings of the 3rd International Symposium on Big Data and Cloud Computing Challenges (ISBCC – 16')*,
Smart Innovation, Systems and Technologies 49, DOI 10.1007/978-3-319-30348-2_1

Hypervisor creates multiple number of Virtual machines to perform the host level operations. Virtual Machines are nothing but the set of files which will inherit the abstract behavior of the physical machine. There are number of Virtual machine migration operations performed on the same cloud environment for continuity of uninterrupted Service offered by the cloud to its customers.

Cloud Computing is a Internet based Technology offers computing resources as a services to its end users support various IT business process. The computing resources are physical infrastructure as a service including compute, network and storage etc., The Cloud computing services are offered to the customers in the different models like Infrastructure as a Service (IaaS), Platform as a Service (PaaS) and Software as a Service (SaaS).

Cloud Deployment models selected by the customers based on their specific business needs. If the customer is more concern about the data and less concern about the cost they opted to go for the private cloud model in the home location. If the customer is less concern about the data security but more concern about the hiring cost then they will go for an public cloud model for their organization.

The third model will provide the intermediate level facility to the customers by offering hybrid infrastructure for their business process.

Virtual Infrastructure is the backbone of the cloud computing business model, this will eliminate the need for setting up the physical infrastructure by customers. Instead of these customers can hire the computing resources as pay as you go model structure.

Virtual machine Manager (Hypervisor) abstract the physical infrastructure to the Virtual Machines running on the Hypervisor. This model allows the Multiple VM's to run multiple guest operation system on the same physical server. Many application's can be deployed using this flexibility of the physical infrastructure. Automated Live VM migration is the one of the technique which is frequently handled by the hypervisor to move the Virtual machine into different physical servers depends on the requirement of the resource availability.

VM level attacks like monitoring and capturing the traffic between the virtual network will lead to the different level of issues on the Co located VMs. VM level challenges like Multitenancy, Colocated VM attack must be Addressed in a Effective way in order to reduce the risk associated with infrastructure level threats in the cloud environment. As an example Cloud Platform like Amazon EC2, Microsoft Azure running multiple number of VMs on the same Environment can lead multiple number of Security issues at the host level. Predominant usage of internet services in the process of Connection establishment between Cloud data center and the Client machine will reflect the untrusted channel communication in public cloud environment. Untrusted Network operations, Untrusted Port group, protocols and interconnecting devices will require maximum attention while setting up the Cloud Environment for running multiple number of VM on the same physical Machine.

Remote Level machine authentication is also part of this data center activity which needs to prove itself as a authorized user to access the Virtual Machine created on the cloud Environment.

Fig. 1 Hosted hypervisor

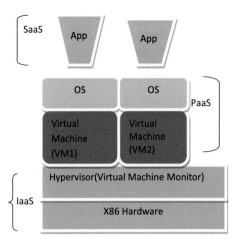

2 Characteristics of VM Operations

Virtual machine is the set of files which inherit the abstract concept from the physical machine and creates the Virtual environment as a mirror image of physical machine. Hypervisor is the software layer which allows the multiple number of OS's run simultaneously by sharing the common resource among themselves from the underlying physical machine.

There are two types of Hypervisor namely (a) Bare metal Hypervisor (b) Hosted Hypervisor that is a software layer which allows the VM level interaction between physical hardware and Virtual Machine created on the same hypervisor (Fig. 1).

From the above diagram it is clear that various level of security enforcement are required on the cloud Infrastructure at Infrastructure level (IaaS), Platform level (PaaS) and Application level (SaaS).

3 Live VM Migration Process

Live VM Migration is the process of migrating the VM from one physical server to another physical server to different location. VM contents like CPU state, Memory pages, VM configuration data etc., migrated to the destination server while VM still in running state.

Two types of Migration strategies are followed in the process of Live Migrations.

1. Pre copy Migration
2. Post copy Migration

In the Pre copy migration process the Running VM contents are copied at a time and all the memory pages of the VM are migrated to destination host, this process was done without disturbing the running application.

In Stop and Copy migration process the currently running VM was stopped for the specific point of time in the source, in subsequent iteration the modified memory pages (Dirty pages) are copied to the destination server, after that VM will be resumed at the destination server.

4 Challenges Related to Co Located VM's

Primary threat for a Virtual machine is due to running multiple VMs on the Same Hypervisor. The attack surface will be very high among the VMs even the malicious VMs can also the part of this Surface. Efficient sharing of resources among the number of VMs will also lead to different issues on this environment. Live Migration of Failed VM from one hypervisor to another hypervisor will carry the sensitive information to the untrusted environment which could have the possibility to attack the target VM and extract the confidential information about the target.

The major challenge in the Live VM migration is the Security issues related to the VM content migration. Secure transfer of VM memory pages must be protected from the unauthorized modification. The VM's are still running state while transferred between different physical servers. There is a possibility of masquerade VM contents into another. Strong Cryptographic Techniques are need to be deployed to avoid this kind of memory modification but key management is the another overhead for this kind of security mechanism. The proposed model trying to eliminate this critical situation by introducing the network level Hashing technique to authenticate the memory page migrations.

Types of Operations on Cloud Virtual Machine (Table 1).

Table 1 VM level operations

S. No.	Type of VM operation	Function
1.	VM live migration	Moving the Live VM from one hypervisor to another hypervisor
2.	VM copy	Copy the contents of VRAM disk into another
3.	VM move	Movement of VM from one hypervisor to another physical server
4.	VM clone	Performing cloning operation to take multiple copies of VM
5.	VM template	Taking the contents of VM in the form of template to produce multiple copies

Fig. 2 Co-located VM
surface attack

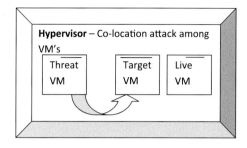

5   Proposed Methodology

Proposed model try to achieve the required level of security by implementing following security polices at VM level.

- Self Destructive nature of VM when there is an unauthorized effect made on the VM.
- Authenticated move or Copy of VM operations at hypervisor level.
- For every VM operation required to prove itself as a authenticated operation on a VM.
- Identification of unauthorized move can be monitored based on the level of traffic flow between two VM's and Virtual Switch.
- If the traffic flow exceed the preset Limit then it will become concluded as a unauthorized move of VM and it will destroy by itself (Figure 2).

Abnormal traffic rate identified between VM's are enforced to apply some set of security policy given below.

6   Behavior of VM Security Policy Rule

Traffic Rate = HIGH − Destroy the VM fail to provide Signature Identity
Traffic Rate = MEDIUM − Signature Identity Check
Traffic Rate = LOW − Continue its normal operation.

6.1   Proposed Algorithm

Let us Consider N-number of Virtual Machines running in an Hypervisor H.

H={VM1,VM2,VM3…VmN}
Th.Value=Max.traffic rate between VM & Vswitch
If(traffic flow rate<=Threshold Value)

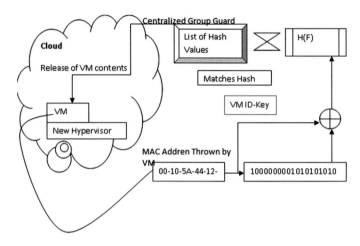

Fig. 3 Proposed architecture

Then
Continue the normal operation
Else if(traffic flow rate>Threshold Value)
Perform Signature Authentication Check
VSigID=GSigID
Enforce the Security policy over VM
Endif.

6.1.1 Vsignature—Authentication Process

Every clustered hypervisor's group generated some hash value signature when it was created as the part of cluster group for the Authentication purpose. This hash based signature used to identify the hypervisor while moving the Live VM's between one Hypervisor to another in the same cluster group (Fig. 3).

This Signature Authentication will provide sufficient security to prevent the Virtual Machine hosted in a attackers hypervisor in unauthorized manner.

Every time the Virtual machine will update the new MAC Address by throwing back to its router where its previously hosted network. The Signature Verification module will now recomputed the hash value based on the MAC address which is thrown by the hypervisor.

7 Implementation and Result Analysis

The above work is simulated using Cloudsim by setting up the two different data center along with multiple number of Virtual machines running on the same hypervisor cloudlet. It is allowed to move a virtual machine from one hypervisor to another hypervisor and the data transfer rates are monitored at the VM and VSwitch level.

$$\text{Capacity of Cloudlet} = \sum_{i=0}^{n} C(i)/Nvm$$

By setting pre Limit condition to the data traffic rate between two VMs and Vswitch identifies the unauthorized behavior of the Virtual machine.

Mathematical Evaluation:

$$Tmr = Tc + Tsd \tag{1}$$

$$T\,\text{service downtime} = Tmr - Tcycles \tag{2}$$

$$Tsd > \text{Est. Th value} \tag{3}$$

Concluded as a abnormal behaviour VM.

RTT/ICMP Packet: (Vswitch)

Total Data transfer = Total no. of RTT peak@/CPU cycles + Data Transfer + Total down time

$$RTT \rightarrow \frac{\text{mean time between Time at first packet sent}}{\text{Time at response}}$$

Let the size of the virtual machine RAM is 'V' the peak rate at which packet's sent 'P' and the network bandwidth available be 'B'.

For an instance, Xen virtualization environment.

The Threshold limit, on the number of RTT peaks/CPU cycles limited to 5.

If RTT > Peaks > 5VM − abnormal

If the maximum amount of data transfers, set to 3 times of Vm RAM.

(1) If amount of data transfer >3 times VRAM

Forcefully stop VM and directed signature authentication

Then, set Th memory page = 50 (or) >50 memory pages

(2) If VM migration amount is <X then, downtime >Th value

Then destroy VM and migration.

→ Data migration during its nth cycle will always less than (or) equal to size of VRAM − V.

$$V * \left(\frac{P}{B} \right) n - 1 < V; \quad n > \,= 1 \tag{4}$$

The RTT peak rate always less than network bandwidth available.

RTTP < B → B

Tc = V/B

$$Tc = [V/RTTP] \tag{5}$$

Total Data migrated during the down time,

$$V\left(\frac{P}{B}\right)n < \Psi V \tag{6}$$

Ψ   Limit data transfer

7.1   Results Analysis

Variation in the IOPS from the above statistical analysis clearly shows that unexpected data traffic between the VM's and Vswitch will lead to an unauthorized move of Virtual machine from the known Hypervisor to target attack surface in the cloud environment (Table 2).

From the estimated values above graph shows a sudden rise of the curve when there is high volume of data transfer at specific point of time (Fig. 4).

7.2   Notations and Symbols Used

See Table 3.

Table 2 Report analysis

S.No.	Preset transfer rate (IOPS)	Starting time (ms)	Completion time (ms)	Actual transfer Rate (IOPS)
1.	3000	4.03	5.08	2800
2.	4000	5.09	6.05	3000
3.	4200	6.15	7.10	7000
4.	3500	7.12	8.10	8000

Fig. 4 Transfer rate result analysis

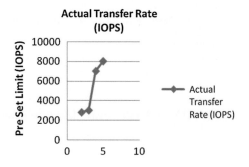

Table 3 Notations and operation

S. No.	Notation used	Operation
1.	VM	Virtual machine
2.	VSwitch	Virtual switch
3.	VRAM	Virtual RAM memory
4.	VsigID	Virtual signature identity
5.	GSigID	Global signature identity

8 Conclusion

The Proposed model tries to reduce the Co-located VM attack on the same hypervisor and enforcing the security polices over the Virtual Machine, when there is an unauthorized move or Copy of Live VMs into malicious hypervisor. Monitoring preset data traffic rate between two VMs and Vswitch node helps to identify uncertainty of VMs at a specific point of time.

Simulated results also shows that proposed model will reduce the risk associated on the VM running on the suspected hypervisor in cloud.

9 Future Enhancements

Proposed model is focused on the live migration of single VM from one hypervisor to another hypervisor, in future case, cluster of VM's running on hypervisor may consider for the secure group VM migration among different hypervisor.

Acknowledgement I dedicate this research paper to my research supervisor Prof Dr. S. Rajalakshmi, Director Advanced Computing Centre, SCSVMV University for the support and guidance throughout my research work.

References

1. Brohi, S.N., Bamiah, M.A., Brohi, M.N., Kamran, R.: Identifying and analyzing security threats to virtualized cloud computing infrastructures. In: Cloud Computing Technologies IEEE (2012)
2. Heen, O., Neumann, C., Montalvo, L., Defrance, S.: Improving the resistance to side-channel attacks on cloud storage services. In: New Technologies, Mobility and Security (NTMS), IEEE (2012)
3. Riquet, D., Grimaud, G., Hauspie, M.: Large-scale coordinated attacks: impact on the cloud security. In: Innovative Mobile and Internet Services in Ubiquitous Computing (IMIS), 2012 Sixth International Conference on IEEE, May 2012
4. Zhang, Y., Li, M., Bai, K., Yu, M., Zang, W.: Incentive compatible moving target defense against VM-colocation attacks in clouds. In: Computer Science Department, Virginia Commonwealth University, IBM T.J. Watson Research Center, USA (2012)

5. Varadharajan, V., Tupakula, U.: On the security of tenant transactions in the cloud. In: Cloud Computing Technology and Science (CloudCom), IEEE, vol. 1 (2013)
6. Bakasa, W., Zvarevashe, K., Karekwaivanane, N.N.: An analysis on the possibilities of covert transfers between virtual machines clustered in cloud computing: survey. Int. J. Innovative Res. Comput. Commun. Eng. 2(6), 4701–4707 (2014)
7. Han, Y., Alpcan, T., Chan, J., Leckie, C.: Security games for virtual machine allocation in cloud computing. In: CCSW'14 Proceedings of the 6th edition of the ACM Workshop on Cloud Computing (2014)
8. Booth, G., Soknacki, A., Somayaji, A.: Cloud security: attacks and current defenses. In: 8th Annual Symposium on Information Assurance (ASIA'13), 4–5 June 2013
9. Hamdi, M.: Security of cloud computing, storage, and networking. In: Ariana, Tunisia in Collaboration Technologies and Systems (CTS), 2012 International Conference on IEEE, May 2012
10. Zhang, N., Li, M., Lou, W., Hou, Y.T.: Mushi: toward multiple level security cloud with strong hardware level isolation. In: Military Communications Conference, IEEE (2012)
11. Zhen, C., Han, F., Cai, J., Jiang, X., et al.: Cloud computing-based forensic analysis for collaborative network security management system. Tsinghua Sci. Technol. 18(1), 40–50 (2013)
12. Alarifi, S., Wolthusen, S.D.: Mitigation of cloud-internal denial of service attacks. In: Service Oriented System Engineering (SOSE), IEEE (2014)

Trust Enabled CARE Resource Broker

Kumar Rangasamy and Thamarai Selvi Somasundaram

Abstract The objective of the grid system is to allow heterogeneous geographically distributed computational resources to be shared, coordinated and utilized for solving large scale problems. Grid users are able to submit and execute tasks on remotely available grid resources. Resource management is a vital part of a grid computing system. The notion of trust is used in resource management of grid computing system. In this work, we apply behavioral trust to the problem of resource selection. Through CARE Resource Broker (CRB), we demonstrate that our behavioral trust scheduling scheme significantly maximize the throughput while maintain the high success rate of task execution. We also analyze results of our behavioral trust based scheduling and rank based scheduling.

Keywords Grid system · Resource management · Scheduling · Behavioral trust

1 Introduction

A grid system is an infrastructure capable of managing the services and resources in a distributed and heterogeneous environment [1]. The computational grids enable the sharing, selection, and aggregation of a wide variety of geographically distributed computational resources and present them as a single, unified resource for solving large-scale compute and data intensive computing applications. A resource broker is an important component of computational grid systems and acts as a

K. Rangasamy (✉)
School of Computing Sciences and Engineering, Vellore Institute
of Technology, Chennai Campus, Chennai, India
e-mail: rangasamykumarme@gmail.com

T.S. Somasundaram
Madras Institute of Technology, Anna University, Chennai, India

© Springer International Publishing Switzerland 2016
V. Vijayakumar and V. Neelanarayanan (eds.), *Proceedings of the 3rd International Symposium on Big Data and Cloud Computing Challenges (ISBCC – 16')*, Smart Innovation, Systems and Technologies 49, DOI 10.1007/978-3-319-30348-2_2

bridge for the users to access grid resources. It leverages users in analyzing the choice of selecting the suitable resource provider for their job submission.

A grid user wants to have the transparent access and negotiates for the resources through the grid resource broker which identifies and allocates the suitable resource providers for job execution. The main function of a grid resource broker is to identify and exemplify the available resources and to select and allocate the most appropriate resources for a given job. The effective allocation of distributed resources requires not only the knowledge about the capabilities of the grid resources but also the assurance that the high availability and the requested capabilities can be fulfilled for successful completion of the job [2]. The resource discovery is one of the biggest challenges in the grid environment. Although, grid resource broker identifies the resource provider according to the suitability of the requested task and it is important to analyze the resource provider and rate them according to the quality of service rendered by them in their earlier performance. The quality of service may be determined based on the past behavior of the resource providers. The matching criteria of a resource provider in terms of credibility, computation power, memory size, resource performance, etc., for a given set of requirements also required for determining the good resource provider. The estimation of the belief on the behavior and reputation results in determining the trust values of the resource providers.

The trust values of the resource providers may be used for grading and ranking them to achieve better quality of service in the grid environment. In this paper, we propose a system that evaluates the trust of a resource provider by obtained by the trust metrics such as credibility and availability. The proposed Trust Management System is generic one and can be easily integrated with any metascheduler(s)/resource broker(s). The proposed system is integrated with the CARE resource broker to evaluate and update the trust value of resource providers. In our pervious work [3], we have considered a simple averaging scheme for the calculation of trust value. In brief, the contributions of this research work are summarized below.

- We propose a trust model to evaluate the trustworthiness of resource provider in the computational grid infrastructure.
- We derive a novel mathematical model for credibility and availability as termed as Trust Scheduling Function (TSF). The credibility is the combinational value of direct and indirect satisfaction degree. The direct satisfaction degree obtained through the job success rate and the indirect satisfaction degree obtained by the feedbacks from the users. We have applied probabilistic approach based on Bayesian inferences. The availability is derived by resource performance and resource busy degree. The computed trust value can be used as input for the grid resource broker to select the good resource provider for the job execution.
- We propose a mechanism to verify the feedbacks from the users after the utilization of resource providers.
- The proposed trust model has been integrated with CARE resource broker. The impact of trust based scheduling versus rank based scheduling has been analyzed.

The rest of the paper is organized as follows: In Sect. 2 gives the definitions of trust. In Sect. 3, presents the related work. In Sect. 4, we discuss about various types of trust in brief. We derive a mathematical formula to compute overall trust value is explained in Sect. 5. The various components of Trust Management System are explained in Sect. 6. The experimental setup made in our research laboratory and the results are explained in Sect. 7. Finally, in Sect. 8, we conclude our research work and outline the future work.

2   Trust Definitions

Several researchers have proposed various definitions for trust. One of the earlier definitions of trust given by Gambetta [4] states "Trust (or, symmetrically, distrust) is a particular level of the subjective probability with which an agent assesses that another agent or group of agents will perform a particular action, both before he can monitor such action (or independently or his capacity ever to be able to monitor it) and in a context in which it affects his own action". This definition reflects that trust is measure of a belief which paves the way for many researchers in exploring the computation of trust in a subjective manner. Castelfranchi and Falcone [5] extended Gambetta's definition that the trustor should have a "theory of mind" of the trustee, which included the estimations of risk and a decision on whether to rely on the other based on the trustor's risk acceptability.

Grandison and Sloman [6] surveyed various definitions of trust. Their definition states "the firm belief in the competence of an entity to act dependably, securely and reliably within a specified context". They define that trust is a combined effect of various attributes which have been considered while defining the trust of a system.

Josang et al. [7] define trust as "the extent to which one party is willing to depend on something or somebody in a given situation with a feeling of a relative security, even though negative consequences are possible". Their definition describes that the trust of an entity depends on the reliability of what other entities observe and also rely that the positive utility results in the good outcome and there is a possibility of risk occurrence by the relative previous entity.

In general, various trust definitions have been proposed by many researchers based on the context of the relationship which establishes among grid entities. In a computational grid environment, we focus on the establishment of trust of a resource provider. Hence, our definition of trust of resource provider states "Trust is a prediction of reliance on the ability and competence of a resource provider based on the commitment of quality of services measured within the specified context of computational grid".

3 Related Work

Trust concept has been addressed at different levels by many researchers. Trust can be broadly classified into Identity trust and Behavioral trust, where Identity trust is concerned with the authentication and authorization of an entity and Behavioral trust deals with entity's trustworthiness based on good or bad performance. This work is focused on evaluating behavioral trust of resource providers in a computational grid.

The notion of "trust management" was introduced by Blaze et al. [8] in their seminal paper. In the computer science literature, Marsh [9] proposed a computational model for trust in the distributed artificial intelligence (DAI) community. Marsh proposes a trust model takes into account direction. It categorized into three types of trust such as Basic trust, General trust and Situation trust.

Abdul-Rahman and Hailes [10] proposed a model for computing the trust for an agent based on the experience and recommendation in a specific context. Trust values are categorized into very trustworthy, trustworthy, untrustworthy and very untrustworthy. Each agent also stores the recommender trust with respect to another agent. The recommender trust values are semantic distances applied for adjusting the recommendation in order to obtain a trust value. They propose a method for evaluating and combining recommendations and updating the trust value. As the model is based on the set theory, each agent has to store all history of past experiences and received recommendations. On a system with a lot of participants and frequent transactions, each agent should have a large storage with this respect. Regarding the network traffic, this is caused by the messages exchanged between agents in order to get reputation information. The authors provide an example of applying the reputation management scheme, but no computational analysis is provided. However, we have given enough computational analysis based on the proposed mathematical model. Moreover, their approach is entirely based on subjective analysis. The researchers [11–14] have attempted to explore the subjective nature of trust metrics. Further, they focus on the reputation or recommendation that depends entirely on the opinion of the entities involved.

Azzedin and Maheswaran [15], view the trust at two different levels viz., direct and overall trust. In this model, the overall trust represents the reputation value earned by the resource provider. This model rates resource provider through various levels A through F, with the assumption that A being the highest trust value and F as the lowest value. It doesn't provide any confined values for those levels. The overall trust value earned by an entity is subjective in nature, depends on the user's view or opinion. Further, this model lacks the truthful assurance of the user's feedback which may lead to decrease or increase the trust level of those resources which they had experienced. In our approach, we concentrate on object nature of trust metrics. They have applied discrete approach in calculating the value of trust. We have applied probabilistic density function to calculate the trust value of the resource providers.

Kamvar et al. [16] designed a Eigen algorithm for P2P system and it is based on the transitive trust. This model has the drawback of storing each other peers trust value locally. We have used trust database to store the trust values of each provider in the computing grid infrastructure. In this way, our system reduces the burden storing trust values locally. However, it does not suggest any method for obtaining these trust values. This model also lacks the dynamic collection of the feedback system. In our work, feedback aggregation is done after the verification of positive feedbacks and negative backs. During the feedback verification, if the feedbacks are found to be faked ones, our system automatically discards it. von Laszewski et al. [17] exploits the beneficial properties of EigenTrust. This work extends the model to allow its usage in grids. They integrate the Trust Management System as part of the QoS management framework, proposing to probabilistically pre-select the resources based on their likelihood to deliver the requested capability and capacity. The authors present the design of the system, but they don't present the experiments in order to prove efficiency of the approach.

Aberer and Despotovic [18] referred to P2P networks, because it fully employs the referral network as being a source for obtaining recommendations. They assume the existence of two interaction contexts: a direct relationship where a destination node performs a task and recommendations when the destination node acts as a recommender of other nodes in the network. Rather than considering the standard P2P architecture, the graph of nodes is built by linking peers who have one the above-mentioned relationships. The standard models usual weight a recommendation by the trustworthiness of the recommender. Instead, they model the ability of a peer to make recommendations, which is different from the peer trustworthiness.

Chen et al. [19] proposed a model for selection and allocation of grid resources based on Trust Management System scheme. The authors have considered trust metrics such as affordability, success rate and bandwidth for the calculation of overall trust value. There is no substantial mathematical derivation has given for the calculation of trust value. The authors have followed the simple averaging scheme for their computation of trust value.

Dessi et al. [20] introduced a concept of Virtual Breeding Environment (VBE). This system has three types of operative contexts namely user operative context, resource operative context and organization operative context. Under these contexts, many trust metrics have been proposed. However, their work lacks in mathematical derivation of each and every metrics. Aforementioned trust metrics are subjective in nature. For instance, the resources are grouped into five reputation levels. The quantification of trust by adopting subjective metrics does not give much accuracy compared to objective metrics that we have adopted in our research work. There is no classification of direct and indirect trust metrics are obtained from the resource providers and the users. We have classified the trust metrics into direct and indirect one. The direct trust metrics are success rate and failure rate of the resource providers. The indirect metrics such as positive feedbacks and negative feedbacks. Moreover, the feedbacks are cross checked by our system.

An effective voting based reputation system has proposed by Marti and Garcia-Molina [21] to facilitate the judicious selection of P2P resources. The major concern in such a system is to isolate the adverse effects of a malicious peer. The simple voting scheme was found to be quite effective towards achieving this goal. Similarly, Damiani et al. [22] also suggested a reputation sharing system to enable reliable usage of remote peer's resources. Their system works by using a distributed polling algorithm. Our system achieves an effective resource selection and allocation of job execution by considering the past behaviours of resource providers.

Sabater and Sierra [23] proposed a model, named REGRET that considers three dimensions of the reputation models such as the individual dimension: which is the direct trust obtained by previous experience with another agent and the social dimension which refers to the trust of an agent in relation with a group and the ontological dimension which reflects the subjective particularities of an individual. In [24] review some works regarding reputation as a method for creating trust from the agent related perspective. They do not categorize the described models, but they try to find how those models are related with some theoretical requirement properties for reputation. The aforementioned works are related to subjective in nature. The quantification of overall trust value is more complex.

Karaoglanoglou and Karatza [25] presented a trust-aware resource discovery mechanism that guarantees satisfying requests with a high value of trustworthiness and in the minimum distance of hops in a Grid system. It is not clear that how trust is calculated for each Virtual Organization (VO) in the Grid system. The authors simply assume random values for all VOs. There is no substantial mathematical background for the calculation of overall trust of VOs. Further, the proposed model is not integrated any metascheduler/resource broker. In [26], the authors have proposed the social-network based reputation ranking algorithm for the peer to peer environment. It is capable of inferring while the calculation of indirect trust ranks of the peers more accurately. Further, the effective measures are still missing in their design and implementation. In [3], the authors mainly considered three trust metrics such as affordability, success rate and bandwidth of the resource providers. The overall trust value can be computed by the simple approach.

In this research work, we propose a trust model that aimed to explore the possibility of obtaining the trust value in an objective fashion and the same has been integrated with CARE resource broker. We compute the reputation of resource providers by means of resource performance and quality of service. We consider the user's opinion in terms of positive and negative feedbacks. The feedbacks are collected from the user after the usage of a particular resource provider can be quantified into a numerical value. We propose a mathematical model to evaluate the trust value of the resource provider. The trust values for each resource provider will be stored in the database for the future use. Each resource provider has its unique resource identity and associated trust value. We have considered two main factors, one reflecting the entity's past experience and the other reflecting the capabilities of resource provider.

4 Types of Trust for Computational Grid

The resource broker is responsible for identifying the suitable resource provider meeting the customer's requirement. It is also responsible for mediating the resource provider and users during grid transactions. Trust in the resource provider's competence, honesty, availability, success rate and reputation will influence the consumer's decision for accessing them. The trust models are categorized into four types such as Service provision trust, Behavioral trust, Identity trust and Reputation trust.

4.1 Service Provision Trust

Service provision trust describes the relying party's trust in service. The trustor trusts the trustee to provide a service that does not involve access to the trustor's resources. This type of trust reflects the resource provider's capability in terms of its computational and connectivity power. The grid resources with a higher computational power are expected to process the task with a reduced amount of execution time with respect to others. The trust metrics are considered in service provision trust such as CPU speed, size of memory, network bandwidth, latency and utilization of CPU, etc.

4.2 Behavioral Trust

Behavioral trust is a measurable trust by the resource broker by its experience and interactions over the resource providers in the grid. It measures the consistency of any grid entity over a period of time in the grid environment. It also helps in the determination of trust acquired by an entity which helps to predict the behavior of that entity in near future. A few examples of behavioral trust metrics are availability, success rate and network speed.

4.3 Identity Trust

Identity trust represents the entity's trust depending on their form of identification to the grid environment. This trust focuses more on the authentication and authorization system which has been adopted in the grid environment. This trust helps to classify the entity's security level. A few examples of identity trust metrics are certificate authority, authentication mechanism, authorization policies.

4.4 Reputation Trust

Reputation trust reflects the trust of a grid entity over a certain period of time based on the remarks made by the grid users. This trust helps to identify the experience of a grid entity directly or indirectly. Reputation based trust is subjective in nature and the value of trust vary from individual to individual. A few examples of reputation trust metrics are feedback, recommendation, grading based on the varieties of jobs handled, etc.

5 Mathematical Model for the Proposed Trust Management System

In this section, we propose a novel trust model of estimating the trustworthiness of the resource provider in the grid infrastructure. Our model has advantages by considering the objective nature of the trust in the computation of trust value. There are various possibilities of failures in the grid environment. The success rate, failure rate and system availability of resource providers are taken into consideration while calculating the overall trust value. Hence, the trustworthiness of the resource provider is the combined effect of the behavioral of current status and the historical data meaning that how it was performed in the past and present. In the following section, we explain the formation of a mathematical model for the proposed system.

Let us take the following scenario. Consider R is a problem, all the possible solutions are representing a set S, set S is called the solution set of the problem. Let us consider $S = (T, -T)$ where T represents a grid node to another grid node of a resource provider is not only credible but also available and $-T$ represents a node set which a grid node to another grid node is not entirely credible and available, is a set of grid nodes in line with the requirements of solutions. For instance, a grid node has very high credibility, but not available, or a node has high availability, but it is not credible, such nodes do not meet the requirement nodes. For the credibility and availability of the grid node, we formulate a trust scheduling function $TSF(i)$ to evaluate the trustworthiness of a resource provider. This scheduling function is evaluated via two functions namely credibility function $C(i)$ and availability function $AVL(i)$. $TSF(i)$ is defined as formula (1).

$$TSF(i) = \alpha \times C(i) + \beta \times AVL(i) \qquad (1)$$

Here α and β are the weights of credibility and availability, sum of α and β is 1. We have assigned equal weights for credibility and availability. If there is a low requirement level for credibility, then we assign the lower value for α and higher value for β. If there is higher requirement of the credibility is needed, then we assign the bigger value for α and lower value for β.

5.1   Credibility

The credibility is an assessment of grid node status and behaviors characterized by the credibility value. This is to make each others satisfaction degree evaluation of the trusted status after interacting information between nodes. We define a double $C(i) = (D(i), I(i))$ to indicate the credibility of the node. Here $D(i)$ is the direct satisfaction degree and $I(i)$ is the indirect satisfaction degree. $C(i)$ is defined as formula (2).

$$C(i) = \{w_1 \times D(i) + w_2 \times I(i)\} \tag{2}$$

Here, w_1 and w_2 are the weights of direct satisfaction degree and indirect satisfaction degree, the sum of w_1 and w_2 is 1. In general, we have more confidence for the direct satisfaction degree and less confidence for the indirect satisfaction degree. Therefore, we usually set the value of w_1 is larger and set the smaller value for w_2.

5.2   Direct Trust Satisfaction

It is the direct evaluation of trusted status of the other node after interacting with each other and comes from the historical record of information. This information is obtained from the job success rate of the resource provider. Direct satisfaction is a kind of reliable information in the derivation of trustworthiness value. The calculation of direct satisfaction degree is done by Bayesian inferences. Two metrics are used in beta distribution to represent the observations are chosen n_s as the number of previous satisfying interactions and n_u as the number of unsatisfying interactions. While computing the values of n_s and n_u, it is assumed that the desired type future interaction is identical to that of previous interactions. By setting $x = n_s + 1$ and $y = n_u + 1$ the estimated value of D(i) is obtained by the expected value of the probability distribution function of the beta distribution.

$$D(i) = E(f(a; x, y)) = \frac{x}{x+y} = \frac{n_s + 1}{(n_s + 1) + (n_u + 1)} = \frac{n_s + 1}{n_s + n_u + 2} \tag{3}$$

The idea of adding 1 each to n_s and n_u (thus 2 to $n_s + n_u$) in formula (3) is that it follows the Laplace's *famous rule of succession* for applying probability to decision making from the history of values.

5.3   Indirect Trust Satisfaction

It is an indirect evaluation of trusted status of the grid node is performed by collecting the feedback from others. The feedback from others could help to find out the quality of the site even without direct interactions. However, feedback is not reliable source of information. By getting feedbacks from the grid users, we compute the indirect trust value of each site. The feedback evaluation is done by the following mechanism. After the usage of the grid resource, the feedback can be collected from the users through web forms. The feedbacks are classified by the user QoS parameter such as a deadline and the recommendation for the future.

The user can give an answer to the question as follows after usage of each grid node.

–   Whether the job has been finished within the deadline or not?
–   Do you recommend to use this node in the future for the other users?

It is obvious that the users could provide both positive feedbacks and negative feedbacks about the grid resource providers. These feedbacks are cross checked and then aggregated. Here P denotes the positive feedbacks and N denotes the negative feedbacks. The positive feedbacks and negative feedbacks are checked by the following scenario. The grid job is through the job submission template to the various resource providers available in the grid infrastructure. The job template consists of job requirements such as CPU, memory, disk size, deadline and etc. After every job execution, the user has to provide the feedback about the resources. Through the user, feedbacks are compared with the scheduler generated execution time of the job.

For instance, the user request 30 min as a deadline parameter to run the job on the resource. If the job is finished on or before 30 min, but the user gives the negative feedback about the resource. We consider this feedback is faked one. So we consider this as a negative feedback. Likewise, user has to provide the feedbacks for recommendation metric. It is possible that the resource provider performs well, but the user can give negative feedback. The grid metascheduler has logs at the end of every job execution. Through the system generated logs our trust model could verify the recommendation parameter and aggregate the negative and positive feedbacks. The mechanism for verifying the feedbacks are hidden to the users of the grid system. After the feedback verification, we categorize the feedbacks and then aggregate using the following formula for the calculation of overall trust.

By setting $x = n_P + 1$ and $y = n_N + 1$ the estimated value of I(i) is obtained by the expected value of the probability distribution function of the beta distribution.

$$I(i) = E(f(a;x,y)) = \frac{x}{x+y} = \frac{n_P + 1}{(n_P + 1) + (n_N + 1)} = \frac{n_P + 1}{n_P + n_N + 2} \qquad (4)$$

5.4   Resource Availability

Resource availability function $AVL(i)$ is defined as the characteristics of resource performance and resource busy degree characterization. It is decided by the resource performance RP(i), resource busy degree RB(i).

The resource performance $RP(i)$ is defined as a formula (5).

$$RP(i) = \sum_{j=1}^{n} (R_{ij} \times p_j) \quad p_j \in (0,1) \tag{5}$$

Here $RP(i)$ is the resource performance number, i is the grid node, R_{ij} is the number, j is the resource attribute number, i grid node, such as a number of CPUs per core, CPU frequency, memory size, hard disk size, network speed, etc., p_j is the weight number, j is the resource attribute. It is a static value and there is no need to update periodically.

The degree of a resource busy is defined as a formula (6) below.

$$RB(i) = \sum_{j=1}^{n} \left(\left(1 - \frac{U_{ij}}{T_{ij}}\right) \times q_j \right) \quad q_j \in (0,1) \tag{6}$$

Here, $RB(i)$ is the busy degree of a number, U_{ij} is the usage number, j is the category resource number, i grid node, such as a CPU, memory, hard disk size, network usage, etc., T_{ij} is the total amount number, q_j is the weight of the resource usage number. The value of $B(i)$ is dynamic in nature. So, this value can be updated periodically to the resource broker in order to take the scheduling decision. Therefore, the resource availability function $AVL(i)$ is the combination of resource performance value and resource busy value.

$$AVL(i) = RP(i) + RB(i) \tag{7}$$

6   Trust Management System

In our earlier work [3, 27], we have proposed a generic life cycle of Trust Management System and its various phases involved in establishing trust across the grid resources. Figure 1, present the proposed Trust Management System has been integrated with CARE resource broker.

The sub components of the Trust Management System are described in this section. The Trust Metrics Identifier represents the trust metrics in a simpler form for the ease of trust value computation. The two main components of Trust Metrics Identifier are Trust Metric Information Collector and Trust Metric Indicator. The Trust Metrics Information Collector (TMIC) retrieves the basic information about trust metrics of every resource provider. It obtains the information from the grid resource broker or the grid middleware depending upon the trust metrics needed. The various metrics which is obtained by TMIC is availability status, job status, CPU and network information of the resource provider.

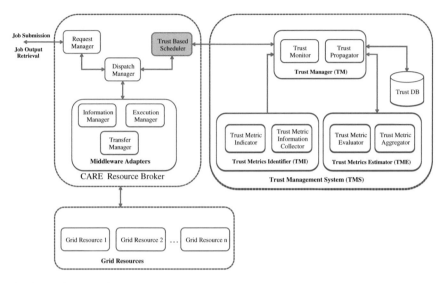

Fig. 1 Architecture of trust management system

Trust Metric Indicator provides the representation of the trust metrics which have been obtained through TMIC. It converts each trust metric into a suitable format for the computational aspect.

The Trust Estimator processes the trust metrics and applies a suitable mathematical formula for the computation of trust value of a resource provider. The two main components of Trust Estimator are Trust Metric Evaluator (TME) and Trust Metric Aggregator (TMA). The Trust Metric Evaluator computes the value of each trust metric considered for the resource provider. It provides the conversion of the trust metrics which is given by the TMI and represents the value of computing resources with respect to each trust metrics. The Trust Metrics Aggregator performs the computation of overall trust of computing resources. It processes each trust metrics, which is provided by the TME and applies a suitable mathematical model to compute the trust value.

The Trust Manager maintains the trust values computed by the Trust Estimator (TE). This maintenance of trust helps in evaluating the past experience of any resource provider in a grid environment. The two main components of Trust Manager are Trust Monitor and Trust Propagator. The Trust Monitor updates the trust value of any computing resources after each job execution. It also keeps logging information of every trust metrics computation. The Trust Propagator communicates with the grid metascheduler. The Trust Propagator retrieves the trust value of the resource provider from the Trust DB, if requested by the metascheduler. The metascheduler can utilize the trust value which is calculated by the Trust Management System and can schedule the jobs accordingly. The Trust DB holds the trust value of all resource providers that are available in the grid environment. It is also responsible for the Trust maintenance of every resource after each task/job completion.

6.1 CARE Resource Broker (CRB)

CARE resource broker [28] is a meta-scheduler that has been developed at the Center for Advanced computing Research Education Laboratory, Madras Institute of Technology. The motivation behind this work is that the application scheduling in grid is a complex task that often fails due to non availability of resources and required execution environment in the resources. The CARE resource broker addresses several scheduling scenarios using the concepts of virtualization. The Virtualization technology offers effective resource management mechanisms such as isolated, secure job scheduling, and utilization of computing resources to the possible extent. However, lack of protocols and services to support virtualization technology in high level grid architecture does not allow management of virtual machines and virtual clusters in grid environment. CRB proposes and implements necessary protocol and services to support creation and management of virtual resources in the physical hosts. Besides, CRB supports Semantic component, Resource leasing and Service Level Agreement (SLA). In addition to the afore-mentioned features of CRB, the proposed trust model that in assists to select good service provider for the reliable job execution. The current form of CARE resource broker provides more features but it lacks in selecting most reliable resource for the job execution. For this reason, we propose and model a novel trust scheme in addition to the aforementioned features.

The conventional CRB is working based on the ranking scheme. The compu-tation of rank is done by considering the CPU and memory of the grid resources. Trust based scheduler makes scheduling decisions for jobs on the available grid resources. The grid jobs can be submitted to Care Resource Broker (CRB) through the job submission portal. The Request Manager component of CRB does the matchmaking process for the submitted jobs against the available resources in the specific interval. By default, the CRB is working based on the conventional (rank) scheme. The new feature is introduced in CRB is trust based scheme. The function of Dispatch Manager is invokes the rank or trust scheme is based on the configu-ration of CRB. If the trust scheme is configured, then the CRB calls the trust based scheduling function. The trust calculation is done by considering both direct and indirect experience. Trust based scheduler is working based on system capability, system performance, direct trust interaction and indirect trust interaction. The algorithm for trust based scheduling is given below.

Algorithm 1 Job submission and Trust based resource scheduling

Step 1   Start the CARE resource broker and Oracle 10g Database and initialize the Number of Jobs (NJ), Trust Scheduling Function (TSF), Credibility (C), Availability (AVL), Direct Trust (DT), In-Direct Trust (IDT), Resource Performance (RP), Resource Busy (RB).

Step 2   Submit 'N' numbers of job requests to CARE resource broker.

Step 3   The Dispatch Manager invokes the Trust Based Scheduler to select the trustworthy resources for job submission.

Step 4   The Trust Based Scheduler invokes the Trust Management System for trust computation.

Step 5   The Trust Management System computes the trust value based on the following trust metrics such as success rate, feedback and availability using the formula (1), (2) & (3) and computed Trust resource list has been sent to Trust Based Scheduler.

For (i=1; i<NJ.Size; i++) {

    For (j=1; j< Matchedresourcelist.Size; j++)

      {

        TSF(j)=C(j) +AVL(j);   - Formula (1)

        C(j)= DT(j)+IDT(j);   - Formula (2)

        AVL(j)=RP(j)+RB(j);   - Formula (3)

        Trustresourcelistcount++;

        Trustresourcelist[j].add (Matchedresourcelist[j]);

      }

}

Step 6   The Trust Based Scheduler selects the most trustworthy resource from trust resource list and it is sent to Dispatch Manager.

Step 7   The Dispatch Manager invokes the Transfer Manager to transfer the executable and input files to selected trustworthy resource.

Step 8   Once the job execution is completed successfully, Dispatch Manager invokes the Transfer Manager to transfer the output files to broker.

Step 9   Then Dispatch Manager updates the job status, which is "Success" or "Failure" in Trust DB through Trust Management System.

Step 10 End

7 Results and Discussion

7.1 Experimental Setup

Figure 2 shows that the experimental testbed is done in our research laboratory for testing the proposed work in real world scenario. The test-bed consists of the Trust Resource Broker named Gridtrustbroker.mit.in and three cluster resources namely Xencluster.care.mit.in, Smscluster.care.mit.in and Centcluster.care.mit.in. This setup is managed by the grid middleware of Globus Toolkit (GT) 4.0.1, Torque-2.0.1 and Sun Grid Engine (SGE) 2.3.0 as Local Resource Manager (LRM), Ganglia 3.0.2 as resource monitoring tool, Network Weather Service (NWS) 2.13 as network monitoring tool and Oracle 10 g as a database for the trust repository. The purpose of using database is reduce the overhead and to store the historical information about the grid resources for a very large environment.

The Xen hypervisor is also installed in the environment because our CARE resource broker supports for the creation of virtual nodes, if there is a shortage of physical nodes. We have submitted and tested grid jobs in this testbed. The setup produces the fruitful results. We have tested 50 grid jobs in this environment initially. In the rank based scheme which is used in gridway metascheduler, the success ratio is 76 %. In trust based scheme, the success ratio is 85 %. Obviously, trust based scheduling is producing better results.

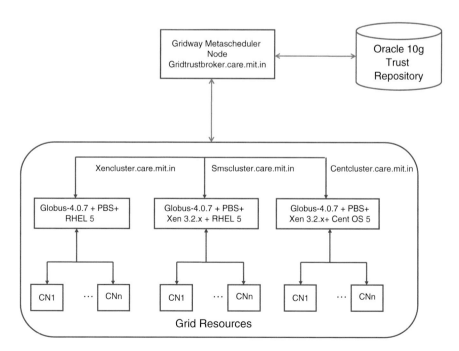

Fig. 2 Experimental setup

Fig. 3 Job success ratio

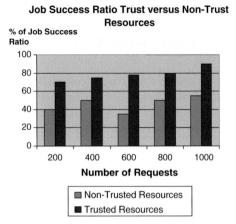

7.2 Simulation Experimental Results and Inferences

The proposed trust model has been simulated using the java based simulator code using a sample of 200, 400, 600, 800, 1000 requests and 200 nodes of grid resources and the requests (jobs) have been submitted to the trust enabled Gridway metascheduler node. The above requests have been tested both with trust and non-trust based model. The percentage of requests handled successfully with respect to the submitted requests is plotted as shown in Fig. 3. The proposed trust model increases the job success ratio, user satisfaction, and utilization of grid resources. The job success ratio in case of trusted resources is gradually increasing but in case of non-trusted resources, the job success ratio shows inconsistent.

The user's satisfaction level is plotted based on the feedback given by the user. The user's satisfaction increases for the trustworthy resources over a period of time whereas the satisfaction level is fluctuating and it is unpredictable for non-trusted resources as shown in Fig. 4.

The resource utilization of the trusted versus non-trusted resources is plotted as shown in Fig. 5. The X-axis represents the number of requests and Y-axis represents the utilization of the resources. The graph is plotted with the sample of 200, 400, 600, 800, 1000 and 1200 requests. The resource utilization of the trusted resources increases with the number of requests and the non-trusted resources utilized in a constant rate.

7.3 Trust Based Scheduling Versus Rank Based Scheduling Algorithm

The proposed Trust Management System has been integrated with the Gridway metascheduler for analyzing the impact of using trust in the selection of the

Fig. 4 Level of user's satisfaction

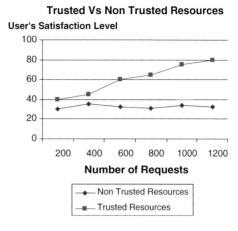

Fig. 5 Resource utilization

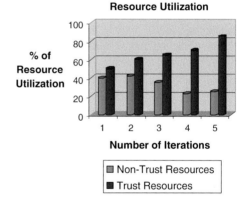

computational resource for the job execution. The Gridway metascheduler enables large scale reliable and efficient sharing of computing resources managed by different LRMS systems. However, Gridway approach follows the resource selection based on the rank mechanism.

The integration of our trust module with Gridway metascheduler is given in the following link: http://www.gridway.org/doku.php?id=ecosystem. In analyzing the impact of the resource selection in the grid environment, a simulator is developed to evaluate the performance of trust based scheduling algorithm with the rank based algorithm. In Fig. 6 represents the comparison of trust based scheduling and rank based scheduling. By default, Gridway uses the rank based scheduling algorithm. The rank based algorithm finds the ranking of computing resources by using the computational capability such as CPU and free RAM available.

The proposed trust model not only considers the computational capability and also takes the past behavior into an account while calculating overall trust. Our trust based scheduling algorithm has been tested using the same resources considered in rank based scheduling. Figure 6 shows that the trust based scheduling scheme

Fig. 6 Rank based
scheduling versus trust based
scheduling

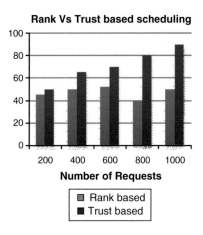

outperforms the ranking mechanism used by the Gridway metascheduler. These results states that the resources are utilized more effectively compare to rank based scheduling. In this way, in a computational grid environment, the throughput of the resources can be maximized by using our approach. The comparison between these two schemes is represented below.

8 Conclusion

In this work, we present the trust enabled CARE by considering the metrics such as direct trust, indirect trust and system availability. The proposed trust model is proficient in choosing reliable, most trustworthy resources which are part of grid environment. The computation of trust value of each resource provider is done through the mathematical model and the measured value of trust is objective in nature. We have tested our trust model with CARE resource broker and observed that the trust based scheduling enables the identification of resource providers yielding increased throughput. We also analyze results of our behavioral trust based scheduling and rank based scheduling. Our trust model is a generic one and the same can be easily integrated with other grid metascheduler(s)/resource broker(s), thus improving the effective resource management of grid infrastructure.

References

1. Arenas, A., Wilson, M., Matthews, B.: On trust management in grids. In: Proceedings of the 1st International Conference on Autonomic Computing and Communication Systems (2007)
2. Zhao, S., Lo, V., Gauthier-Dickey, C.: Result verification and trust-based scheduling in peer-to-peer grids. In: Proceedings of the Fifth IEEE International Conference on Peer-to-Peer Computing (2005)

3. Thamarai Selvi, S., Balakrishnan, P., Kumar, R., Rajdendar, K.: Trust based grid scheduling algorithm for commercial grids, vol. 1, pp. 545–5581, ICCIMA (2007)
4. Gambetta, D.: Can we trust trust? In: Gambetta, D. (ed.) Trust: Making and Breaking Cooperative Relations, pp. 213–237 (1988)
5. Castelfranchi, C., Falcone, R.: Principles of trust for MAS: cognitive anatomy, social importance and quantification. In: Proceedings of the Third International Conference on Multi-agent Systems. IEEE C.S., Los Alamitos (1998)
6. Grandison, T., Sloman M.: Specifying and analysing trust for internet applications. In: The Second IFIP Conference on E-Commerce, E-Business, E-Government (I3E), Lisbon, Portugal (2002)
7. Josang, A., Ismail, R., Boyd, C.: A survey of trust and reputation systems for online service provision. Decis. Support Syst. **43**, 618–644 (2005)
8. Blaze, M., Feigenbaum, J., Lacy, J.: Decentralized trust management. In: IEEE Conference on Security and Privacy (1996)
9. Marsh, S.: Formalizing trust as a computational concept. PhD thesis, University of Stirling (1994)
10. Abdul-Rahman, A., Hailes, S.: Supporting trust in virtual communities. In: HICSS '00: Proceedings of the 33rd Hawaii International Conference on System Sciences, vol. 6, p. 6007, Washington, DC, USA, IEEE Computer Society (2000)
11. Huynh, D., Jennings, R., Shadbolt, R.: Developing an integrated trust and reputation model for open multi-agent system (2004)
12. Dingledine, R., Mathewson, N., Syverson, P.: Reputation in P2P anonymity systems. In: Proceedings of the First Workshop on Economics of P2P systems (2003)
13. Zhou, R., Hwang, K.: Trust overlay networks for global reputation aggregation in P2P grid computing. In: IEEE International Parallel and Distributed Processing Symposium (IPDPS-2006), Rhodes Island, Grace (2006)
14. Resnick, P., Zeckhauser, R., Friedman, E., Kuwabara, K.: Reputation systems. Commun. ACM **43**(12), 45–48 (2001)
15. Azzedin, F., Maheswaran, M.: Integrating trust in grid computing systems. In: Proceedings of the International Conference on Parallel Processing (2002)
16. Kamvar, S.D., Schlosser, M.T., Garcia-Molina, H.: The Eigen Trust algorithm for reputation management in P2P networks (2004)
17. von Laszewski, G., Alunkal, B.E., Veljkovic, I.: Towards reputable grids. Scalable Comput. Pract. Experience **6**(3), 95–106 (2005)
18. Aberer, K., Despotovic, Z.: Managing trust in a peer-2-peer information system. In: CIKM '01: Proceedings of the Tenth International Conference on Information and Knowledge Management. pp. 310–317, New York, USA, ACM Press (2001)
19. Chen, C., Li-ze, G., Xin-xin, N., Yi-xian, Y.: An approach for resource selection and allocation in grid based on trust management system. In: First International Conference on Future Information Networks (2009)
20. Dessi, N., Pes, B., Fugini, M.G.: A distributed trust and reputation framework for scientific grids. In: Third International Conference on Research Challenges in Information Science (2009)
21. Marti, S., Garcia-Molina, H.: Limited reputation sharing in P2P systems. In: EC '04 Proceedings of the 5th ACM Conference on Electronic Commerce, New York, USA
22. Damiani, E., Vimercati, D.C., Paraboschi, S., Samarati, P., Violante, F.: A reputation-based approach for choosing reliable resources in peer-to-peer networks. In: 9th ACM Conference on Computer, ACM Press (2002)
23. Sabater, J., Sierra, C.: Regret: a reputation model for gregarious societies. In: Fourth Workshop on Deception, Fraud and Trust in Agent Societies, ACM Press (2001)
24. Sabater, J., Sierra, C.: Review on computational trust and reputation models. Artif. Intell. Revision **24**(1), 33–60 (2005)

25. Karaoglanoglou, K., Karatza, H.: Resource discovery in a grid system: directing requests to trustworthy virtual organizations based on global trust values. J. Syst. Softw. **84**(3), 465–478 (2011)
26. Wang, Y.F., Nakao, A.: Poisonedwater: an improved approach for accurate reputation ranking in P2P networks. Future Gener. Comput. Syst. **26**(8), 1317–1326 (2010)
27. Thamarai Selvi, S., Kumar, R., Balachandar, R.A., Balakrishnan, P., Rajendar, K., SwarnaPandian, J.S., Kannan, G., Rajiv, R. Prasath, C.A.: A framework for trust management system in computational grids. In: National Conference on Advanced Computing (NCAC) (2007)
28. Somasundaram, T.S., Amarnath, B.R., Kumar, R., Balakrishnan, P., Rajendar, K., Kannan, G., Rajiv, R., Mahendran, E., Rajesh, B.G., Madusudhanan, B.: CARE resource broker: a framework for scheduling and supporting virtual resource management. Future Gener. Comput. Syst. (2009)
29. Huedo, E., Montero, R.S., Llorente, I.M.: A modular meta-scheduling architecture for interfacing with pre-WS and WS Grid resource management services. Future Gener. Comput. Syst. **23**, 252–261 (2007)

An Initiation for Testing the Security of a Cloud Service Provider

D.M. Ajay and E. Umamaheswari

Abstract If Security and Privacy are the biggest concerns in cloud, how to find which Cloud Service Provider is safe? This paper proposes an idea to test the security of a Cloud Service Provider, in a way helping new customers adopting cloud, independently select their service provider. This testing will create a healthy competition among the service providers to satisfy their Service Level Agreement and improve their Quality of Service and trustworthiness.

Keywords CSP—Cloud Service Provider · SaaS—Software-as-a-Service · PaaS—Platform-as-a-Service · IaaS—Infrastructure-as-a-Service · TaaS—Testing-as-a-Service

1 Introduction

The idea of cloud computing was first coined by Professor. John McCarthy, MIT at the MIT's centennial celebration in 1961 [1]. He quoted "Computing may someday be organized as a public utility just as the telephone system is a public utility. Each subscriber needs to pay only for the capacity he actually uses, but he has access to all programming languages characteristic of a very large system. Certain subscribers might offer service to other subscribers. The computer utility could become the basis of a new and important industry". His words presciently describe a phenomenon sweeping the computer industry and internet today: Cloud Computing.

D.M. Ajay (✉) · E. Umamaheswari
School of Computer Science and Engineering,
Vellore Institute of Technology University, Chennai Campus,
Vandaloor-Kelambakkam Road, Chennai 600127, Tamil Nadu, India
e-mail: dm.ajay2015@vit.ac.in

E. Umamaheswari
e-mail: umamaheswari.e@vit.ac.in

© Springer International Publishing Switzerland 2016
V. Vijayakumar and V. Neelanarayanan (eds.), *Proceedings of the 3rd International Symposium on Big Data and Cloud Computing Challenges (ISBCC – 16')*,
Smart Innovation, Systems and Technologies 49, DOI 10.1007/978-3-319-30348-2_3

1.1 Characteristics of Cloud Computing

- Companies, individuals, and even governments instead of owning their own computer systems, will be able to share computing resources on a common computing infrastructure.
- The common computing infrastructure consists of interchangeable parts providing computation, communications, and data storage.
- Any of the component malfunctions or need to update, the programs and data of that component will automatically move to other components.
- This model is cheaper to operate, since both the hardware infrastructure and administrative staff can be utilized much more efficiently.
- The companies, end users can scale up and scale down their resources at any time, in no time, saving a lot of cost and time.

1.2 Predictions and Market Estimates for Cloud Computing

- Global Cloud Computing Market Forecast 2015–2020 expects, the global cloud computing market to grow at a 30 % Compound Annual Growth Rate (CAGR) reaching $270 billion in 2020 [2].
- 90 % of the worldwide mobile data traffic is expected to be accounted in cloud applications by 2019 [3].
- Software-as-a-Service workloads will raise to 59 % by 2018 from 41 % in 2013 [4] (Fig. 1).

The second section will explain the importance of security in Cloud Computing, major attacks and data breaches in recent times. The third section will explain the security standards in Cloud Computing defined by various Organizations. The fourth section will explain about Cloud Testing. The fifth section will explain the proposed idea for testing the security of a Cloud Service Provider.

2 Importance of Security in Cloud Computing

Cloud Computing with all its benefits and computing trend for the future, has some major areas of concern among which security is an important factor. Security is the one of the biggest barrier for the IT industry to switch to cloud. While an organization or enterprise adapts to cloud, the crucial data and important file management and protection responsibility switches to the service provider. Therefore, it is very crucial for the enterprise or organization to test the security, prevention capability, protection strategy and recovery standards of a cloud service provider before switching to cloud (Fig. 2).

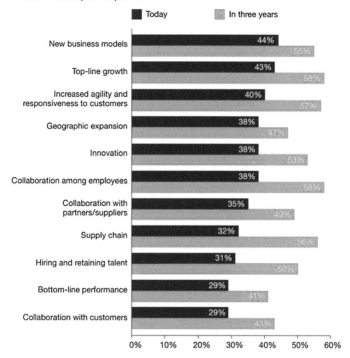

Fig. 1 2015—three years predictions for, impact of cloud computing on business [3]

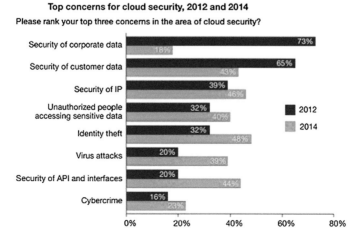

Fig. 2 Top concerns for cloud security, 2012 and 2014

The magnitude of attacks and data breaches in recent times explain the importance of security aspect in cloud computing.

- The HealthCare industry is the biggest sufferer of data breaches. 7,787,832 records of Community Health System were breached in 2014. Overall Health care Industry had about 495 breaches in total, and the total records breached are 21.2 million, at a total loss of $4.1 billion [6, 7].
- Educational Institutions are the second attacked industry accounting 27 % of data breaches [8]. Privacy Rights ClearingHouse Reported, 17 % of all reported data breaches involve higher education institutions [9]. In 2014, 30 institutions had suffered data breaches. In 2015, 9 institutions suffered attacks including big names like Harvard, University of California Los Angeles, and Penn State University.
- In the Retail Industry retailers like Target, Home Depot, Neiman Marcus, Dairy Queen suffered data breaches. About 64,668,672 customer records were breached in 2014.
- In the Financial Sector 1,182,492 records of NASDAQ was breached in 2014.
- Even Government Organizations suffered security breaches. About 6,473,879 US Post Office, State Department records were breached in 2014 (Table 1).

Table 1 Loss as a result of hacking cloud environment [10, 11]

Year	Country	Company	Attack type	Impact
Jul 2011	China	Alibaba	XSS payload	Goods ordered by customers were never delivered, resulting in financial loss estimated $1.94 million
Dec 2013	USA	Target	Credentials was stolen by hackers from a third party HVAC company through phishing mails	70 million customers' credit card information were compromised
April 2014	USA	Home Depot	Custom-built malware	Compromised 53 million emails and 56 million credit card or debit card information
Sep 2014	Global	iCloud	By using a Brute force service technique called "iBrute", Hackers hacked the accounts of major celebrities	Major celebrities' private photographs were publicized
Nov 2014	Global	Sony	Brazen cyber-attack	Hacking copies of unreleased Sony films, information about employees and their families, intercompany e-mails
Jun 2015	Russia, Iran	Kaspersky Lab	Duqu 2.0	Advanced attack on Kaspersky internal networks

3 Cloud Computing Standards

It is extremely important for Enterprises or Organizations to carefully evaluate the security and data protection standards followed by a cloud service provider, before switching to cloud. Many International Organizations have defined several Cloud Computing standards and security standards, guidelines and procedures for increasing the security of Cloud Computing services. Some of them are [12]

1. Cloud Standards Customer Council (CSCC).
2. The European Telecommunications Standards Institute (ETSI).
3. Distributed Management Task Force (DMTF).
4. European Union Agency for Network and Information Security (ENISA).
5. Global Inter—Cloud Technology Forum (GICF).
6. ISO/IEC JTC 1.
7. National Institute of Standards and Technology (NIST).
8. International Telecommunications Union (ITU).
9. Open Cloud Consortium (OCC).
10. Storage Networking Industry Association (SNIA).
11. OpenCloud Connect.
12. Object Management Group (OMG).
13. Organization for the Advancement of Structured Information Standards (OASIS).
14. The Open Group.
15. Open Grid Forum (OGF).
16. Association for Retail Technology Standards (ARTS).
17. Enterprise Cloud Leadership Council (ECLC).

3.1 Cloud Computing Security Threats

Cloud Security Alliance (CSA) in March 2010, had listed seven items of cloud computing security threats [13].

1. **Abuse and Nefarious Use of Cloud Computing**: Traditionally PaaS service providers suffer most from this kind of attacks. But, recent studies states that attackers have also started targeting IaaS providers. Since the registration process in IaaS providers are often easy, anyone with a valid credit card can register and avail the services immediately. Using this spammers, malicious code writers perform their illegal activities with immunity. Stricter initial registration and validation processes, introspecting the traffic of the customer network, improved fraud monitoring of credit cards will help in handling this threat in a more efficient way.

2. **Insecure Interfaces and APIs**: Provisioning, Monitoring, Management, Orchestration of services to the customers are provided by exposing a set of software interfaces or APIs. APIs are responsible for security and availability of general cloud services. Auditing the security model of cloud provider interfaces, implementing strong authentication and access controls with encrypted transmission will help in handling this threat.
3. **Malicious Insiders**: There is little or no standards followed in hiring and practices for cloud employees. The malicious insider with the level of access granted, can harvest confidential data and also gain complete control over the cloud services, with minimal risk of detection. This security threat can be handled by following standards in hiring cloud employees, Insiders' requirement conditions and processes should be transparent to the customers.
4. **Shared Technology Issues**: Sharing infrastructure, scalability are the biggest advantages in Cloud Computing. In this infrastructure, the underlying components aren't designed to offer strong isolation among users. Virtualization, hypervisor acts as mediator between guest operating system and the physical compute resources. However, flaws have been exhibited by hypervisors which enabled the guest operating system to gain control of inappropriate levels or control the underlying platform. In-depth defence strategy is recommended which includes computing, storage, and monitoring and network security enforcement. Strong compartmentalization among individual customers should be employed in such a way that no individual customer impacts other tenant's operations running on the same cloud provider.
5. **Data Loss or Leakage**: Deletion and alteration of records, loss of encoding key are some of the ways to compromise data. The architectural or operational characteristics of cloud environment increases the data leakage threat.

   (a) Implementing strong API access control.
   (b) Backup and retention strategies should be specified to the provider in the contract.
   (c) Strong key generation, management, storage and destruction practices should be implemented.

6. **Account or Service Hijacking**: Attacker gains access to cloud service provider's credentials and can manipulate data, redirect customers to illegitimate sites, return falsified information. Detection of unauthorized activity by employing proactive monitoring system and prohibition of sharing of account credentials between users and services are ways in which this threat can be handled.
7. **Unknown Risk Profile**: Details like internal security procedures, patching, auditing, internal security procedures are not often clearly stated. Infrastructure details should be fully/partially disclosed, Alerting and monitoring on necessary information are the measures to handle this threat.

4   Cloud Testing

From Sect. 2, we can infer security is one of the biggest concerns in Cloud Computing. To prevent and overcome the security issue, it's important to test the cloud.

4.1   Types of Cloud Testing

There are three different types of cloud-based testing [14].

1. **Testing a Software-as-a-Service in a Cloud**
   Software-as-a-Service testing comprises of validating SaaS applications with data integration, business workflow, compliance, application/network security, availability, performance, multi-tenancy, multi-browser compatibility, and live upgrade testing and disaster recovery. The focus of SaaS testing is on the core components of application, infrastructure, and network.
2. **Testing as a Service (TaaS)**
   Testing as a Service can be defined as automated software testing as a cloud-based service. Companies prefer to reduce costs, improve quality, and speed up the process of the applications. TaaS is an important paradigm which is the latest trend in cloud services. TaaS provides Testing services on demand on clouds any time and all the time.
3. **Testing a Cloud**

   - **Testing of Cloud**:
     Cloud service providers and end users are interested in carrying this type of testing. This testing validates the quality of a cloud environment from an external view based on the provided cloud service features and specified capabilities.
   - **Testing inside a Cloud**:
     Cloud vendors can perform this type of test. Access to internal infrastructure is necessary to perform this testing. This testing checks the quality, security, management and monitoring capabilities of a cloud from an internal view based on the internal infrastructure and specified cloud capabilities.

5   Cloud Security Testing

The biggest barrier or major factor of concern for users or organizations that are thinking of shifting to cloud services, is finding the best secured Cloud Service Provider from the market. Users or Organizations concerned about privacy and data security in CSP, find the above as the major problem. This makes testing the security measures, procedures followed by a Cloud Service Provider vital. The motivation of this paper is to help new customers to cloud, find the most secured and reliable CSP in terms of trust and security. In this work, a new idea is being proposed to test the security of a Cloud Service Provider. There has been large number of attacks and data breaches in recent times, which has caused huge data and revenue loss. Some of the most important are mentioned in Sect. 2. Each attack has taught new lessons and paved the way for new strategies and standards to avoid, handle those attacks with minimal loss. Cloud Computing standards have been defined by many International Organizations as mentioned in Sect. 3. These Organizations periodically update the latest cloud standards [12]. Theoretically lots of attacks have been predicted, practically new attacks have evolved and used to breach data and security, making security the biggest concern in cloud services. The idea proposed is to collect the major attacks on security of Cloud services and find out best strategies that are followed to prevent those attacks. The best strategies for preventing those attacks can be determined by studying the standards proposed by International Organizations, Procedures followed by Cloud Service Providers, and solutions put forth by Researchers'. A single attack can have many number of efficient strategies and the best among them can be determined by trial and error method. The best strategies in handling an attack should be updated frequently by observing the new trends and if any new strategy is found more efficient than the current one, it should be updated and also checked whether the Cloud Service Providers have updated it or not. Cloud Service Providers that follow the best strategy and keeps updating time to time are considered more secure, according to this proposed idea (Fig. 3).

Fig. 3 Process flow for testing security of a cloud service provider

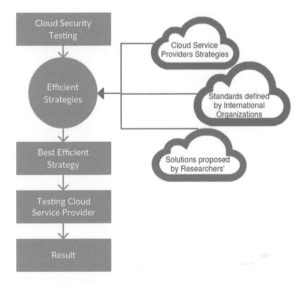

6 Conclusion and Future Work

This paper proposed a new idea to test the security of a Cloud Service Provider. This idea will take into consideration all the theoretically predicted and practically evolved attacks used to breach security of cloud services. Then the best strategies to prevent and handle those threats will be determined and Cloud Service Providers using those strategies will be considered more secure.

The future work will focus on implementing this idea in a more effective way, taking into consideration all the pros and cons.

References

1. Garfinkel, S.L., Abelson, H.: Architects of the Information Society: Thirty-Five Years of the Laboratory for Computer Science at MIT. In: Abelson, H. (ed.) The MIT Press, Cambridge (1999)
2. Global Cloud Computing Market Forecast 2015–2020, Tabular analysis. January 2014, http://www.marketresearchmedia.com/?p=839
3. Louis Columbus, 55 % of enterprises predict cloud computing will enable new business models in three years. http://www.forbes.com/sites/louiscolumbus/2015/06/08/55-of-enterprises-predict-cloud-computing-will-enable-new-business-models-in-three-years/
4. Cisco Global Cloud Index: Forecast and Methodology, 2014–2019, White Paper Cisco
5. Perspecsys, 2014 Data breaches. http://perspecsys.com/2014-data-breaches-cloud/
6. Bill kleyman, March 2015, Security Breaches, data loss, outages: the bad side of cloud. http://www.datacenterknowledge.com/archives/2015/03/16/security-breaches-data-loss-outages-the-bad-side-of-cloud/
7. HITRUST Report—U.S. healthcare data breach trends. https://hitrustalliance.net/breach-reports/
8. 2015 Data Breaches. http://www.cloudhesive.com/2015-data-breaches/
9. Privacy Rights Clearing House. https://www.privacyrights.org/
10. Cloud security breaches still the stuff of IT nightmares. http://searchcloudcomputing.techtarget.com/feature/Cloud-security-breaches-still-the-stuff-of-IT-nightmares
11. Mayayise, T.O., Osunmakinde, I.O.: A compliant assurance model for assessing the trustworthiness of cloud-based e-commerce systems. In: IEEE (2013)
12. Cloud Standards. http://cloud-standards.org/wiki/index.php?title=Main_Page
13. Cloud Security Alliance. Top threats to cloud computing V1.0. https://cloudsecurityalliance.org/topthreats/csathreats.v1.0.pdf
14. Nasiri, R., Hosseini, S.: A case study for a novel framework for cloud testing. In: IEEE (2014)

Security and Privacy in Bigdata Learning Analytics

An Affordable and Modular Solution

Jeremie Seanosky, Daniel Jacques, Vive Kumar and Kinshuk

Abstract In a growing world of bigdata learning analytics, tremendous quantities of data streams are collected and analyzed by various analytics solutions. These data are crucial in providing the most accurate and reliable analysis results, but at the same time they constitute a risk and challenge from a security standpoint. As fire needs fuel to burn, so do hacking attacks need data in order to be "successful". Data is the fuel for hackers, and as we protect wood from fire sources, so do we need to protect data from hackers. Learning analytics is all about data. This paper discusses a modular, affordable security model that can be implemented in any learning analytics platform to provide total privacy of learners' data through encryption mechanisms and security policies and principles at the network level.

Keywords Bigdata · Learning analytics · Analytics · Security · Privacy

1 Introduction

It is becoming obvious that the future and success of learning analytics and analytics in general reside in "bigdata" [1]. By this, it is meant that more and more data will need to be collected, or sensed, in order to provide always more accurate and reliable analytics results, as these additional data provide much context for each learning interaction. Fulfilling the goal of learning analytics, advanced analyses on these data will help learners improve their learning skills through an individualistic, learner-centered approach, which creates a learning environment where each and every student is cared about as opposed to the traditional class where some students may be left behind due to the work load on the teacher. However, the successful adoption and implementation of these fabulous ideas will necessitate properly

J. Seanosky (✉) · D. Jacques · V. Kumar · Kinshuk
School of Computing and Information Systems, Athabasca University,
Athabasca, Canada
e-mail: jeremie@rsdv.ca

© Springer International Publishing Switzerland 2016   43
V. Vijayakumar and V. Neelanarayanan (eds.), *Proceedings of the 3rd International Symposium on Big Data and Cloud Computing Challenges (ISBCC – 16'),*
Smart Innovation, Systems and Technologies 49, DOI 10.1007/978-3-319-30348-2_4

addressing the security question, as this new growing global dataset of learners' interactions may become a high-value target.

Ensuring total security, privacy, and protection of students' data is the ultimate goal for success in learning analytics. However, a proper equilibrium, or the "Golden Mean", needs to be achieved between two security extremes for optimal efficiency of such a learning analytics system. One side of the pendulum to be careful of is an over-complexification of advanced security layers. Such a scenario will arise if very advanced, complex security solutions are implemented properly or improperly that result in possible data bottlenecks, data losses, and ultimately higher costs. Data bottlenecks will arise from security mechanisms that overload the system and cause the intake of learners' interaction data to slow down to a trickle. Data losses may also be a result of levels of security that are too high and thus may mistakenly reject valid data packets as "malicious", thus creating an incomplete dataset and possibly inaccurate analysis results. Sometimes, such high-level security mechanisms do not cover all possible system flaws, as they tend to be more focused on network-level security while the application layer may have implementation flaws. In addition, this over-complex security may induce a false feeling of quasi-invincibility, which may lead to other consequences. The other side of the pendulum to be aware of is laxity with regards to security and the naïve belief that hackers are not interested in that kind of data. Neither side of the pendulum is appropriate, and the latter creates openness for hackers to break into such a system, leak personal and confidential data, and thus cause harm to students, institutions, companies, etc.

The security model discussed in this paper arose from the need for a production-level security layer in building a learning analytics solution targeting the writing and coding learning domains to be used in various academic institutions worldwide [2–5]. With several institutions showing interest in the product, pilot experiments started. It became clear that the security question needed to be addressed before this system could be deployed in production mode. Different security strategies were explored and an affordable, modular security model that can be implemented in any learning analytics system and provide an acceptable level of security for students' learning interaction data has been developed.

The above-mentioned bigdata learning analytics platform provides the following features to learners and teachers:

- It captures a wide variety of observations of learners at different times and frequencies from a widerange of learning contexts and learning domains.
- It analyzes and transforms those observations into insights by means of different analytics solutions such as competence and confidence analytics, metacompetence assessment, causal inferencing, clustering techniques, predictive analysis, and personalization and adaptation with regards to the learning contents.
- It observes the impact that learning analytics feedback has on each student's learning progress.

For a practical understanding of how the system works, Fig. 1 shows the technical architecture of this system.

As a complement to Figs. 1 and 2 demonstrates this same learning analytics architecture from a data flow perspective, indicating where the data come from, through what analytics apparatus they pass, and what is the expected outcome.

OpenACRE is currently an operational system provided to students in different institutions and this research paper aims at providing a solution to the data security aspect in such a learning analytics system.

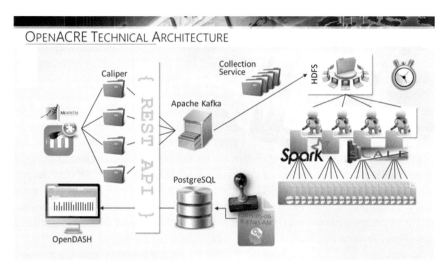

Fig. 1 OpenACRE technical architecture

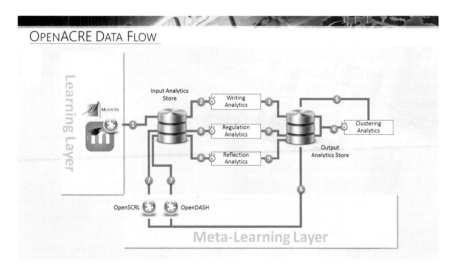

Fig. 2 OpenACRE data flow

Learning analytics is the process of analyzing learning interactions, or activities, by learners in any learning environment and perform analytics on those data in order to assess the learning competence of the learners. Different variants of such systems exist that use numerous techniques to analyze various aspects of learning interactions from learners, but fundamentally, the goal of any such learning analytics system is to provide helpful feedback to learners as to how and where to improve their learning process.

As obvious as it may seem, almost every online environment is under constant threat from hackers and ill-willed people who are intent on destroying others' work, stealing confidential information, and harming people's lives. Learning analytics systems are no exception. The following describes a possible disastrous scenario in which students' learning data would be hacked into and leaked to the world and the ensuing consequences.

Students' data from an unsecure learning analytics environment could be rather easily hacked into and released publicly. Unveiling of this information, which should be private between the teacher and student, could potentially cause a student to be denied a job at a company and thus harm his/her career. Furthermore, the simple fact that this information belongs to the student and is confidential as far as the student permits, that the student has been assured that his/her data were secure, and that then he/she suddenly finds out that a breach occurred is in and of itself totally unacceptable. This creates a breach of trust towards the academic institution and the system and cancels any good that the learning analytics solution might do or have done for students. Such scenarios absolutely need to be prevented.

Several security solutions exist these days for almost every imaginable possible scenario from advanced encryption techniques to firewall solutions, which solutions can be very expensive and complex to setup and maintain. However, protection needs to be addressed in two layers of any learning analytics system: (1) the network and (2) application layers. A learning analytics system can boast a very secure network apparatus, but flaws in the application layer (the code of the system) could expose vitally important data to hackers.

On the other hand, it may be justifiably argued that some of the security measures introduced in this paper are overboard, unnecessary, and incur useless overhead if only K-12 and university students' data are considered. However, the security and privacy model discussed in this paper has been designed from ground up not only with academic institutions in mind, but also and importantly with industry based on previous experience with building a training system for operators in the oil industry [6]. When dealing with training of employees in the industry, whether it be software companies, medical institutions, oil and gas industries, etc., employee and company data are extremely sensitive, and thus much higher security is required than with academic institutions. One should thus keep in mind the application of learning analytics in the industry and the higher risks involved despite the fact that this paper uses an academic learning analytics system as its use case to better convey the concepts of the described security and privacy model.

Given the aforementioned facts, this paper proposes a simple, secure, and affordable learning analytics security layer that could be implemented in any learning analytics system, be it for academic institutions or industries, and thus provide sufficient security as per the system's requirements.

2 Proposed Solution: Security, Affordability, Modularity (SAM) Model

2.1 Introduction

Given these factors to consider and given the risks incurred by the learning analytics world, this research has come up with a solution named the SAM Model. SAM stands for Security, Affordability, and Modularity and is a model specifying how to practically implement security into a new or existing learning analytics system while incurring minimal or no costs. Modularity means that this model provides a basic, no-cost security model that different institutions can add upon with their respective security layers as they see fit. By "affordability", it is intended to reach a wider range of institutions and share a great learning experience not only with large, well-off institutions but also with less-favored institutions in under-developed or developing countries around the world. These less-favored institutions oftentimes are those who are most in need of higher-quality education, but who generally cannot afford it. The SAM Model strives to offer the most affordable security solution for these institutions.

The underlying, fundamental principle of the SAM Model is centered around the idea of making all of the data marshalled by the learning analytics system "completely meaningless" to any hacker. This is considered the first layer of the SAM Model upon which are built the other features to protect the data through encryption and a secure network architecture. The SAM Model relies on the use of good practices, common sense, and open-source software to achieve a reasonably high level of security without involving costly, security solutions.

2.2 Privacy and Anonymity

In any learning environment, there are key elements of the data that constitute a threat to the privacy of students and institutions, and that are of interest to hackers, namely PII, or Personally Identifiable Information. PII is any information shared by students to the system that can be used to identify a student as a person. This includes username, first and last names, phone number, email address, home address, social network IDs, age, gender, etc. These personal details are in no way required for accurate, reliable analysis results. This constitutes the first layer of the SAM Model whereby all PII is removed from students' learning interaction data. To achieve this, user accounts are created by the learning analytics system and the user ID is then assigned to the student.

The SAM Model defines a user profile as an entity with four tokens, which are called SII (System-Identifiable Information). SII means that these tokens can only be used by the system to identify a student as a system entity, not a person.

The four SII tokens are:

1. Institution ID (unique identifier per institution)
2. Learner ID (institution-unique identifier per learner)
3. PIN Number (password-like token private to each student)
4. System ID (unique private key per learner entity in the system; used for decryption of each student's data)

These tokens contribute to making the notion of student or learner abstract and anonymous to any outsider to the system, thus mitigating the possible damage from a hacking attack.

As per the SAM Model, mapping between the student ID provided by the institution and the learner ID created by the learning analytics system should not be done by the learning analytics system and stored by the system. As shown in Fig. 3, it is the responsibility of the institution to do this mapping if they desire so and keep that information on their side. It should be noted, however, that most institutions use meaningless numerical or alphanumerical character strings as Student IDs, which in and of itself does not constitute an appreciable risk with regards to the disclosing of students' personal information. However, if a learning analytics system would go about doing this mapping and storing this mapping on the server in the same place as the rest of the system, that would constitute a risk in the following manner. A hacker could break into the learning analytics system and find the LearnerID $\rightarrow$ StudentID mapping. Then if the hacker knows which institution these students belong to, he could hack into the institution's system where the

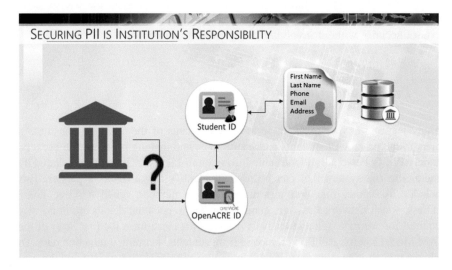

Fig. 3 LearnerID $\rightarrow$ StudentID mapping

personal information of students is stored associated with their StudentID and thus bridge the gap between the data and their owner. However, in such a case, data would still be encrypted as per the section below and would require some effort on the hacker's part to decrypt them. In addition, each institution is identified with a unique, meaningless Institution ID that cannot trace back to the institution itself. Therefore, the SAM Model proposes a network-level security layer that can avert hacking into the core system itself. This shall be discussed below in this research document.

2.3   Encryption of Data

Included in the Security part of the SAM Model is a learners' data encryption layer which programmatically encrypts all of the data in transit between the learning environment (client-side) and the data analytics engine (server-side). The encryption mechanism is presented in a visual manner in Fig. 4.

In the SAM Model, the encryption mechanism works as follows. Upon the successful creation of a learner account in the learning analytics system, an encryption key pair is generated by the system for that particular learner, and the private key is stored as part of that learner's private profile information. For security's sake, this private key is kept exclusively private to the system forever and not even the learner gets access to it. However, the public key, which is also stored in the system as part of that learner's private profile information is passed to the client-side during the authentication handshake process when the learner successfully authenticates into the learning tool from where the learning interaction data are

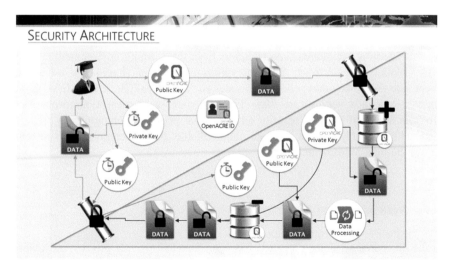

Fig. 4 Encryption mechanism

captured. From that point onwards, the learning tool uses the authenticated learner's public key to send encrypted data packets to the backend where these are stored in their encrypted form.

Upon reception of encrypted learning interaction data packets into the backend system, the data analytics engine retrieves these data and decrypts them using the appropriate learner's private key which resides exclusively on the server. Only then is the data analytics engine able to process and analyze these data and make sense of them.

When the data analytics engine has completed analyzing the learning data, the results of the analysis need to be encrypted again, which operation is performed using that learner's public key that also resides both on the server and on the client (only when the client requests it through authentication). These encrypted analysis results are then stored on the backend waiting to be requested by the visualization (dashboard) tools.

The final step in the encryption process occurs when learners want to view their learning analytics results, that is how well their learning performances are in the learning domain at hand. In this scenario where the user requests to view his/her data, the encryption process needs to be reversed. Therefore, upon successfully authenticating in the dashboard tool, a session-based encryption key pair is generated by the client learning tool on the client-side. The private key is kept on the client side and the public key is sent to the server as part of the transaction in which results are requested for displaying in the dashboard visualizations. Upon receiving the public key from that session, the server-side uses it to encrypt the analytics results and send these data encrypted to the client for viewing in the dashboard. When the client (dashboard) receives the encrypted data, it uses its private key to decrypt the data and show them in an intelligible manner to the user in the dashboard. Given that this encryption pair is session-based, when the student logs out of the dashboard tool or the session expires, the key pair is invalidated and destroyed so that no one else can retrieve the key pair and decipher learners' results.

It is correct to say that the analytics results are more in need of protection than the raw learning interaction data captured from the learning environment. In the SAM Model specifications, this encryption mechanism should use advanced client-side public-key encryption techniques such as OpenPGP or RSA in order to encrypt data in a very secure way. The aforementioned encryption mechanism provides a second layer of security on top of the existing anonymity and privacy layer which already renders the data unidentifiable. Now encryption aims at rendering the data unreadable.

2.4   Network-Level Security

In the preceding sections, emphasis has been directed towards privacy and anonymity as the first protection layer and then encryption of the data. These two layers are application-level security mechanisms. However, these mechanisms alone do not protect the data against the most important threats. This section proposes a secure network architecture or topology for hosting the different components of a typical learning analytics system, which components could be summarized as follows:

1. Learning Sensors—sensors tracking learning interactions from various learning environments
2. Ingestion Service—web service taking in learning data packets and passing them on to the server for further processing
3. Queueing System—the queue serving as a buffer zone to temporarily store learning data and avoid any data loss
4. Collection Service—internal service taking data from the queue and persisting them in a database
5. Input Analytics Store—database system storing raw learning interaction data before they are processed and analyzed
6. Data Processor—parallel-processing engine that performs custom-built analyses on the raw learning interaction data and persists the results in an output data store
7. Analytics Results Store—database system storing the results of the processing done by the Data Processor on the learning interaction data
8. Results Service—web service providing the analytics results data back to the client for visualization in the dashboard component

This shows how data are flowing through a typical learning analytics system, and from there, possible attack points, flaws, or high-risk areas can be discerned. Based on the current system architecture, vulnerable, high-risk areas have been identified and a proposed secure network architecture design when deploying learning analytics systems ensued. This security-centric architecture is generic and can be applied to any such system.

Figure 5 shows the components described above from a network and security perspective.

As this architecture clearly demonstrates, three layers are concerned:

1. Client-side sending out and receiving data
2. Middleware securely bridging the gap between client-side and private network hosting the system's core
3. Private server network hosting the learning analytics system's core

This architecture stems from a principle of isolation of the at-risk components of the system in order to minimize and even eradicate any unnecessary exposure. That is why the whole system has been broken down into three network layers that interact securely together.

The most sensitive, at-risk component of a learning analytics system is the system's core where the data processing is done and where all of the data is stored. This is the safe, or vault, of the learning analytics store. Therefore, higher protection is required.

This secure-centric system architecture starts with the client-side where Learning Sensors and Dashboards reside. This is the layer with which learners interact

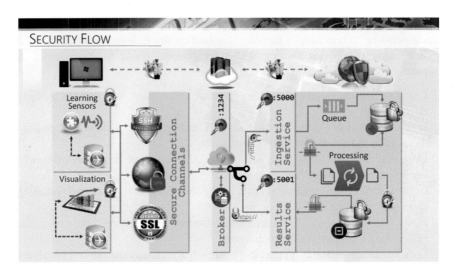

Fig. 5 Learning analytics system network-level security

sending their learning interaction data via the Learning Sensors and viewing their performance reports in the Visualization layer. This information travelling back and forth between the server and the client needs to be protected, which is achieved by the afore-mentioned first layer of client-side encryption.

Then comes the Secure Connection Channels. The raw data, even though they are client-side encrypted, need to travel in a secure environment where they cannot be sniffed. To achieve that, at least three possible and well-known solutions exist, namely SSL, VPN, and SSH.

SSL (Secure Sockets Layer) is the default secure channel for conveying information between applications and server-side components using the concept of signed certificates. However, SSL has vulnerabilities that sometimes get uncovered and can be exploited by hackers to sniff in the traffic [7–9]. Still, it is a good security layer. Also, SSL interception techniques could possibly make in-transit data vulnerable to exposure [10].

VPN (Virtual Private Network) is a more secure approach towards protecting data to and from the server. Most VPN solutions tend to be costly solutions, though reliable, free VPN solutions exist that can be used to provide more advanced security of the data such as SoftEther VPN [11, 12] and OpenVPN [13] that are worth considering. According to the SAM Model, modularity is there to allow the adding of new or more advanced security features on top of the existing default security setting. VPN is one such option that can provide more security, though it is less user-friendly as learners have to setup a VPN connection in order to work with the system. Again, depending on the case and the level of security required, VPN can or cannot be considered.

SSH (Secure Shell) tunneling, or port forwarding, could be used with a secure SSH key pair to establish a very secure connection to the server and send data securely. Furthermore, with this approach, as with the VPN solution, the risk of opening up server ports for client connections is mitigated since connections are made via SSH and not directly to a specific port. Though no security mechanism is absolutely 100 % secure and invincible, these security mechanisms offer high-level protection against sniffing of learners' data.

These three possible secure connection channels are options that can be used as per the SAM Model to ensure a double encryption layer for the data. This means that data are doubly encrypted while in transit. First of all, the client-side encrypts the data and sends them out, but then the connection channel such as SSL, VPN, and SSH provide a very strong second layer of encryption over these already-encrypted data. Therefore, even if someone would breach SSL, VPN, or SSH, the data would still be encrypted, and they would have to try to decrypt them.

Those data are now encrypted and passed onto the middleware layer called the "Broker". This broker plays a critical role in the whole architecture, as it bridges an important gap between the client-side sending encrypted data and the system core totally protected in a private network, which otherwise would be completely inaccessible. This Broker is a single server hosting a web service that listens on a single port and accepts only predefined types of data packets and otherwise rejects non-compliant data types. This Broker does not decrypt any of the data, but only relays them to the protected core. The Broker may also provide a temporary storage of data in transit in the case where the core is unreachable.

Finally, the most important part of the learning analytics system, the core, is hosted in a totally private network of servers hosting the different components of the system. This network comprises the two critical web services provided by the learning analytics system, namely the Ingestion and Result Services. These web services are unreachable from the web as they reside in a private network. This isolation of the core component protects it better due to its greater risk level as compared to the other components.

In order to send the learning data to the core and retrieve performance results, data pass through the Broker, which in turn relays those encrypted data to and from the respective web services on the private network. The key here is that the Broker server is on the same physical network as the core servers but is reachable from the internet only through a specific port and accepts only a predefined structured type of data. The other servers hosting the core are private and have no public IP address through which they could be web-reachable. Using this approach, all of the raw and processed data are protected against exposure by preventing any external access to the servers storing these data.

However, one possible flaw remains through the Broker. One can argue that someone breaking into the Broker's web-reachable server could then potentially gain unauthorized access to the internal, private network to which the Broker is connected. That is truly a possibility, but it is possible to mitigate this threat by isolating the Broker and private network so that only strict standards-adhering communications via the web services can be accomplished between the Broker and private core network. For instance, SSH access to any server within the private network could be denied to the Broker server, thus ensuring that no one hacking into the Broker could penetrate into the private network's servers.

In summary, the SAM Model, as laid out above, is a learning analytics architecture design that if implemented can ensure a proper balance between protection and productivity in the learning analytics world.

3 Conclusion

As the learning analytics world grows, security concerns with respect to the marshalled data will become increasingly important. It is therefore of utmost importance to address these issues in the beginning and propose concrete plans to fix possible current and future issues. In addition, proliferation of such systems on a worldwide scale will demand from them that they be as low-cost as possible in order to reach out to those most in need of high-quality education. This is what this research strives to do by designing a secure, affordable, modular learning analytics architecture based on an actual learning analytics system pilot-tested in several institutions as of now. The SAM Model proposed in this research paper will raise security awareness in the learning analytics world and provide a basic, no-cost solution to remedy to security issues. This will also serve as the basis upon which to build new learning analytics security models as well as expand on the SAM Model in order to always provide better protection of learners' data and thus build a trust relationship between the learner and the system providing help to the learner.

The SAM Model serves as a model, or specification, that different learning analytics platforms can implement with different variants depending on their respective context. The SAM Model should be considered a draft specification, or work in progress, that needs to be refined, reviewed, critiqued, validated, and approved by the general consensus of the learning analytics world.

Areas to be considered for future improvement and research include, though not restricted to:

1. Security provisions against DNS Hijacking
2. Secure authentication mechanisms (e.g. Central Authentication Service) to ensure the learner's credentials cannot be intercepted

Finally, researchers may also be interested in pursuing the creating and development of an HTTPA-like protocol where learners have more control over their data, thus providing for greater transparency as to who is doing what with their data [14–17].

References

1. Korfiatis, N.: Big data for enhanced learning analytics: a case for large-scale comparative assessments. In: Garoufallou, E., Greenberg, J. (ed.) Metadata and Semantics Research, vol. 390, pp. 225–233. Springer International Publishing, (ISBN: 978-3-319-03436-2). (2013). Retrieved from http://dx.doi.org/10.1007/978-3-319-03437-9_23

2. Seanosky, J., Boulanger, D., Kumar, V.: Unfolding learning analytics for big data. In: Emerging Issues in Smart Learning, pp. 377–384. Springer, Berlin (2015)
3. Boulanger, D., Seanosky, J., Kumar, V., Panneerselvam, K., Somasundaram, T.S.: Smart learning analytics. In: Emerging Issues in Smart Learning, pp. 289–296. Springer, Berlin (2015)
4. Kumar, V., Boulanger, D., Seanosky, J., Panneerselvam, K., Somasundaram, T.S.: Competence analytics. J. Comput. Educ. 1(4), 251–270 (2014)
5. Boulanger, D., Seanosky, J., Pinnell, C., Bell, J., Kumar, V.: SCALE: a competence analytics framework. In: State-of-the-Art and Future Directions of Smart Learning, pp. 19–30. Springer Singapore (2016)
6. Boulanger, D., Seanosky, J., Baddeley, M., Kumar, V., Kinshuk.: Learning analytics in the energy Industry: measuring competences in emergency procedures. In: 2014 IEEE Sixth International Conference on Technology for Education (T4E), pp. 148–155. (2014)
7. Wagner, D., Schneier, B.: Analysis of the SSL 3.0 protocol. In: The Second USENIX Workshop on Electronic Commerce Proceedings, pp. 29–40 (1996)
8. Sarkar, P.G., Fitzgerald, S.: Attacks on SSL a comprehensive study of beast, crime, time, breach, lucky 13 and RC4 biases (2013). Internet: https://www.isecpartners.com/media/106031/ssl_attacks_survey.pdf (2014)
9. Prado, A., Harris, N., Gluck, Y.: SSL, gone in 30 seconds. Breach attack (2013). Internet: http://news.asis.io/sites/default/files/US-13-Prado-SSL-Gone-in-30-seconds-A-BREACH-beyond-CRIME-Slides_0.pdf
10. Jarmoc, J., Unit, D.S.C.T.: SSL/TLS interception proxies and transitive trust. Black Hat Europe (2012). Internet: http://docs.huihoo.com/blackhat/europe-2012/bh-eu-12-Jarmoc-SSL_TLS_Interception-WP.pdf
11. SoftEther VPN. https://www.softether.org/
12. Nobori, D.: Design and Implementation of SoftEther VPN. Unpublished master's thesis, Department of Computer Science, Graduate School of Systems and Information Engineering, University of Tsukuba, Japan (2011). Retrieved from https://www.softether.org/@api/deki/files/399/=SoftEtherVPN.pdf
13. OpenVPN. https://openvpn.net/index.php/open-source.html
14. Senevitane, O., Kagal, L.: Addressing data reuse issues at the protocol level. In 2011 IEEE International Symposium on Policies for Distributed Systems and Networks (POLICY), pp. 141–144 (2011)
15. Seneviratne, O., Kagal, L.: HTTPa: accountable HTTP. In: IAB/w3C internet privacy workshop (2010)
16. Seneviratne, O.W.: Augmenting the web with accountability. In: Proceedings of the 21st International Conference Companion on World Wide Web, pp. 185–190. ACM (2012)
17. Seneviratne, O., Kagal, L.: Enabling privacy through transparency. In: 2014 Twelfth Annual International Conference on Privacy, Security and Trust (PST), pp. 121–128. IEEE (2014)

Detection of XML Signature Wrapping Attack Using Node Counting

Abhinav Nath Gupta and P. Santhi Thilagam

Abstract In context of web service security, several standards are defined to secure exchanges of SOAP messages in web service environment. Prominent among these security standards is the digital signature. SOAP messages are signed partially or fully before being transmitted. But recent researches has shown that even signed messages are vulnerable to interception and manipulation of content. We refer to these types of attacks as XML signature wrapping attacks. In this paper, an approach is proposed to detect the XML signature wrapping attacks on signed web service requests using node counting. We detect XML signature wrapping attacks by calculating the frequency of each node in web service request. Experiments show that the proposed solution is computationally less expensive and has better performance in securing the exchange of SOAP messages.

Keywords XML digital signature · XML signature wrapping · Web services

1   Introduction

Web Service is generally used to describe web resources that are accessed by the software applications rather than users. Web services are a set of functionalities designed to work in collaboration to complete a task. Web services standardized the business applications and due to the code reusability and interoperability feature provided, the business environment is falling for web services. Due to extensive usage of web services in business scenario it gives enough importance to Web Services to raise the security concerns.

Making Web Services secure means making SOAP messages secure and keeping them secure wherever they go [1]. The group of security standards in WS-Security is used to secure exchanges of SOAP messages in Web Service environment. However, despite all of these security mechanisms, certain attacks on

A.N. Gupta (✉) · P. Santhi Thilagam
Department of Computer Science and Engineering, NITK Surathkal,
Mangalore 575025, India

© Springer International Publishing Switzerland 2016
V. Vijayakumar and V. Neelanarayanan (eds.), *Proceedings of the 3rd International Symposium on Big Data and Cloud Computing Challenges (ISBCC – 16')*,
Smart Innovation, Systems and Technologies 49, DOI 10.1007/978-3-319-30348-2_5

SOAP messages may still occur and lead to significant security faults [2]. Illustrated that the SOAP message, protected by an XML Digital Signature as specified in WS-Security, can be modified without invalidating the signature. These kind of attacks are called XML Signature Wrapping Attacks. They can happen because XML Digital Signature assigned to an object in an XML document does not depend on the location of the object in the document.

Moreover, SOAP extensibility model by default has a very less restriction of the presence of headers and elements inside soap message and hence an unrecognized security header or soap header can be present inside soap message. All of these features along with vulnerabilities of XML Digital Signature gives away a way for performing wrapping attacks on SOAP messages.

Different solutions have been proposed to solve this problem. For example, in Ref. [3] the authors proposed an inline approach that uses the structure information of the SOAP message by adding a new header element called SOAP Account. In Ref. [4] the authors extended the inline approach by considering not only structure but depth and parent child relationships in soap message. However, none of these solutions could properly detect wrapping attacks. Moreover, not much attention has been laid on how to recover from the attack.

Section 2 describes the related work done on the same topic. Section 3 gives the details of system design. Section 4 describes the experimental setup and performance evaluation against other popular approaches to counter xml signature wrapping attack. Section 5 is conclusion followed by references considered for the study.

2 Related Work

To secure the exchange of XML document over the web, to maintain the authenticity and integrity of the exchange, XML digital signature is deployed [2]. XML Digital signature provides mechanism to encrypt an XML document partially or fully thereby securing the content of the XML document. In Fig. 1 [3] the structure of signed XML document is given.

The major vulnerability of xml digital signatures which leads to xml signature wrapping attack is xml processing is done twice when xml digital signature is present: once for the validation, and once for application use. Issue is that, for each case, validation and application, different approach is used to access xml data. XML Signature validation finds the signature element and use the references id inside to locate the signed element. The application parser instead analyze the message thoroughly to find the data application is interested in. Generally the results are same, but in the signature wrapping case, the attacker replaces the original signed element by a fake and relocating the original element inside the soap message from its original place.

SOAP message is depicted in Fig. 1 [3]. The figure shows the function of deleting the user present in the SOAP message body. Authentication and integration

Fig. 1 Example of XML signature applied on the SOAP body

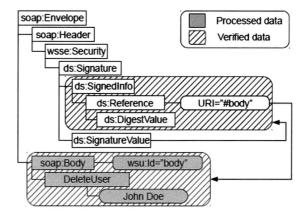

of the SOAP body is done with the help of XML signature. XML signature is present in SOAP header consisting two elements <SignedInfo> which is the identification pointing to the SOAP body and digest value computed over cited element and <SignatureValue>. The authentication for <SignedInfo> is provided by computing the signature value and assigning it to <SignatureValue>. The end-user first searches for the cited element given in <SignedInfo>. Than the digest value over the cited element is computed and comparison is done with the value given in the <DigestValue>. Than the signature is verified using <SignedInfo>. At the end, function defined in the SOAP body is executed. This was first observed by McIntosh and Austel [2], Example of the XML Signature Wrapping attack is shown in Fig. 2 [3]. In this example an attacker moves the original body of the SOAP inside header. Then he creates a body of the SOAP new id and invoke different function. Since the Signature is not altered just relocated and the concerning parts

Fig. 2 Example of XML signature wrapping attack on the SOAP body

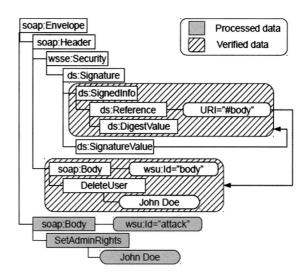

are also unchanged, the security logic will be able to do the verification of integrity and authenticity. This new SOAP body is taken as the input by the business logics.

XML Signature Wrapping attack is a newly discovered attack and as explained above, this attack exploits the loophole in processing of XML signature, which is designed for the purpose of authentication and integrity of the request message, to inject malicious data inside the request. Only a handful of papers are available to provide insight of this attack and even less counter measures. McIntosh and Austel [2] show ways to defend against wrapping attacks by laying certain message exchange policies for both sender and receiver. These policies have to be hardcoded into the application. But due to this the advantages of service oriented architectures is lost as the web services no longer remain independent.

Another popular countermeasure is XML Schema validation [3]. However, performing schema validation in the Web Services framework is not suitable, since it could adversely affect the performance of the service. Furthermore xml schema validation doesn't provide good security against the xml signature wrapping attack.

Second category of proposed countermeasures is called the inline approach and was presented in [3]. This approach fixes the relative location of the signed element so that any movement or alteration is detected. This approach works by adding a header element called SOAP account for each signed element containing its number of child element, parent elements, depth from SOAP header and Envelope.

But this idea has some disadvantages, first among those is its not being a standard of xml. Secondly, attacker could alter the message content while still being validated successfully. That means that inline approach cannot prevent against signature wrapping attacks in general.

Other approaches like schema hardening [3], attaching XPath expressions and ontology based approaches in [3]. But All these approaches have to be hardcoded into the application itself, and it means loss of flexibility and independence of service oriented architecture. All the approaches defined above are proactive approaches and demand a good understanding of the system from both the client side and service provider side.

3   System Design

The System is divided into 3 main modules, interception module, detection module, and logging module. First module is basically the interception module which intercepts the incoming digitally signed SOAP requests and forward it to the detection module.

Detection module then applies the algorithm given below to detect the presence of the XML signature wrapping attacks by analysing the request thoroughly. If detected, the request is denied and log is generated via logging module else the request is forwarded to intended recipient and another log is generated of successful forwarding of the request.

The Algorithm to detect the XML signature wrapping is

Input: InputStream of SOAP Envelope

1. See if <Signature> is child of <Header> element using//Signature//Reference
   [@URI] xpath expression, if present move to step 2 else there is no attack.
2. If found, extract the URI attached with the Reference attribute using string api in
   java, and check if it begins with #, if it does move to step 3 else match the URI
   against the URI Signatures.
3. If the URI begins with #, check the SOAP Header for optional header elements
   using//*[contains(@mustUnderstand,′0′)]/*/*xpath expression. It also extract all
   the children of the optional header element and save it in a NodeList data
   structure from java xpath api.
4. Extract all the children of SOAP Body element including itself
   using/Envelope/Body/*and save them in another NodeList.
5. Now compare NodeList from above two steps by counting the frequency of
   child nodes, if a single element is common in both list there is XML Signature
   Wrapping attack, else no attack.

Output: Boolean value notifying the presence of attack vector.

This concludes the algorithm to detect the XML signature wrapping attack.

4 Implementation and Performance Evaluation

We have implemented 3 services as per the specifications provided by a popular
benchmark for web services named TPC-APP, and deployed them on Apache
Tomcat using Axis2 service engine.

Test Environment machine composed of Hardware: 100 Mbps Ethernet card,
4 GB memory, Intel core i5 cpu with clock speed of 2.40 GHz, and Software:
Windows 7 Home Basic Edition, Java2 Standard Edition jdk 1.7.0. All the
implementation including web services and WS-IDS is in java.

Colored Petri Net (CPN) tools version 4.0 are used for the simulation of XML
signature wrapping attack in Lab Environment. Using the CPN tools we have
simulated 100 web services request messages which contains 20 malicious request
of namespace injection, 20 each of both id based and xpath based signature
wrapping attacks and rest of the request are all valid. Several attacking cases
including Replay Attack, Redirection Attack and Multiple Header Attacks is simulated inside the header and body of service requests along with attacks specific to
XML signature wrapping. We have also considered the specific case of XML
signature wrapping attacks and also sub types of the attack itself like namespace
injection or id based wrapping. The system starts by intercepting the incoming
SOAP requests and analysing them according to three approaches, i.e., WS
Security, SOAP Account and Node Counting to detect the XML signature wrapping attack. The results are presented in Table 1.

Table 1 Performance of node counting against other detection approaches

XML signature wrapping attack detection technique	False positive	True positive
WS security	8	52
Node counting	0	60
SOAP account	16	44

The results presented in Table 1 makes it is easy to recognize that our approach has bettered the other approaches. WS-Security approaches show the lowest attacks detection because as we mentioned in Sect. 2, XML Digital Signature has limitations to protect SOAP message from wrapping-attacks.

The next, SOAP Account approaches show the second best result. This is because with this approach, analysis of SOAP account, a header for each referenced element contains the number of child nodes of that element, would decrease the performance of the approach. Results may vary on different configurations and more rigorous testing, but while comparing Node Counting approach with other approaches it is noticed that Node Counting is only approach independent of any prior interaction to web service client or web service provider of any kind. In case of SOAP Account approach a separate header element for each of referenced element has to be added inside SOAP Header which has to be analysed at the receiving end in order to detect any XML Signature attack, a computationally expensive task.

We lose the independency and flexibility of web services as it is not a standard yet. Same is the case with WS-Security approach in which an xpath expression is included to locate the referenced element, again expensive in terms of computation cost. Also it has to be noted that even though approach bettered other approaches, it cannot detect 100 % of attacks. It is because slight variations in the behaviour of the attack may happen or new attack may appear.

Aside from detection rate of the XML signature wrapping attack, there is one more factor to consider and that is time taken to analyse the service request to detect the presence of XML signature wrapping attack. Node counting approach performs better than WS Security and SOAP Account as it takes less time to analyse the service request to detect the presence of attack as shown in Table 2.

The reason for this time difference is node counting approach directly analysis the received service request without referencing or dereferencing other extensions to the original request like the case in SOAP Account and WS Security Approach.

Table 2 Timed analysis of XML signature wrapping attack detection

XML signature wrapping attack detection technique	Time taken in analysis (ms)
WS security	1200
Node counting	450
SOAP account	700

5 Conclusions

In this paper, we have studied about web services and attacks on web services, specifically about the XML signature wrapping attack. We have studied various vulnerabilities of xpath and XML digital signatures and also considered various scenarios in which those vulnerabilities are exploited to mount signature wrapping attack on web services. We have also studied popular proposed mechanism to counter the XML signature wrapping attacks and there shortcomings in catering the XML signature wrapping attack.

In this paper, we proposed a mechanism based on node counting to combat with XML Signature Wrapping Attacks. Experiments showed that the proposed solutions have better performance in securing the exchange of SOAP message comparing to other methods.

Thus, we believe that our approach can protect SOAP message from wrapping-attacks and therefore, bring a reasonable protection to entire Web Service environment. Our current method may cause a reduction of the effectiveness when slight variations in the behaviors of the attack happen or when new attacks appear.

References

1. Bhargavan, K., Fournet, C., Gordon, A.D.: Verifying policy based security for web services. In: Proceedings of the 11th ACM conference on Computer and communications security, pp. 268–277 (2004)
2. McIntosh, M., Austel, P.: XML signature element wrapping attacks and countermeasures. In: Proceedings of the 2005 Workshop on Secure Web Services, pp. 20–27 (2005)
3. Jensen, M., Meyer, C., Somorovsky J., Schwenk, J.: On the effectiveness of XML schema validation for countering XML signature wrapping attacks. In: 1st International Workshop on Securing Services on the Cloud (IWSSC), pp. 7–13 (2007)
4. Benameur, A., Kadir, F.A., Fenet, S.: XML rewriting attacks: existing solutions and their limitations. IADIS Appl. Comput. **812**(1), 4181–4190 (2008)
5. Bhargavan, K., Fournet, C., Gordon, A.D., O'Shea, G.: An advisor for web services security policies. In: Proceedings of the 2005 Workshop on Secure Web Services, pp. 1–9 (2005)
6. Bartel, M., Boyer, J., Fox, B., LaMacchia, B., Simon, E.: XML-signature syntax and processing. W3C Recomm. **12** (2002)
7. Gajek, S., Liao, L., Schwenk, J.: Breaking and fixing the inline approach. In: Proceedings of the 2007 ACM Workshop on Secure Web Services, pp. 37–43 (2007)
8. Nasridinov, A., Byun, J.-Y., Park, Y.-H.: UNWRAP: An approach on wrapping attack tolerant SOAP messages. In: Second International Conference on Cloud and Green Computing (CGC), pp. 794–798 (2012)

Secure and Efficient Distance Effect Routing Algorithm for Mobility (SE_DREAM) in MANETs

H.J. Shanthi and E.A. Mary Anita

Abstract The aim of this paper is to provide security against Denial of Service attack for position based routing in MANET. In the position based routing, the message forwarding area is restricted by using the position information of the destination node because routing overhead is lower than ad hoc routing protocols. The presence of misbehaving nodes inside the forwarding zone is the hindrance to provide reliable data transmission. To subdue this, we propose Secure and Efficient Distance Effect Routing Algorithm for Mobility (SE_DREAM). In this protocol, the misbehaving nodes inside the forwarding zone are detected by analyzing the traffic flow between each pair of nodes in the network. If the traffic flow is abnormal, then that node is added into the malicious list. The malicious node is revoked from forwarding zone before forwarding the message towards destination position. The proposed algorithm outperforms existing geographic routing algorithm in the presence of malicious nodes in the network.

Keywords Mobile ad hoc network · Denial of service attack · Distance effect routing algorithm for mobility · Misbehaving nodes

1 Introduction

Mobile Ad hoc Network is consists of number of mobile terminals with no infrastructure and centralized control. The mobile nodes in the ad hoc network change its position more frequently. So, providing an efficient route between the mobile modes is the ultimate goal of the Mobile Ad hoc Network. The route should be discovered with less overhead and less bandwidth consumption [1]. The Geographic routing in MANET can achieve these requirements.

H.J. Shanthi (✉)
AMET University, Chennai, India
e-mail: shanthi_harold@yahoo.co.in

E.A. Mary Anita
S.A. Engineering College, Chennai, India

© Springer International Publishing Switzerland 2016   65
V. Vijayakumar and V. Neelanarayanan (eds.), *Proceedings of the 3rd International Symposium on Big Data and Cloud Computing Challenges (ISBCC – 16')*,
Smart Innovation, Systems and Technologies 49, DOI 10.1007/978-3-319-30348-2_6

1.1 Geographic Routing

In geographic routing, the nodes know their location by using any positioning system like Geographical Positioning System (GPS). The node estimates the position of its immediate neighbors and the destination node to forward a data packet to the destination. Even though the node in MANET changes its location dynamically, a node can estimate its neighbor's location easily in geographical routing [2, 3].

Many routing protocols of MANET uses the topology information to discover the route in the dynamic environment. The geographic or position based routing overcomes some of the disadvantages of topology based routing using some additional information. The geographic routing forwards the data packet towards the position of the destination node. There are two types of geographic routing such as proactive and reactive routing. In proactive routing, each node maintains the location table. The location table is updated by transmitting location packets among the nodes. So, the node will get the location information from its location table in case of proactive routing. The best example of proactive geographic routing is Distance Effect Routing Algorithm for Mobility (DREAM) [4]. In reactive routing, the node will estimate the position of its neighbor at the time of route discovery by using location services like GPS [1]. This kind of protocol does not maintain the location table. The best example of reactive routing is Location Aided Routing (LAR).

1.2 Security of Geographic Routing

Trusting the neighbor nodes in the geographical routing plays an important role as all the nodes forward the data via its immediate neighbors. The forwarder nodes are selected based on the location information. So, the attackers try to hack the location information. So, we need to provide the security to the location information exchanged between the nodes. The attacker may change the message or drop some packets if it has been in the forwarding zone. Hence, the routing protocol without security cannot provide better performance in the presence of malicious nodes [5].

In this paper, we propose a Secure and Efficient Distance Effect Routing Algorithm for Mobility (SE_DREAM) to ensure the reliable communication in MANET in the presence of misbehaving node in the network. The misbehaving nodes inside the forwarding zone are detected by analyzing the traffic flow in the network. According to the traffic flow analyzed, the nodes are categorized and exclude from the forwarding zone.

The rest of the paper is organized as follows: Sect. 2 discusses various works have done related to our proposed work. In Sect. 3, the process proposed work is explained in a prolix manner. Section 4 presents the results obtained by implementing the proposed idea in the Network Simulator.

2   Related Works

Malgi et al. [6] have proposed an Anonymous Position-based security aware routing protocol (APSAR) to maintain the anonymity protection from malicious nodes. This protocol has provided the anonymity protection for source, destination and the route by dividing the network area into several zones.

Rao et al. [7] have proposed secure geographical routing using Adaptive Position Update technique. In this protocol, the source node discovers the secure geographical route by using group signature when it wants to transmit the data. The Adaptive position update technique is used to update the position of the nodes according to their mobility patterns dynamically.

Carter et al. [8] have proposed Secure Position Aided Ad hoc Routing (SPAAR) which protects the position information to provide security for position based routing. A node verifies its one hop neighbors before adding them in the route. So, the unauthorized users cannot participate in the route as well as route discovery process.

Pathak et al. [9] analyze the well known geographic routing protocol Greedy Perimeter Stateless Routing protocol with Adaptive Position Update (APU) technique for position update. In this paper, the author provides the security by encrypting the data packets during transmission. The RC4 algorithm is used for encryption.

Ranjini et al. [10] have proposed security efficient routing for highly dynamic MANETs. The authors said that the Position based Routing is the best solution for delivering data packets in the highly dynamic environment. But the position based routing did not check whether the intermediate nodes in the route are secure or not. The security efficient routing gives the solution for this problem.

Malgi et al. [11] have proposed SC_LARDAR (Security Certificate Location Aided Routing Protocol with Dynamic Adaptation of Request Zone) protocol which is a new location based ad hoc routing protocol. SC-LARDAR concentrates on black hole attack in MANETs. The main advantages of this protocol are reduction in flooding RREQ packets and reduction of power consumption. But certificate based security scheme consumes extra memory to store keys. But our proposed method uses the traffic analyses method to detect the abnormal traffic flow in the network. The traffic analyses method is applied only to the nodes inside the forwarding zone towards the destination. So, the proposed method can reduce the overhead, memory usage and energy consumption to detect the misbehaving nodes in the forwarding area.

3   Proposed Work

DREAM is a proactive routing protocol because each mobile node maintains location table. The location table is updated by transmitting location packets (LP) to nearby nodes in higher frequency and to faraway nodes in lower frequency.

Fig. 1 Data forwarding via
nodes in the forwarding zone
except malicious node

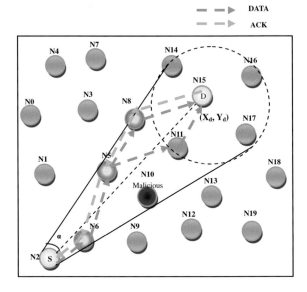

The mobile node in ad hoc network is not necessary to maintain up to date location
information for far away mobile nodes. Each location packet consists of
co-ordinates of the source node, source node's speed and the time in which the LP
was transmitted. Upon receiving the LP packet each node updates its location table.

Consider the network with 20 mobile nodes as shown in Fig. 1. Let us consider
node N2 as the source node and the node N15 as the destination node. Each and
every node broadcast Location Packets (LP) to its neighbors. The Location Packet
contains the co-ordinate of the source node and the time at which the LP was
transmitted. Each mobile node maintains Location Table (LT) and updates the
Location table by using LP it received. The source node calculates the expected
zone by using the Eq. (1). In Fig. 1, the dotted circle around the destination node
N15 is the expected zone. The nodes N11, N14, N16 and N17 are the nodes
available inside the circle around the destination node. Then source node defines its
forwarding zone by calculating minimum angle α as shown in Fig. 1. The minimum
angle is calculated by using Eq. (2). The forwarding zone is the area enclosed by
angle α vertex is at source N2 and whose sides are tangent to the circle around the
destination N15. The nodes N5, N6, N8, and N10 are present in the forwarding
zone.

The malicious node inside the network is discovered by analyzing the traffic flow
between each pair of nodes. If the detected malicious node is present inside the
forwarding zone in the sense that should be remove from the forwarding zone list.
Assume that, the node N10 is the malicious node in our network as shown in Fig. 1.
The source node sends the data packet for destination in the forwarding zone. When
the destination receives the data packet successfully, it returns as ACK packet to the
source node N2.

3.1 Process

1. Source first calculates a circle around the most recent location for destination D, using the last known speed.

$$R = V_{max} * (t_1 - t_0) \text{ Centered at } X_d, Y_d \tag{1}$$

2. Source node S defines its forwarding zone to be the region enclosed by an angle α, whose vertex is at S and whose sides are tangent to the circle calculated for D.

$$Angle\ \alpha = \arcsin(R \div d_{SD}) \tag{2}$$

3. Each of these neighbors then computes their own forwarding zone, based on their own location tables.
4. In our proposed work, we are going to filter out the malicious node present in the forwarding zone by using following method.

3.2 Traffic Matrix Construction and Traffic Analysis

In Mobile ad hoc network, individual mobile node dynamically come into and goes out of communication range of another individual mobile node. To simplify the analysis, we assume that mobile node will stay in communication range of other nodes for certain time interval.

Assume that there are altogether N mobile nodes in network denoted by m_i (i = 1, 2, …, N). Considering two individual mobile node m_i and $m_j (i \neq j)$, the process of m_i sending data to m_j can be divided as follows.

First m_j discover nodes those are present inside the communication range (Neighbor list), in which the probability of containing m_i is denoted by P_{ji}^s. Then assume that the data transfer rate from m_j to m_i is T_{ji} and assume that probability of m_i data receiving rate for the data transmitted by node m_j is P_{ji}^r. Finally m_i begins transferring data to m_j with the flow throughput B_{ij}.

Therefore, the traffic from mobile node m_i to m_j is

$$TF_{ij} = P_{ji}^s * T_{ji} * B_{ij}. \tag{3}$$

Now, we will go through every parameter. The probability of not containing m_i in the neighbor list of m_j is given by following equations,

Consider that, the total number of neighbor nodes of m_j is denoted by L_j and the total number of mobile nodes in the network is denoted by N. Then the probability of not getting m_i inside the communication range of m_j is given by

$$\overline{P_{ji}^s} = L_j/N \tag{4}$$

$$P_{ji}^s = 1 - \overline{P_{ji}^s}. \tag{5}$$

The data transmission rate is the amount of data transmitted from one mobile node to another in a given time. It can be viewed as speed of travel of a given amount of data from one node to another node. The data transfer rate from m_j to m_i is T_{ji} is given by

$$T_{ji} = Amount\ of\ data\ transferred/Time \tag{6}$$

Probability of mobile node m_i's data receiving rate for the data transmitted by node m_j is given by following equation.

Consider that the Service capacity is denoted by S_i (Maximum amount of data it can receive) and Demand is denoted by D_i (Actual size of the data node m_i received)

$$P_{ji}^r = S_i/D_i \tag{7}$$

Throughput is the rate at which the data can be transmitted. The flow throughput B_{ij} is given by,

$$B_{ij} = Amount\ of\ data/Transmission\ Time \tag{8}$$

3.3 Threshold Traffic Value

The threshold traffic value is calculated by analyzing the traffic of source node S. The following is the pseudo code to calculate the threshold traffic value.

```
Proc ThresholdValue (S, NL(s))
        List trafficValues;
        NL (n) denotes neighbor list of node n;
        FNL denotes Forwarding node list;
        for each node  n ∈ NL(s)
                Append TF(s, n) to trafficValues;
        End for
        Sorted value [] = Sorted value of list
trafficValues in descending order;
        mt=sorted value [0];
        return mt;
End Proc
```

Then compare the calculated threshold value with the traffic matrix of the nodes in the forwarding zone. If the traffic value of node m_i and m_j value is greater than threshold value in the sense, the node m_j send the revocation message against the node m_i to all its neighbors. The revocation message is transmitted with the signature in the following way:

The misbehavior node revocation is comprised of three sub process.

- Misbehavior Notification
- Revocation Generation
- Revocation Notification.

3.4 Misbehavior Notification

After the detection of node m_i's misbehavior, the mobile node m_j notify the misbehavior detection of node m_i to entire neighbor nodes. The notification message consists of ID of the node m_i along with time stamp value $\{ID_i, T_j\}$

3.5 Revocation Generation

Whenever a particular node got a misbehavior notification, the message will be dropped if the node m_j itself has revoked already.

The node m_i is diagnosed as malicious node when the number of misbehavior notification message reaches the predefined threshold value.

The neighbor nods construct the revocation message in the following way:

$$\operatorname{Re} v_j = H(ID_i) \times d \qquad (9)$$

where 'd' denotes the shared key among the mobile nodes in the ad hoc networks

d is the highest prime number. Each and every mobile node has its own public key pub (i), which is generated by $pub(i) = r \times d$. $r \in Z$ is an random integer generated by each mobile node. Then the value $d_{pub} = H(pub(j)) * d$, which is used to verify the signature appended with the revocation message.

3.5.1 Revocation Verification

Upon receipt of revocation message, each neighbor node verifies it by checking whether the equation $H(pub(j)) * \operatorname{Re} v_j = d_{pub} * H(ID_i)$ holds. If the equation holds, the particular node m_i is revoked from forwarder zone list.

5. Now the forwarding zone is free from malicious nodes. So, each neighbors of source node in the forwarding zone forwards the data accordingly.
6. When destination D receives a data packet, D returns an ACK packet.
7. If source node S does not receive an ACK packet within a time out period, then source resorts to a recovery procedure.

4 SE_DREAM Algorithm

In this section, we will describe how to transmit the data securely and efficiently in the ad hoc network. For that, we introduce SE_DREAM algorithm for eliminate misbehaving nodes in the forwarding area to provide reliable communication in the ad hoc network. Algorithm 1 depicts the pseudo code of SE_DREAM algorithm. The parameters used in our algorithm are described in Table 1.

Table 1 Parameter description used in Algorithm 1

Parameter	Description
S	Source node intended to transmit the data
D	Destination node receives the data from S
d_{ij}	Distance between the node 'i' and 'j'
CR_i	The communication range of the node i
V_{max}	The maximum speed of the destination node
t_0	Time at which transmission starts
t_1	Time at which transmission ends
TF_{ij}	Traffic flow between node 'i' and 'j'
$NL\ (i)$	Neighbor list of node i
mL	List name used to store the malicious node's ID
X_d	X co-ordinate of the destination node
Y_d	Y co-ordinate f the destination node
N_A	List of ad hoc nodes in our network
F_Z	Forwarding zone
$Re\ v_j$	The revocation message sent by node j
α	Minimum angle used to define the forwarding zone
R	The radius of the circle around the destination

Require: S, d_{ij}, CR_i, V_{max}, t_1, t_0, TF_{ij}, NL (i), mL
Ensure: The position of the destination node is X_d, Y_d
Event: Node S a data packet to send to the destination D in time interval from t_0 to t_1
/* Steps

1. Source S calculates a circle around the most recent location for destination D
2. The radius of circle is R;
3. $R = V_{max} * (t_1 - t_0)$
4. Source node S defines its forwarding zone
5. $Angle \alpha = \arcsin(R \div d_{SD})$
6. F_Z =List of nodes inside the forwarding zone
7. //call the function ThresoldValue
8. T_h =ThresoldValue(S, NL(S));
9. For each node $i \in N_A$ do
10. For each node $j \in N_A$ do
11. If $TF_{ij} > T_h$ then
12. Node j send revocation message Rev_j to NL (j)
13. If Rev_j is valid
14. Add node i into mL
15. End if
16. End if
17. End for
18. End for
19. //Remove malicious node from forwarding zone list
20. For each node $n \in F_z$ do
21. If $n \in mL$ then
22. Remove n from F_z
23. End if
24. End for
25. //Forward the data packet
26. fn=Source node S
27. While $fn \neq D$ do
28. For each $n1 \in F_z$ do
29. For each $n2 \in NL(n) \in F_z$ do
30. Forward the data packet to n
31. If $n1 = D$ then fn=D
32. Else fn=n
33. End for
34. End for
35. End while
36. If D receives data packet then
37. Send ACK back to S
38. Else
39. Update location table
40. Go to step 1
41. End if

Initially, the source node S calculates the circle around the destination node. The radius of the circle is calculated in line 3. The line 4–6 describes how to calculate the forwarding zone. The threshold value is calculated by calling the procedure ThresholdValue. It has two arguments such as source node ID and the neighbor list of source S. It returns the threshold traffic value. The malicious node is detected by analyzing the traffic matrix values from line 9 to 18. From line 20 to 24, the malicious nodes are removed from the forwarding zone. The data is forwarded to destination via forwarding zone by the source node is given by line 26–35. If the destination receives the data packet successfully, it will send the acknowledgement back to the source node else the source node update the location table and do the process from line 1.

5 Simulation Results

5.1 Simulation Model and Parameters

The detailed simulation parameters are listed in Table 2.

5.2 Performance Metrics

The NS2 Simulator is mainly used in the research field of networks and communication. The performance evaluated by using the network parameter packet delivery ratio, packet loss ratio, end to end delay, routing overhead and throughput.

Table 2 Simulation parameters

Parameter type	Parameter value
Simulation time	60 ms
Simulation area	1500 × 1000 m
Number of nodes	50
Mobility Speed	3, 4, …, 10 m/ms
Path loss model	Two ray ground
Channel bandwidth	2 Mbps
MAC protocol	802.11
Transmission range	250 m
Traffic model	CBR

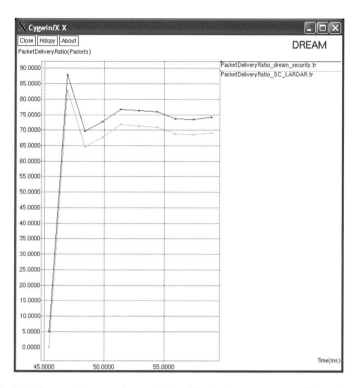

Fig. 2 Packet delivery ratio comparison analysis with existing secure geographic routing protocol SC_LARDAR

The performance of our proposed scheme is evaluated by comparing the proposed scheme SE_DREAM with the existing location based secure and reactive routing protocol SC_LARDAR.

Figure 2 gives the comparison analysis output of the proposed scheme SE_DREAM with the existing approach SC_LARDAR. The proposed scheme outperforms than SC_LARDAR. In our context, the packet transmission starts at time 45.0 ms and 100 packets have to transmit for each and every 0.05 ms. Initially 87 packets have been delivered to the destination successfully remaining packets have lost due to link breakage and channel fading only as we have remove malicious nodes from forwarder list. In Fig. 3, the packet loss ratio has plotted. The packet delivery rate varies during the runtime because of frequent link failure, channel fading and presence of malicious node in MANETs. When the simulation end, our proposed scheme has delivered 74 packets out of 100 packets to the

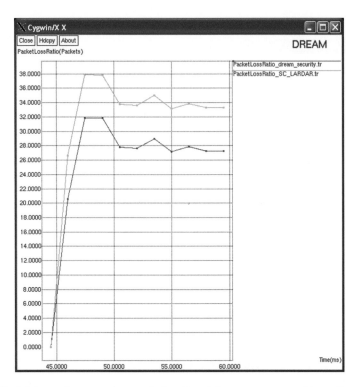

Fig. 3 Packet loss ratio comparison analysis with existing secure geographic routing protocol SC_LARDAR

destination successfully on the other hand the existing scheme has delivered 69 packets to the destination successfully. Thus our proposed scheme provides higher performance than the existing one in terms of packet delivery ratio. Higher the Packet delivery ratio indicates that the high performance of the network.

The packet loss ratio of the proposed scheme is compared with the existing approach SC_LARDAR. The packet loss ratio of the proposed scheme is lower than the SC_LARDAR as shown in Fig. 3. The Packet delivery ratio is indirectly proportional to the packet loss ratio. Lower the Packet loss ratio indicates that the high performance of the network.

Figure 4 shows that the proposed scheme leads to less delay when compared with the existing approach SC_LARDAR because of existing scheme provides certificate based security. In our context, delay refers to the time taken by the source

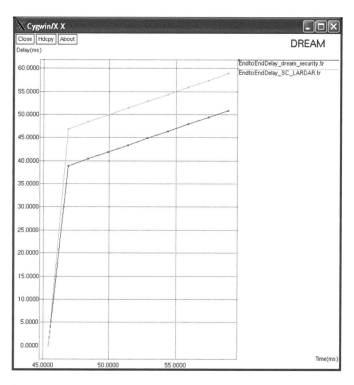

Fig. 4 End to end delay comparison analysis with existing secure geographic routing protocol SC_LARDAR

node to deliver the data successfully to the intended destination. In our simulation, our proposed scheme takes 38.0 ms to deliver 87 packets successfully whereas the existing scheme have taken 46.0 ms to deliver 82 packets successfully. The simulation result shown in Fig. 6 proves that the routing overhead in SC_LARDAR is lower than the proposed scheme. Our security scheme has to update the location table periodically by broadcasting Location Packets as the topology of the network changes dynamically. In the existing scheme, the location table is not maintained by the node. On the fly, it discover the by exchanging location information. The routing overhead is slightly increases during the runtime in Fig. 5. The route discovery process starts from 0.0 ms. At simulation end time, our proposed scheme uses 5524 packets to discover the route whereas the existing scheme uses only 5000 packets to discover the route.

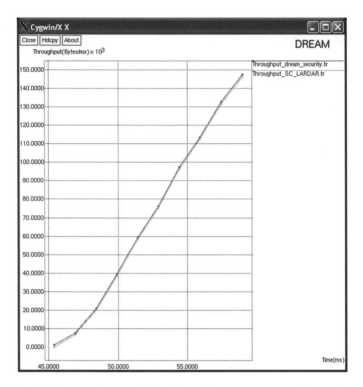

Fig. 5 Throughput comparison analysis with existing secure geographic routing protocol SC_LARDAR

   The comparative analysis of the proposed scheme with the existing approach SC_LARDAR is given by the graph shown in Fig. 5. Tracing the values from Fig. 5, we can say that our proposed scheme has delivered 91418 bytes successfully per unit time on the other hand the existing scheme has delivered 90320 bytes successfully per unit time. As a result, the proposed routing algorithm is can able to guarantee reliable communication in the highly dynamic environment.

6   Conclusion

Providing reliable communication in the Mobile Ad hoc Network in the presence of misbehaving nodes is the challenging one. This paper proposes a Secure and Efficient Distance Effect Routing Algorithm for Mobility (SE_DREAM) scheme to enable reliable communication and optimal communication overhead by detecting malicious nodes in the network. The attacks described in this paper focus on DoS attacks from outside nodes in the forwarding zone of DREAM. This proposed scheme parry the malicious node in the forwarding zone while forwarding the data

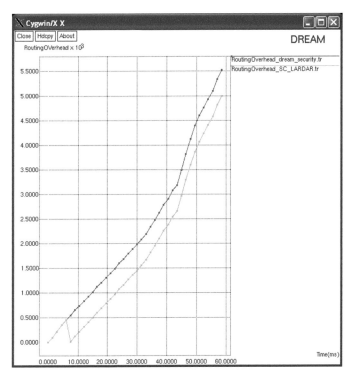

Fig. 6 Routing overhead comparison analysis with existing secure geographic routing protocol SC_LARDAR

packets towards the destination position. The performance metrics of the proposed scheme is obtained by using the simulator NS2. The performance of the proposed scheme is evaluated by comparing with the existing secure geographic routing protocol Secure Position Aided Ad hoc Routing. Our proposed scheme outperforms than the existing scheme in terms of throughput and overhead.

References

1. Song, J.-H., Wong, V.W.S., Leung, V.C.M.: A framework of secure location service for position-based ad hoc routing. In: Proceedings of the 1st ACM International Workshop on Performance Evaluation of Wireless Ad Hoc, Sensor, and Ubiquitous Networks, Venezia, Italy, pp. 99–106 (2004)
2. Song, J.H., Wong, V.W., Leung, V.C.: Secure position-based routing protocol for mobile ad hoc networks. Ad Hoc Netw. J. **5**(1), 76–86 (2007)
3. Sharma, S., Singh, S.: A survey of routing protocols and geographic routing protocol using GPS in manet. J. Glob. Res. Comput. Sci. **3**(12), (December 2012)
4. Shanthi, H.J., Mary Anita, E.A.: Performance analysis of black hole attacks in geographical routing MANET. Int. J. Eng. Technol. (IJET)7 **6**(5), (2014)

 5. Shanthi, H.J., Mary Anita, E.A.: Heuristic approach of supervised learning for intrusion detection. Ind. J. Sci. Technol. **7**(S6), 11–14 (2014)
 6. Malgi, P., Ambawade, D.: APSAR: anonymous position base security aware routing protocol for MANETs. Int. J. Comput. Appl. (0975–8887), (2013)
 7. Rao A.S.L., Sunitha, K.V.N.: Secure geographical routing in MANET using the adaptive position update. Int. J. Adv. Comput. Res. **4**(16), 3 (2014). ISSN (print): 2249-7277 ISSN (online): 2277-7970
 8. Carter, S., Yasinsac, A.: Secure Position Aided Ad hoc Routing. U.S. Army Research Laboratory and the U.S. Army Research Office Under Grant Number DAAD19-02-1-0235
 9. Pathak, V., Yao, D., Iftode, L.: Securing geographical routing in mobile ad-hoc networks. The NSF Grant CNS-0520123
10. Ranjini, S.S., Let, G.S.: Security-efficient routing for highly dynamic MANETS. Int. J. Eng. Adv. Technol. (IJEAT) **2**(4), (2013). ISSN 2249-8958
11. Anamika, Tyagi, K.: Secure approach for location aided routing in mobile ad hoc network Int. J. Comput. Appl. (0975-8887) **101**(8), (2014)
12. Khaleel, T.A., Ahmed, M.Y.: The enhancement of routing security in mobile ad-hoc networks. Int. J. Comput. Appl. (0975–888) **48**(16), (2012)
13. Shanthini, A.V., Mathan Kumar, M.: Security for geographic routing in mobile adhoc networks using RC4 algorithm. Int. J. Innov. Res. Sci. Eng. Technol. **3**(3), (2014) (An ISO 3297: 2007 Certified Organization)
14. Wadhwa, D., Deepika, V.K., Tyag, R.K.: A review of comparison of geographic routing protocols in mobile adhoc network. Adv. Electr. Electr. Eng. **4**(1), 51–58 (2014). ISSN 2231–1297
15. Sachan, G., Sharma, D.K., Tyagi, K., Prasad, A.: Enhanced energy aware geographic routing protocol in MANET: a review. Int. J. Mod. Eng. Res. (IJMER) **3**(2), 845–848 (2013)
16. Pyati, D., Rekha S.: High secured location-based efficient routing protocols in MANET's. Int. J. Recent Dev. Eng. Technol. **2**(4), (2014). ISSN 2347-6435 (online)
17. Gowda, S.R., Hiremath, P.S.: Review of security approaches in routing protocol in mobile adhoc network. IJCSI Int. J. Comput. Sci. Iss. **10**(1), 2 (2013)
18. Xu, K., Shen, M., Cui, Y., Ye, M., Zhong, Y.: A model approach to the estimation of peer-to-peer traffic matrices. IEEE Trans. Parallel Distrib. Syst. **25**, 5 (2014)
19. Li, L.-C., Liu, R.-S.: Securing cluster-based ad hoc networks with distributed authorities. IEEE Trans. Wirel. Commun. **9**(10), (2010)

Intrusion Detection Model Using Chi Square Feature Selection and Modified Naïve Bayes Classifier

I. Sumaiya Thaseen and Ch. Aswani Kumar

Abstract There is a constant rise in the number of hacking and intrusion incidents day by day due to the alarming growth of internet. Intrusion Detection Systems (IDS) monitors network activities to protect the system from cyber attacks. Anomaly detection models identify the deviations from normal behavior and classify them as anomalies. In this paper we propose an intrusion detection model using Linear Discriminant Analysis (LDA), chi square feature selection and modified Naïve Bayesian classification. LDA is one of the extensively used dimensionality reduction technique to remove noisy attributes from the network dataset. As there are many attributes in the network data, chi square feature selection is deployed in an efficient manner to identify the optimal feature set that increases the accuracy of the model. The optimal subset is then used by the modified Naïve Bayesian classifier for identifying the normal traffic and different attacks in the data set. Experimental analysis have been performed on the widely used NSL-KDD datasets. The results indicate that they hybrid model produces better accuracy and lower false alarm rate in comparison to the traditional approaches.

Keywords Accuracy · Chi square · Feature selection · Intrusion detection system · Linear discriminant analysis · Modified Naïve Bayes

1 Introduction

Intrusion Detection Systems (IDS) are built to defend networks from various viruses and attacks. IDS deploys efficient classification techniques to discriminate normal behavior from abnormal behavior in network data. The major challenge for building IDS is the management of large scale data clustering and classification.

I.S. Thaseen (✉)
School of Computing Science and Engineering, VIT University, Chennai, India
e-mail: sumaiyathaseen@gmail.com

Ch.A. Kumar
School of Information Technology and Engineering, VIT University, Vellore, India
e-mail: aswanis@gmail.com

© Springer International Publishing Switzerland 2016   81
V. Vijayakumar and V. Neelanarayanan (eds.), *Proceedings of the 3rd International Symposium on Big Data and Cloud Computing Challenges (ISBCC – 16'),*
Smart Innovation, Systems and Technologies 49, DOI 10.1007/978-3-319-30348-2_7

Many of the existing IDS face difficulty in dealing with new types of attacks or changing environments. Hence machine learning approaches are used for intrusion detection such as support vector machine [1], genetic algorithm [2], neural network [3], K-nearest neighbor [4], rough set theory [5], Naïve Bayes [6] and decision tree [7]. These approaches create an opportunity for IDS to learn the behavior automatically by analyzing the data trails of activities. The major steps involved in intrusion detection using machine learning techniques are

(i)   Capturing packets from the network
(ii) Extracting the optimal set of features that describe a network connection
(iii) Learning a model that describes the behaviour of normal and abnormal activities
(iv) Identifying the intrusions using learned models.

In our paper, we consider step (i) has been already completed and data is available.

The important contributions of this paper are as follows:

1. Hybrid model including a rank based feature selection integrated with modified Naïve Bayes classifier to minimize the computational task involved in classification.
2. Minimizing the training data and also allowing new data to be added to the training phase dynamically.

2   Related Work

IDS is considered as a crucial building block for any network system. Malicious attacks result in more adverse effect on the network than before and hence there is a need for an effective solution to recognize such attacks more effectively. Naïve Bayes is one of the classification approaches that identifies the class label in a faster manner due to the less complexity involved in the model. Fast prediction is also one of the major reasons for the selection of Bayesian approach by many researchers. Virendra et al. [7] developed a new integrated model that ensembles Naïve Bayes and Decision Table techniques. Dewan et al. [8] proposed an algorithm for adaptive network intrusion detection using Naïve Bayesian classifier integrated with decision tree. A balanced data set is deployed to the integrated model to reduce false positives for different types of network attacks and the model also eliminates redundant attributes as well as contradictory examples from training data to avoid building a complex detection model. The model also addresses limitation of data mining techniques such as handling continuous attribute, dealing with missing attribute values, and reducing noise in training data. Due to the large volumes of security audit data as well as the complex and dynamic properties of intrusion behaviors, several data mining based intrusion detection techniques have been applied to network-based traffic data and host-based data in the last decades.

Many of the hybrid intrusion detection models used feature selection and dimensionality reduction techniques to remove the noisy features and unimportant features which decreases the classification accuracy. Bhavesh et al. [9] built a hybrid model using LDA (Latent Dirichlet Allocation) and genetic algorithm (GA). LDA is used for generation of optimal attribute subset and GA performs the calculation of initial score of data items and then performs the different operations like breeding, fitness evaluation using objective function and filtering to produce new generation. Saurabh et al. [10] proposed an attribute vitality based reduction method in conjunction with a Naïve Bayes classifier for classification. Thaseen et al. [11, 12] proposed a novel model for intrusion detection by integrating PCA and Support Vector Machine (SVM) after optimizing the kernel parameter using variance of samples belonging to same and different class. This variance plays a key role in identifying the optimal kernel parameter to be deployed in the model for training. Hence this method resulted in a better classification accuracy.

Hence from the literature it is evident that dimensionality reduction along with optimal feature selection yields a better performance for the classifier model. In our approach, we develop a model by integrating dimensionality reduction with optimal subset selection using rank based feature selection and thereby predicting the class label using modified Naïve Bayesian approach.

3 Background

3.1 Normalization

Normalization is the process of altering the data to tumble in a lesser range such as $[-1, 1]$ or $[0.0, 1.0]$. Normalization is mostly required for classification algorithms. There are many methods for normalizing data such as min-max normalization, z-score and normalization by decimal scaling. In this paper, we deploy z-score technique as it is based on the mean and standard deviation of the attribute.

$$d^1 = \frac{d - mean\,(f)}{std\,(f)} \tag{1}$$

where,
mean (f)   sum of all values of a specific feature 'f'
std (f)   standard deviation of all values of a specific feature 'f'

3.2 Linear Discriminant Analysis

In many applications, the data set is noisy and has columns with single values. Hence a pre-processing stage is required to filter the unwanted data. Linear

Discriminant Analysis has been used in most of the applications to extract the relevant attribute set. LDA is used to produce a new subset of uncorrelated features from a set of correlated ones. The resultant data can be expressed in a less dimensional space with more powerful data representation. There has been a wide usage of PCA also for dimensionality reduction in many fields like face recognition, image compression and also for intrusion detection. The advantages of LDA in comparison with PCA are specified as follows.

(i) LDA handles both within-class and between class discrimination whereas PCA does not consider the within class variance [13].
(ii) Dimensionality reduction is performed by preserving class discriminatory information as high as possible [14].
(iii) LDA performs better than PCA when number of samples in a dataset is large [15].

3.3 Feature Selection

Feature selection is the process of retrieving a subset of relevant features for building a model. The major reasons for using feature selection are:

- Simplify the models for easier interpretation by researchers,
- Reduced training times,
- Reduction of over fitting to improve generalization.

The most widely used feature selection techniques are mutual information and chi-square as they improve the overall accuracy of the classifier. In our paper, we have deployed chi square feature selection technique.

3.3.1 Chi-Square Feature Selection

Chi-squared feature selection is a commonly used approach in many applications [16]. In this approach, the feature is evaluated by calculating the chi-squared statistic value relative to the class variable. The hypothesis is of the assumption that two features are not related and it is verified by chi-square formula:

$$\chi^2(f,c) = \frac{N * (WZ - YX)^2}{(W+X)(W+Z)(W+X)(Y+Z)} \tag{2}$$

where,
W No. of times feature 't' and class label 'c' co-occurs
X No. of times 't' appears without 'c'
Y No. of times 'c' appears without 't'

Z   No. of times neither '*c*' nor '*t*' appears
N   Total number of records.

3.4 Modified Naïve Bayes (MNB) Classifier

Naïve Bayes technique utilizes the Bayesian probability model for prediction of class label [17]. Naïve Bayes assumes that the probability of one feature has a null effect on the probability of the other feature which is considered as the hypothesis. Naïve Bayes classifier makes 2n! distinct assumptions for n attributes It is specified as given below:

$$P(C|Y) = \frac{P(Y|C)P(C)}{P(Y)}$$

$$i.e \ (C|Y) = P(y_1|C) * P(y_2|C) * \cdots * P(y_n|C) * P(C)$$

(3)

where,
$P(c|y)$ is the posterior probability of class for the analyst attribute
$P(c)$ is the prior probability
$P(y)$ is the prior probability of predictor.

The assumption in Naïve Bayes is that the attributes are conditionally independent.

$$P(Y|C_i) = \prod_{k=1}^{n} P(y_k|C_i) = P(y_1|C_i) * P(y_2|C_i) * \cdots P(y_n|C_i)$$

(4)

This results in reduction of computational cost. If the attribute is categorical, $P(Y_k|C_i)$ is the number of tuples in C_i having value y_k for A_k divided by $|C_{i,D}|$. Number of tuples of C_i in class D. If the attribute is continuous in nature, $P(y_k|C_i)$ is evaluated based on guassian distribution with mean μ and standard deviation σ.

$$g(x, \mu, \sigma) = \frac{1}{\sqrt{2\pi\sigma^2}} e^{-\frac{(x-\mu)^2}{2\sigma^2}}$$

(5)

and

$$P(X_K|C_i) = g(x_k, \mu_c, \sigma_{c_i})$$

(6)

Naïve Bayes assumes normal distribution for the numerical attributes and also considers that all features are independent for a given class.

In this paper, we will use modified Naïve Bayes as the experimental results show that this will increase the performance.

According to modified Naïve Bayes, we do not attempt to multiply individual probabilities but rather we will perform an summation to produce final probability.

Therefore, Eq. (3) will be modified as

$$P(Y|C_k) = \sum_{1 < k < n} P(y_k|C_k) \tag{7}$$

4 Proposed Intrusion Detection Model

The proposed intrusion detection model is built integrating dimensionality reduction, optimal subset selection and classification techniques. The entire process is divided into three stages. (i) Preprocessing of the data by normalization and dimensionality reduction using LDA. (ii) optimal subset selection using chi square feature selection and (iii) modified naïve bayes classifier for predicting the normal and abnormal data samples. In stage 1, the entire data set is sent to a preprocessor which normalizes the data using z-score normalization and dimensionality reduction is performed using LDA which removes noisy attributes from data. In stage 2, the genetic algorithm is deployed to identify the best optimal subset for classification. The features with maximum chi square statistic are supplied to the modified naïve Bayesian classifier for classification.

4.1 Proposed Algorithm

Input: Attributes of dataset.
Output: Class label prediction, accuracy
Hybrid Approach

Step 1:   Z-Score Normalization and LDA.
Step 2:   Remove symbolic features and normalize the continuous features within the range [0, 1] using z-score

  (i)   Calculate the mean (μ) and standard deviation (σ) of every continuous feature.
  (ii) Re calculate the value of every attribute using z-score technique.

Step 3:   Remove noisy attributes using LDA

  (i)   Calculate the within class (S_w) variance for normal and attack data and between class variance (S_B).
  (ii) Compute the eigen vector and sort the feature according to highest variance.

Step 4:   Optimal subset selection using chi square feature selection

      (i)   Determine the chi square statistic on every feature

     (ii)   Select the optimal feature subset of the individuals which are having a maximum chi square value.

Step 5:   Modified Naïve Bayes Classifier

      (i)   Predict the class label using probability estimation of normal class and abnormal class.

     (ii)   If the probability (normal) > probability (abnormal) assign class label to normal other wise to abnormal.

5   Performance Analysis

We have used an Intel Core CPU @2.30 GHz with Matlab R2013a for conducting experiments. The proposed model has been evaluated with NSL-KDD dataset containing 33,300 records [18]. Z-score normalization is performed as a part of preprocessing to transform all attribute values into the normalized range [0, 1]. In this paper, the various performance metrics evaluated are detection rate and false alarm rate which are the major metrics used to analyze the performance of intrusion detection (Fig. 1).

In this paper, the metrics used are

$$\text{Detection rate} = \frac{TN}{TN + FP} \tag{8}$$

$$\text{False-Alarm-Rate} = \frac{FN}{TN + FP} \tag{9}$$

where, TP indicates True positives, TN specifies True Negatives, FP indicates False Positives and FN specifies False Negatives. A confusion matrix is generated to identify the actual and predicted classification accuracy of every data category.

5.1   Study of NSL-KDD Dataset

The NSL-KDD data set contains five divisions of network traffic such as normal, denial of service, unauthorized access to local (U2R), Remote to Local (R2L) and probe After preprocessing on the NSL-KDD dataset, dimensionality reduction using LDA is performed resulting in removal of symbolic features and features with

Fig. 1 Proposed intrusion
detection model

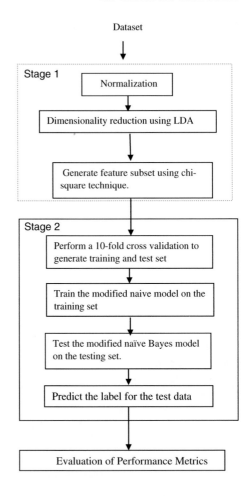

less variance. Table 1 shows the 31 attributes selected after dimensionality reduction and 22 optimal attribute subset selected after obtaining the maximum chi square statistic value on every attribute. A ten-fold cross validation is performed to generate different folds of training and test data. Classification using modified naïve bayesian is performed to predict the class label. The entire data is split into training and test set. The training set contains nearly 15,000 records and the test set contains 18,800 records. Table 2 shows the confusion matrix format generated for naïve bayes classifier for each of the individual attack type and normal data for the test set The confusion matrix specifies the information about actual and predicted classifications for the network traffic obtained from the modified naïve Bayesian classifier. This matrix helps us to predict how many samples are classified correctly in each category and how many are misclassified.

Table 1 Features selected after dimensionality reduction and optimal subset selection using chi-square

31 features selected after dimensionality reduction by LDA	Service, Dst_bytes, dst_host_diff_srv_rate, diff_srv_rate, flag, dst_host_serror_rate, dst_host_srv_count, same_srv_rate, count, dst_host_same_srv_rate, dst_host_srv_serror_rate, serror_rate, src_bytes, dst_host_srv_diff_host_rate, srv_serror_rate, dst_host_same_src_port_rate, logged_in, dst_host_count, hot, dst_host_rerror_rate, srv_count, duration, srv_diff_host_rate, dst_host_srv_rerror_rate, rerror_rate, protocol_type, srv_rerror_rate, is_guest_login, srv_count, num_compromised, num_failed_logins
22 features selected after chi-square test	Service, Dst_bytes, dst_host_diff_srv_rate, flag, dst_host_serror_rate, dst_host_srv_count, same_srv_rate, dst_host_same_srv_rate, serror_rate, src_bytes, dst_host_srv_diff_host_rate, hot, dst_host_rerror_rate duration, srv_diff_host_rate, dst_host_srv_rerror_rate, rerror_rate protocol_type, srv_rerror_rate, is_guest_login, srv_count, num_compromised

Table 2 Confusion matrix of proposed model on NSL-KDD data set using MNB classifier

	Probe	DoS	U2R	R2L	Normal
Probe	1837	0	0	0	31
DoS	2	6495	0	0	19
U2R	0	0	11	0	10
R2L	0	1	1	1275	19
Normal	29	29	0	0	8521

5.2 Discussion

The accuracy and false alarm rate are the two major factors that specify the cost task of an intrusion detection model. The performance of the proposed model has been analyzed on NSL-KDD dataset and the results are shown in Table 3. The results also show that the feature selection plays an important role in minimizing the computational complexity of the model. The accuracy is obtained from the average accuracy of all class labels present in the respective datasets. The major challenge in any intrusion detection model is selection of unbalanced training set for multiple classification task and selection of critical features which can distinguish the attackers and normal users with their network behavior. Both the challenges are met in our proposed model as the dataset we deployed in our model are very huge and imbalanced in nature. The chi square feature selection plays a major role in ranking the feature and identifying the class label. The minority attacks in NSL-KDD dataset namely U2R and R2L are detected with a good classification accuracy of 91 and 99 % respectively. In contrast to other classification techniques such as decision

Table 3 Performance of proposed model on NSL-KDD dataset

Metrics	NSL-KDD dataset
Detection rate	97.78
False alarm rate	2.46
Data preparation time	4.20 s
Training and testing time	10235 s

Table 4 Comparison of proposed model with other Naïve Bayesian techniques

Models	Detection rate	False alarm rate
Traditional Naïve Bayes [13]	70.01	29.99
Naïve Bayes with feature selection [20]	75.78	24.21
Discriminative multi nomial Naïve Bayes [19]	94.84	5.16
Proposed model	96.80	4.1

tree, nearest neighbor and genetic programming no retraining of the classifier is required in the proposed model for any new datasets arriving in the model. In addition as the training and testing stages are deployed as two modules independently, the process of training and testing can also be deployed parallel in any intrusion detection model. This is considered as an essential factor for real time intrusion detection. We have also illustrated in Table 4 that the proposed model produces better results in comparison to the other existing models in terms of accuracy and false alarm rate.

6   Conclusion

In this paper we develop and evaluate an intrusion detection model using LDA, chi square and modified naïve bayesian classification. LDA is one of the widely practiced dimensionality reduction technique to remove noisy attributes from the network dataset. As there are many attributes in the network data, optimal subset of attributes are identified by calculating the chi square statistic and obtaining the maximum test value attributes. The optimal subset is then used by the modified naïve bayesian classifier for identifying the normal and different attacks in the data set. With the experimental analysis performed using the NSL-KDD dataset. It is evident that the majority attacks can be classified accurately which is very crucial for any intrusion detection model whereas the minority attacks and normal data accuracy is slightly less due to the less number of training samples that illustrate their behavior. Our model focusses on building a defense mechanism to identify the majority attacks with maximum accuracy.

References

1. Hussein, S.M., Ali, F.H.M., Kasiran, Z.: Evaluation effectiveness of hybrid IDS using snort with Naïve Bayes to detect attacks. In: 2012 Second International Conference on Digital Information and Communication Technology and its applications (DICTAP), pp. 256–260, May 2012
2. Pu, W., Jun-qing, W.: Intrusion detection system with the data mining technologies. In: 2011 IEEE 3rd International Conference on Communication Software and Networks, pp. 490–492
3. Muda, Z., Yassin, W., Sulaiman, M.N., Udzir, N.I.: Intrusion detection based on K-Means clustering and Naïve Bayes classification. In: 2011 7th IEEE International Conference on IT in Asia (CITA), pp. 278–284
4. Panda, M., Patra, M.: Ensemble rule based classifiers for detecting network intrusion detection. In: 2009 International Conference on Advances in Recent Technology in Communication and Computing, ArtCom '09, pp. 19–22
5. Karthick, R.R., Hattiwale, V.P., Ravindran, B.: Adaptive network intrusion detection system using a hybrid approach. 2012 IEEE Fourth International Conference on Communication Systems and Networks (COMSNETS), pp. 1–7, Jan 2012
6. Muda, Z., Yassin, W., Sulaiman, M.N., Udzir, N.I.: A K-mean and Naïve Bayes learning approach for better intrusion detection. Inf. Technol. J. **10**(3), 648–655 (2011)
7. Barot, V., Toshniwal, D.: A new data mining based hybrid network intrusion detection model. In: 2012 IEEE International Conference on Data Science & Engineering (ICDSE), pp. 52–57
8. Ferid, D.M.D., Harbi, N.: Combining Naïve Bayes and decision tree for adaptive intrusion detection. Int. J. Netw. Secur. Appl. (IJNSA) **2**, 189–196 (2010)
9. Kasliwal, B., Bhatia, S., Saini, S., Thaseen, I.S., Kumar, C.A.: A hybrid anomaly detection model using G-LDA. In: 2014 IEEE International Advance Computing Conference, pp. 288–293
10. Mukherjee, S., Sharma, N.: Intrusion detection using Naïve Bayes classifier with feature reduction. C3IT-2012 Procedia Technol **4**, 119–128 (2012)
11. Thaseen, I.S., Kumar, Ch.A.: Improving accuracy of intrusion detection model using PCA and optimized SVM. CIT J. Comput. Inf. Technol. (Accepted manuscript, 2015)
12. Thaseen, I.S., Kumar, Ch.A.: Intrusion detection model using fusion of PCA and optimized SVM. In: 2014 International Conference on Computing and Informatics (IC3I), pp. 879-884, 27–29 Nov 2014
13. Martinez, A., Kak, A.: PCA versus LDA. IEEE Trans. Pattern Anal. Mach. Intell. **23**(2), 228–233 (2001)
14. Baek, K., Draper, B., Beveridge, J.R., She, K.: PCA vs. ICA: A comparison on the FERET data set. In: Proceedings of the fourth international conference on computer vision, pattern recognition and image processing, Durham, NC, USA, pp. 824–827, 8–14 Mar 2002
15. Delac, Kresimir, Grgic, Mislav, Grgic, Sonja: Independent comparative study of PCA, ICA and LDA on the FERET data set. Int. J. Imaging Syst. Technol. **15**(5), 252–260 (2005)
16. Liu, H., Setiono, R.: Chi2: feature selection and discretization of numeric attributes. In: Proceedings of IEEE 7th International Conference on Tools with Artificial Intelligence, pp. 358–391, 1995
17. Panda, M., Patra, M.R.: Network intrusion detection using Naïve Bayes. Int. J. Comput. Sci. Netw. Secur.**7**(12), (2007)
18. KDD. (1999). Available at http://kdd.ics.uci.edu/databases/-kddcup99/kddcup99.html
19. Panda, M., Abraham, A., Patra, M.R.: Discriminative multi nomial Naïve Bayes for network intrusion detection. In: 2010 Sixth International Conference on Information Assurance and Security, pp. 5–10
20. Tavallaee, M., Bagheri, E., Lu, W., Ghorbani, A.A.: A detailed analysis of the KDDCUP'1999 dataset. In: IEEE Symposium on Computational Intelligence in Security and Defence Application, 2009

An Intelligent Detection System for SQL Attacks on Web IDS in a Real-Time Application

K.G. Maheswari and R. Anita

Abstract Web application plays an important role in individual life as well as in any country's development. Web application has gone through a rapid growth in the recent years and their adaptation is moving faster than that was expected few years ago. Web based applications constitute various types of attacks, in that SQL injection is the worst threat which exploits the most web based applications. It is done by injecting the SQL statements as an input string to gain an unauthorized access to a database. However, in previous system the injection gives an access to some unauthorized users because the development of different approaches to prevent SQL injection still remains an alarming threat to web application. To address this problem an extensive review on different types of SQL injection attacks are presented in this paper. The web intrusion detection system is focused in this paper for threat detection and prevention by using renowned datasets. The strength and weakness of the entire range of SQL injection is estimated by addressing with mathematical models.

Keywords Web IDS · SQL · Attacks · Dataset · Protocol

1 Introduction

Intrusion detection is the software application to monitor the flaws in the network. It searches for the security violations, unauthorized access, identify information leakage and other malicious programs. IDS can also take some steps to deny access to intruders. Web applications are most widely used for providing services to the

K.G. Maheswari (✉)
Department of MCA, Institute of Road and Transport Technology,
Anna University, Erode, Tamil Nadu, India
e-mail: kgmaheswari@gmail.com

R. Anita
Department of EEE, Institute of Road and Transport Technology,
Anna University, Erode, Tamil Nadu, India
e-mail: anita_irtt@yahoo.co.in

© Springer International Publishing Switzerland 2016   93
V. Vijayakumar and V. Neelanarayanan (eds.), *Proceedings of the 3rd International Symposium on Big Data and Cloud Computing Challenges (ISBCC – 16')*,
Smart Innovation, Systems and Technologies 49, DOI 10.1007/978-3-319-30348-2_8

user like online shopping. Online banking and many other applications is designed in perspective of users [1]. Web sites and web applications are rapidly growing; however the complex business applications are now delivered over the web i.e. (HTTP). The web applications increased with the web hacking activity and also it get attacked by the internet worms. So there is a big question mark in damage done and firewall utility.

The web pages involved in drive-by-download attacks typically include malicious JavaScript code [2]. This code is usually obfuscated and it finger prints the visitors browser, identifies the vulnerabilities in browser itself. Hence the Web attacks such as URL interpretation attack, Input validation attacks, Session hijacking, Buffer overflow attack, XSS attack etc. The web IDS system is used to reduce the injection of these attacks and also it reduces the analysis time and increases the efficiency of the system.

2 SQL Attacks

The objective of the SQL injection attack is to shaft the database system into running harmful code that can reveal confidential information. Hence the SQL injection attacks are an unauthorized access of database [2, 3]. The attack repository is built to evaluate the classification scheme by collecting SQL injection attacks from various references. Table 1 explains the various types of attacks with the help of their scenarios.

The SQL injection attack is categorized in different names with different case depending on the system scenarios [3]. The attack is broadly classified based the orders of vulnerability types as first order injection attack, second order injection attacks and lateral injection attacks. It is also classified on the bases of blind and against database.

Table 1 Types of SQL attacks with their scenarios

Types of attacks	Working methods	Scenarios
Tautologies	SQL injection codes are injected into one or more conditional statement so that they are always evaluated to be true	SELECT * from users where id = '101' OR '1' = '1'
Logically incorrect queries	Using error messages rejected by the database to find useful data facilitating injected of the backend database	SELECT * from users where name = 'Lucia01'
Union queries	Injected query is joined with a safe query using the keyword UNION in order to get information related to other tables	SELECT salary from employee where empid = "union select * from employee"
Stored procedure	Many databases have built in stored procedures. The attackers execute these built in functions	Update department Set dept = 'abc'; SHUTDOWN;—where dept = 'aaa'

3 Related Works

Panda and Ramani [4] presented an approach which protects the web applications against SQL injection attacks. Some predefined methods are discussed and applied in the database to avoid attack in the web application login phase. Indrani Balasundaram et al. proposed a hybrid encryption to prevent the SQL injection where they used AES encryption and RSA Cryptosystem to make the authentication. Hasan Kadham et al. proposed mixed cryptography to encrypt database.

   Denial of service attacks are large scale co-operative attacks on the networking structure. It disables servers/victims from providing services to its clients. These attacks adversely affect the network badly. Therefore they must be detected on time. Introduces a classification technique which plays a significant role in detecting such intrusions but it takes significant classification time due to large number of features. It reduces its efficiency. So in order to improve efficiency or to reduce classification time, researcher provides relevant set of features for detection of DOS attacks. For this purpose, researcher is using NSL KDD dataset and analysis is performed on orange canvas V2.6.1 data mining tool.

4 Web Intrusion Detection System

Web Intrusion detection system is security programs to decide whether events and activities occurring in a web applications or network are legitimate [5]. The objective of web IDS is to identify intrusions with low false alarm rate and high detection rate while consuming minor properties. There are abundant issues in out dated web IDS including regular updating, low detection capability to unfamiliar attacks.

   In order to protect web applications from SQL injection attacks, there is a great need of a mechanism to detect and exactly identify SQL injection attack (Fig. 1).

   The SQL injection is the injection of threat by means of vulnerable code. This web attacks is analyzed and the results are shown in Fig. 2. The main aim of web

```
1339916791.542641 q5z39ByqpB1 192.168.245.1 51279 192.168.245.128 80 tcp
HTTP::SQL_URI_Injection_Attack SQL Attack from 192.168.245.1 to destination:
192.168.245.128 with Attack
string /dvwa/vulnerabilities/sqli/?id=1'&Submit=Submit - 192.168.245.1
192.168.245.128 80 - bro
Notice::ACTION_LOG 6 3600.000000
1339916811.243118 8CtiEfN7jG9 192.168.245.1 51284 192.168.245.128 80 tcp
HTTP::SQL_URI_Injection_Attack SQL Attack from 192.168.245.1 to
destination: 192.168.245.128 with Attack
string /dvwa/vulnerabilities/sqli/?id=1'+or+'1'='1&Submit=Submit - 192.168.245.1
192.168.245.128 80 -
bro Notice::ACTION_LOG 6 3600.000000 Γ
```

Fig. 1 SQL injection notice log

Fig. 2 Fuzzy featured set
versus fuzzy sets

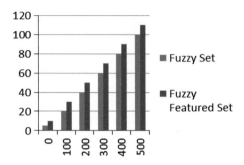

IDS is to detect the web attacks, especially the database corruption attack. For identifying these attacks the hybrid feature relevance algorithm is used by collaborating mRMR and by calculating information gain ratio. After feature selection process of the renowned datasets, the classification of attack data is organized by fuzzy cognitive map technique. The fuzzy system will generate the rule for prediction, which is used for the detection of wrong query.

4.1 Temporal Classification

The Temporal FCM Based Classifier is constructed using Fuzzy Temporal featured Set approach [6]. Generalization of featured set is done using higher detection approximation operators. Fuzzy decision table is a decision table with fuzzy attributes. The simple higher detection approximation operators of a fuzzy set is defined in for every $A = F(U)$, $F(U)$ is the fuzzy power set the residuated implicator, R describes a fuzzy similarity relation and At(u) corresponds to activation function. The tstart and tend defines the starting time and the ending time of the temporal concept. The higher detection approximation operation using generalization of fuzzy featured sets is

$$R \vartheta \alpha(A)(x) = \inf \vartheta (R(x,u), \alpha)$$
$$At (u) \leq \alpha$$
$$^{\wedge}\inf \vartheta (R(x,u), At (u)), \forall \times \in U \qquad (1)$$
$$At(u) > \alpha$$
$$\forall \, tstart < t < tend$$

The higher detection approximation hypothesis and fuzzy decision table are used to obtain fuzzy decision classes [7] for constructing the classifier. The higher detection approximations are used to define the decision classes. By taking a subset of attributes into consideration, the higher detection approximations are designed. The elementary sets will be the outcome of this stage from which the objects are classified into the decision classes. The decision classes are fixed as 0 and 1, which

Table 2 Fuzzy decision table

a/c	C_1	C_2	C_3	C_4	Dc
A_1	100	009	013	019	1
A_2	007	264	011	215	1
A_3	009	064	241	321	0
A_4	100	016	158	009	0

is defined in Table 2. The value 0 and 1 indicates that an attack belongs to category that web application is or not affected by threat.

5 Computational Results and Analysis

In order to illustrate the feasibility and effectiveness of the algorithm, we construct the detection mechanism by means of temporal data. Figure 2 explains the comparison analysis of fuzzy sets and fuzzy featured sets. The analysis of data is maintained by SPSS which originates the combination of rule generation for the prediction analysis, the data set is analyzed and managed by SPSS, the result is obtained as follows:

The SQL detection accuracy is managed by the Chi Square test by means of SPSS v20. The result of the test is analysed by means of significant value shown in Table 3 and Curve fit is statistically subjected by a graph in Figs. 3 and 4.

Table 3 Chi-Square tests

	Value	df	Asymp. Sig. (2-sided)
Pearson Chi-Square	0.804[a]	2	0.669
Likelihood ratio	0.814	2	0.666
Linear-by-linear association	0.119	1	0.730
N of valid cases	15		

[a]Poission-distributed random variable

Fig. 3 SQL injection in some duration

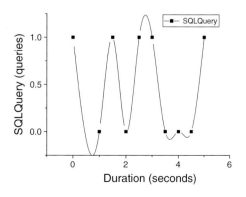

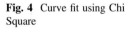

Fig. 4 Curve fit using Chi
Square

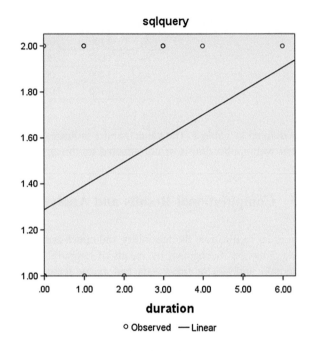

Table 4 SQL detection
accuracy

SQL attacks	Unprotected server	Protected server
Use of tautologies	Not detected	Detected
Additional SQL stmt	Not detected	Detected
Valid user input of (')	Query allowed	False positive

The value accuracy shows that it is greater than 0.5. This shows that the detection of SQL injection can be done by analyses of protocols in renowned Data. We subjected both the protected and the unprotected database instances to different types of SQL attacks namely tautology, piggy back and SQL statements, which is analysed in Table 4.

6 Conclusions

SQL injection is a common techniques hacker which employees to attack underlying databases. These attacks reshape the SQL queries, thus altering the behaviour of the program for the befit of the hackers. In our technique the automatic scanner tool is introduced for detecting the SQL injection attacks based on the WEB intrusion detection system. This IDS model captures the different types of injection attacks which results in the various performance metrics affecting database access. The feature selection methods which were newly proposed and tested with public data have shown improved performance in terms of the feature size and the classification accuracy. In particular, the result of temporal classification method demonstrated the most powerful and stable performance over all the other methods considered. As part of the future work, we plan to extend the implementation of the proposed architecture. We also try to implement and explore the possible functionalities to avoid explicit instrumentation of source code.

References

1. Ruse, M., Sarkar, T., Basu. S.: Analysis and detection of SQL injection vulnerabilities via automatic test case generation of programs. In: Proceedings of 10th Annual International Symposium on Applications and the Internet, pp. 31–37, 2010
2. Tajpour, A., Masrom, M., Heydari, M.Z., Ibrahim, S.: SQL injection detection and prevention tools assessment. In: Proceedings of the 3rd IEEE International Conference on Computer Science and Information Technology (ICCSIT'10), pp. 518–522, 9–11 July 2010
3. Bhoria, M.P., Garg, K.: Determining feature set of Dos attacks. Int. J. Adv. Res. Comput Sci. Softw. Eng. 3(5), (2013)
4. Panda, S., Ramani, S.: Protection of web application against SQL injection attacks. Int. J. Mod. Eng. Res. (IJMER) 3, 166–168 (2013) ISSN 2249-6645
5. Beniwal, S., Arora, J.: Classification and feature selection techniques in data mining. Int. J. Eng. Res. Technol. 1(6) (2012)
6. Zhao, S., Tsang, E.C.C., Chen, D., Wang, X.: Building a rule-based classifier—a fuzzy-rough set approach. IEEE Trans. Knowl. Data Eng. 22(5), 624–638 (2010)
7. Acampora, G., Loia, V.: On the temporal granularity in fuzzy cognitive maps. IEEE Trans. Fuzzy. Syst. (June 2011)

Real-Time Scene Change Detection and Entropy Based Semi-Fragile Watermarking Scheme for Surveillance Video Authentication

V.M. Manikandan and V. Masilamani

Abstract This paper proposes a semi-fragile video watermarking scheme suitable for data authentication of surveillance video. To authenticate any video it is not mandatory to watermark all the frames within it. Instead of that watermarking a few frames such that modification of those frames will lead to change the semantic is also sufficient. A novel scene change detection algorithm is proposed to detect the major scene changes in video and a few selected frames in each scene have been used for watermarking. The scene change is identified by a combined measure of weighted block based mean absolute frame differences and count of edge pixels. To watermark a specific frame an authentication code has been generated from the highest informative block based on the discrete cosine transform (DCT) coefficients. The authentication code is further inserted into the less informative blocks of same frame by a new watermarking algorithm based on pixel intensity swapping within a four neighborhood region. Information within a block of frame is measured in terms of entropy. Experimental study has done on video from standard video data set shows good performance. The comparative study of performance with existing technique has been done.

1 Introduction

Nowadays video surveillance is very much popular for monitoring the daily activities in places like banks, hospitals, retail shops etc. The advances in digital video technology and communication networks made that video surveillance

---

V.M. Manikandan (✉) · V. Masilamani
Computer Engineering, IIITD & M Kancheepuram, Chennai 600127, Tamilnadu, India
e-mail: vmkmanikandan@gmail.com

V. Masilamani
e-mail: masila@iiitdm.ac.in

© Springer International Publishing Switzerland 2016
V. Vijayakumar and V. Neelanarayanan (eds.), *Proceedings of the 3rd International Symposium on Big Data and Cloud Computing Challenges (ISBCC – 16')*,
Smart Innovation, Systems and Technologies 49, DOI 10.1007/978-3-319-30348-2_9

systems are more cost effective and flexible in use. The major advantage of video surveillance is that authorized persons can go through the recorded video to observe the activities happened in past and can be used as a proof or evidence to find crime, robbery or any other unusual incidents. Most of the organizations drastically reduced their manpower resources for security purpose by the replacement of surveillance video camera. The man power can be restricted to the locations where physical intervention is required and remaining security can be provided by the installation of surveillance video [1, 2]. Surveillance video can be efficiently kept in cloud storage provided by any reputed cloud storage provider with affordable cost [3]. Unfortunately the networking environment becomes unsafe due to the malicious activities by fraud people. Especially in the case of surveillance video, it may transmit from different locations to the cloud storage through internet [4, 5]. In between this, malicious people may be able to access the video content and can modify it according to their wish.

Video watermarking technique provides a very promising solution to ensure the authenticity of a video. In video watermarking unique information which is known as watermark will embed into video sequences, in future it can be used as proof of copy right or owner identification information or for data authentication. In video watermarking, watermark can be either directly inserted in raw video data or inserted during encoding process or implemented after compressing the video data [6, 7]. Based on the survival capability watermark against signal processing attacks, image or video watermarking schemes is categorized into robust, fragile and semi-fragile watermarking schemes [8–13]. Robustness, invisibility and real-time processing are the major three challenges in video watermarking. Invisibility and real-time performance are important for a watermarking scheme to be used for surveillance video data authentication [12]. Scene change detection is a crucial part in video watermarking algorithms and different methods are already proposed for this [14–18]. Real-time performance and stability of scene change detection is the main parameter to evaluate a scene change detection algorithm.

This research work contributes two key algorithms. Firstly a reliable scene change detection technique with real-time performance, secondly a semi-fragile watermarking scheme for video authentication. The proposed watermarking scheme is evaluated by peak signal to noise ratio (PSNR). The survival capability of watermark against noise addition, filtering, histogram equalization and contrast stretching has been shown experimentally with surveillance video data set.

2 Preliminaries

For better readability of the paper, a few essential topics are revisited in this section.

Entropy Entropy is a statistical measure of randomness that can be used to characterize the texture of the input image.

$$Entropy = - \sum_{k=0}^{255} \Pr(rk).Log_2\Pr(rk) \tag{1}$$

Here rk is the number of pixels with pixel intensity k and $\Pr(rk)$ is the probability of intensity rk be present in an image [19].

Mean Absolute Frame Difference (MAFD) Mean absolute frame difference between two image blocks IB_1 and IB_2 of size $S \times S$ can be computed as follows.

$$MAFD = \frac{1}{S^2} \sum_{i=1}^{S} \sum_{j=1}^{S} |IB_2(i,j) - IB_1(i,j)| \tag{2}$$

3 Proposed Scheme

Overview of the proposed watermarking scheme is shown in Fig. 1. Video frames from surveillance video will be analyzed to identify scene change detection and the frames with scene changes only will goes to watermarking module. For both scene change detection and watermarking we have proposed new algorithms. Proposed video authentication scheme has three major modules.

(1) Scene change detection (2) Authentication code generation and insertion (3) Watermark extraction and data authentication

3.1 Scene Change Detection

Algorithm 1: Scene change detection

Input : A video file consist of *VN* number of frames, *Video=(F₁, F₂,... F_VN)*
Output : Array of frame numbers indicating scene change, say *SceneID*
Step 1 : PrevFrame=F₁; C=2;
Step 2 : Convert *PrevFrame* to grayscale of size *RxC* pixels, say *PrevFrameG*;
Step 3 : While *C≤VN*
Step 4 :   Convert frame *Cᵗʰ* frame *F_c* to grayscale of size *RxC* pixels, say *CurrentFrameG*;
Step 5 :   Divide both *PrevFrameG* and *CurrentFrameG* into non-overlapping blocks of size *SxS*.
Step 6 :   Consider blocks with same index on *PrevFrameG* and *CurrentFrameG*, find the weighted sum (*Ws*) of absolute difference in number of edge pixels and mean absolute frame difference after normalization.
Step 7 :   If Ws>Th   // Th: predefined threshold value
Step 8 :   SceneID[k]=C; k=k+1; PrevFrame=CurrentFrameG;
Step 9 :   EndIf
Step 10:   C=C+1;
Step 11: EndWhile;
Step 12:Return *SceneID*;

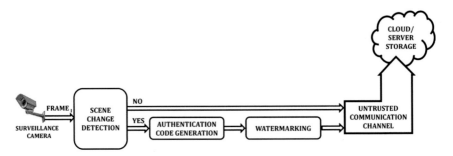

Fig. 1 Overview of proposed watermarking scheme

A video will consist of a sequence of frames, say $(F_1, F_2, ..., F_{VN})$. Practically it is not possible to watermark all the frames within a video file mainly due to time constraint. One more reason to avoid watermarking all video frames is that the visual quality of video will get degraded. In our proposed scheme, from the incoming video frame sequence major scene changes will identify by a novel algorithm. It is basically finding the difference between a reference frame with an incoming frame in terms of weighted block mean absolute frame differences and count of edge pixels. Initially the first frame will be considered as the reference frame and when next frames are coming for processing, the scene change detection algorithm will find a measure of dissimilarity with the current frame and the reference frame. Firstly both reference frame (*PrevFrame*) and current frame (*CurrentFrame*) need to be converted into gray scale images of size $R \times C$ and divide into non-overlapping blocks of size $S \times S$. The blocks with same index on *CurrentFrame* and *PrevFrame* will be considered and for each block we will get a mean absolute frame difference. Mean absolute frame difference (MAFD) between two blocks in different frames which are in the same index can be computed by Eq. (2). Next find the number edge pixels in each block of *CurrentFrame*, *PrevFrame* and find the absolute difference between these two. Both the measures are normalized between 0–1 before going for further computation. In our method we have assigned equal weight for both measures as *0.5*. If the sum of weighted measures is exceeding a threshold (*Th*), it shows a scene change from *PrevFrame* to *CurrentFrame*.

3.2 Authentication Code Generation and Insertion

Algorithm 2 describes the steps for watermarking a RGB video frame.

Algorithm 2: Watermark insertion

Input   : RGB video frame need to be watermarked, Say *F*
Output : Watermarked video frame *Fw*
Step 1 : Extract the blue color plane from *F*, say *BPlane* and divide it into non-overlapping blocks of size *SxS* and find the entropy of each block.
Step 2 : Generate authentication code from *N* blocks with highest entropy by comparing DCT coefficient values
Step 3 : For insertion of authentication code, blocks with least entropy will get preference. Consider one block *B* at a time until all the watermark bit get inserted.
Step 4 :   Based on *key* find a pixel positions in *B* and to insert *0* as watermark swap the centre pixel intensity value with minimum pixel intensity among the 4-neighbourhood and to insert *1* bit as watermark swap the centre pixel intensity value with maximum value among the 4-neighbourhood.
Step 5 : Combine watermarked *BPlane* with the unmodified red and green color planes of video frame to generate watermarked video frame *Fw*.

Firstly, separate red, green and blue color planes from the identified frames which need to be watermarked. Assume that *BPlane* represent the blue color plane extracted from the frame that only will consider for authentication code generation and watermarking. The key idea behind the authentication code generation is that an image frame will divide into non-overlapping blocks of size $S \times S$. After finding the entropy of blocks in *BPlane*, *N* blocks will be considered in the descending order of their entropy value for authentication code generation. The authentication codes generated from different blocks need to be concatenated for the final authentication code for a video frame. DCT values of *BPlane* blocks need to be analyzed for authentication code generation. To generate the authentication code from the DCT coefficient matrix *D* of size $S \times S$, compare the values in $D\,(I, J)$ and $D\,(I, J + 1)$ here $1 \le I \le S$ and $1 \le J \le S - 1$. If $D\,(I, J) > D\,(I, J + 1)$ then *1* will produce as code and otherwise *0*. For *N* number of blocks the code will generate and concatenate one by one in predefined order. This will produce final authentication code for the frame. To insert the watermark we propose a new method which will ensure that the entropy value of a block will not get updated after inserting the authentication. Watermark insertion is dependent on the key value (*key*) and the insertion will start with the block that has least entropy. The *key* is a sequence of index pairs with the restriction that any two pair of index should not have a common 4-neighborhood.

3.3 Watermark Extraction and Data Authentication

Algorithm 3: Video frame authentication

Input : Watermarked RGB video frame need to be watermarked, Say *Fw'*
Output : Similarity measure which indicate the authentication status
Step 1 : Extract blue color plane from *Fw'*, say *BPlane'* and divide it into non-overlapping blocks of size *SxS*, find the entropy of each block
Step 2 : Generate authentication code from *N* blocks with highest entropy by comparing DCT coefficient values
Step 3 : Consider blocks in their descending order of entropy. One block *B* at a time, until all the [*N*S*(S-1)*] watermark bits are extracted
Step 4 : Based on *key* find a pixel positions in *B* and compare the pixel intensity at centre position with pixel intensity of 4-neigbourhoods and if the value at centre pixel intensity is minimum among all extract watermark bit as *0* and if it is maximum extract *1* as watermark bit.
Step 5 : Determine the authenticity of video frame by comparing generated authentication code with extracted watermark bit sequence.

$$Similarity = \frac{Number\ of\ bits\ extracted\ correctly}{Total\ number\ of\ bits\ in\ authenication\ code} \tag{3}$$

4 Experimental Study

We have done experimental study for both scene change detection algorithm and watermarking scheme with surveillance video data set downloaded from VIRAT [20].

4.1 Analysis of Scene Change Detection Algorithm

Performance of the proposed scene change detection algorithm has been evaluated with two parameters.

Average success rate It can be measured as ratio of number of successfully detected scene changes and actual number of scenes in the video. The threshold to find to weight difference between consecutive frames can be adjusted to capture a very small change in the scene also. Practically speaking we prefer to select a threshold value between 15 and 20 which will be able to find the normal scene changes.

Stability Sensitivity to the noise in the video stream. To evaluate the stability of proposed scheme we have added heavy amount of salt and pepper noise and gaussian noise (noise density = 0.1) to the video frames and tried to identified the scene changes and we observed the stability of scene change detection algorithm.

Fig. 2 Block size versus
PSNR

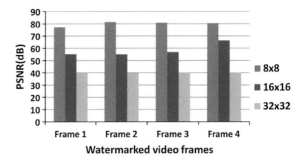

Fig. 3 Comparison of PSNR

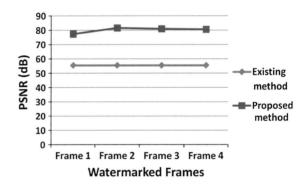

4.2 Analysis of Proposed Watermarking Algorithm

The visual quality of the proposed video watermarking scheme is measured by Peak
Signal to Noise ratio (PSNR) between the original video frame and watermarked
video frame. The modification on watermarked frame purely dependent on the
length of the authentication code. Because to insert a long authentication code
probably we need to modify many pixel values from the selected frame. From a
single block of size $S \times S$, an authentication code of length $S * (S - 1)$ will be
generated. We took $R = 512$, $C = 512$ and $S = 8$ for all the experimental study.
The PSNR value obtained for varying block sizes is shown in Fig. 2 and note that
all PSNR values are measured in decibel (dB). The number of blocks selected for
authentication code generation is fixed as 2. We have achieved higher PSNR value
when block size is set as 8×8. Robustness and visual quality of watermarked
frames are considered for comparative study. The comparative study has been
conducted against [12], which proposed with the objective for surveillance video
authentication. The PSNR value obtained for existing method and proposed method
after watermark insertion (same number of watermark bits) is shown in Fig. 3.

The similarity of extracted watermark and inserted watermark obtained for
existing method [19] and proposed method after common signal processing operations are shown in Table 1.

Table 1 Comparison between proposed and existing method

Operations	Salt & pepper	Speckle noise		Average filtering		Gaussian filtering		Histogram equalization	Contrast stretching
Parameters	Noise density = 0.1	Noise density = 0.1	Noise density = 0.1	3 × 3	5 × 5	3 × 3	5 × 5	–	–
Existing method [12]	0.47	0.42	0.47	0.54	0.46	0.52	0.48	0.43	0.47
Proposed method	0.50	0.42	0.38	0.54	0.49	0.47	0.51	0.43	0.52

5 Conclusion

This research work proposed a novel scene change detection algorithm that can be used for real-time applications and a watermarking scheme for data authentication of video data. Scene change detection is distinguished from traditional algorithm we have introduced weighted block based measure on absolute mean frame difference and edge pixel's count. The watermark information will be generated from the video frames itself based on the DCT coefficient values of selected blocks. Entropy has been used as the criteria for selection blocks for watermark generation and watermark insertion. To reduce the visual quality degradation watermarking has been done in blocks with less entropy. The interesting thing is that proposed watermarking scheme will not change the entropy of a particular block or entire image. Experimental study shows that the proposed scheme works better than the existing methods.

References

1. Ko, T.: A survey on behavior analysis in video surveillance for home-land security applications. In: Applied Imagery Pattern Recognition Workshop (IEEE), pp. 1–8 (2008)
2. Birnstill, P., Ren, D., Beyerer, J.: A user study on anonymization techniques for smart video surveillance. In: 2015 12th IEEE International Conference on Advanced Video and Signal Based Surveillance (AVSS). IEEE, pp. 1–6 (2015)
3. Rodriguez-Silva, D., Adkinson-Orellana, L., Gonz'lez-Castano, F., Gonz'lez-Martinez, D., et al.: Video surveillance based on cloud storage. In: Fifth International Conference on Cloud Computing (CLOUD), IEEE, pp. 991–992 (2012)
4. Subashini, S., Kavitha, V.: A survey on security issues in ser-vice delivery models of cloud computing. J. Network. Comput. Appl. **34**(1), 1–11 (2011)
5. Wang, H., Wu, S., Chen, M., Wang, W.: Security protection be-tween users and the mobile media cloud. Commun. Mag. IEEE **52**(3), 73–79 (2014)
6. Chang, X., Wang, W., Zhao, J., Zhang, L.: A survey of digital video watermarking. In: Seventh International Conference on Natural Computation, pp. 61–65 (2011)
7. Rao, Y.R., Prathapani, N.: Robust video watermarking algorithms based on svd transform. In: 2014 International Conference on Information Communication and Embedded Systems, pp. 1–5 (2014)
8. Lee, S.-J., Jung, S.-H.: A survey of watermarking techniques applied to multimedia. In Proceedings of IEEE International Sym-posium on Industrial Electronics, vol. 1, pp. 272–277 (2001)
9. Jayamalar, T., Radha, V.: Survey on digital video watermarking techniques and attacks on watermarks. Int. J. Eng. Sci. Technol. **2**(12), 6963–6967 (2010)
10. Bianchi, T., Piva, A.: Secure watermarking for multimedia content protection: A review of its benefits and open issues. IEEE. Sign. Proces. Mag. **30**(2), 87–96 (2013)
11. Wang, C.-C., Hsu, Y.-C.: Fragile watermarking scheme for h. 264 video authentication. Opt. Eng. **49**(2), 027003–027003 (2010)
12. Xu, D., Zhang, J., Pang, B.: A digital watermarking scheme used for authentication of surveillance video. In: 2010 International Conference on Computational Intelligence and Security, pp. 654–658 (2010)

13. Fallahpour, M., Shirmohammadi, S., Semsarzadeh, M., Zhao, J.: Tampering detection in compressed digital video using water-marking. IEEE Trans. Instrum. Meas. **63**(5), 1057–1072 (2014)
14. Radke, R.J., Andra, S., Al-Kofahi, O., Roysam, B.: Image change detection algorithms: a systematic survey. IEEE Trans. Image Process. **14**(3), 294–307 (2005)
15. Bruzzone, L., Prieto, D.F.: Automatic analysis of the difference image for unsupervised change detection. IEEE Trans. Geosci. Rem. Sens. **38**(3), 1171–1182 (2000)
16. Del Fabro, M., Böszörmenyi, L.: State-of-the-art and future challenges in video scene detection: a survey. Multimedia Syst. **19**(5), 427–454 (2013)
17. Reddy, B., Jadhav, A.: Comparison of scene change detection algorithms for videos. In 2015 IEEE International Conference on Advanced Computing and Communication Technologies (ACCT), pp. 84–89 (2015)
18. Mas, J., Fernandez, G.: Video Shot Boundary Detection Based on Color Histogram. Notebook Papers TRECVID2003, Gaithersburg, Maryland, NIST (2003)
19. Gonzalez, R.C.: Digital Image Processing. Pearson Education India (2009)
20. Video data set.: Available: http://www.viratdata.org

Secure Video Watermarking Technique Using Cohen-Daubechies-Feauveau Wavelet and Gollman Cascade Filtered Feedback Carry Shift Register (F-FCSR)

Ayesha Shaik and V. Masilamani

Abstract The enormous growth of digital data in the form of digital videos has brought many issues for the copyright holders. Due to advanced technologies available, the legal data can be distributed illegally and modified easily, which creates a problem for the owners of the content and may lead to financial degradation of that organization. These problems arise as unauthorized users can also exchange the copyrighted content illegally over peer-to-peer networks without any legal issues. Hence, authentication of digital data and identifying authorized user is very essential. So, in this paper a secure watermarking technique for video has been presented for user identification and data authentication. In this paper Cohen-Daubechies-Feauveau (CDF) 9/7 lifting wavelet has been used and it is observed that the results are promising. The watermarking has been performed on LH5 and HL5 sweatbands that consists of the most significant features of the video. In order to authenticate the video, watermarking the most significant information is important. The reason is that, any modification done to the watermarked video will modify the significant information. Gollman cascade filtered-feedback carry shift register (F-FCSR) is used for generating secure random numbers that will be used for watermarking.

Keywords Watermarking · Secure · Video · CDF97 · F-FCSR · Gollman · Cascade

A. Shaik (✉) · V. Masilamani
Indian Institute of Information Technology Design & Manufacturing Kancheepuram, Chennai 600127, India
e-mail: ayeshanoormd@gmail.com

V. Masilamani
e-mail: masila@iiitdm.ac.in

© Springer International Publishing Switzerland 2016   111
V. Vijayakumar and V. Neelanarayanan (eds.), *Proceedings of the 3rd International Symposium on Big Data and Cloud Computing Challenges (ISBCC – 16')*,
Smart Innovation, Systems and Technologies 49, DOI 10.1007/978-3-319-30348-2_10

1 Introduction

These days the rapid growth in the amount of digital data is increasing the security and authenticity problems. Data authentication is very essential for many applications in the networked environment. The attackers can easily modify the content, so the authorized user needs to make sure that the data has not been modified. Watermarking is a technique of inserting data called watermark into the digital data, this watermark will hold the information about the content provider or authorized owner for proving ownership. Recently, many watermarking techniques have been proposed by researchers in this area to solve these security and authentication issues. For authenticating digital video, a very few watermarking techniques are available. The recent developments in multimedia technologies made the unauthorized user to produce illegal copies easily, which is a threat for the content provider or copyright holder. The software implementation of 3D-DWT (DWT) watermarking technique for video using a secret key is proposed in [1]. The applications of video watermarking can be listed as copy control, broadcast monitoring, fingerprinting, video authentication and copyright protection. The purpose of these applications is preventing unauthorized copying, identifying the video that is being broadcast, tracing un-authorized user, verifying the data has been altered or not and proving ownership respectively [2].

A semi-fragile compressive sensing watermarking algorithm has been proposed in [3]. In this technique, for sensing I frame's discrete cosine transform (DCT) coefficients, the compressive sensing watermark data are generated from the block compressive sensing measurements. Another video watermarking method that uses scene change detection and adjustment of pixel values is proposed in [4]. In [5], the macroblock's and frame's indices will represent the watermark signals and embedded into non zero quantized DCT values of the blocks. Various types of attacks on the watermark has been detailed and discussed in [6]. Usually, most of the video watermarking techniques will apply DCT or haar wavelets for watermarking. In our work, CDF 9/7 wavelet is used for video watermarking since it has highest compression efficiency compared to DCT and conventional haar wavelets. CDF 9/7 wavelet is an effective biorthogonal and a secure random number generator known as Gollman cascade feedback carries shift register is used. The proposed video watermarking is secure, reduces visual quality degradation and suitable for data authentication. To achieve this, less number of coefficients from a significant portion of the video are used for watermarking.

The rest of the paper is organized as follows. Section 2 provides literature survey. Section 3 explains proposed watermarking technique in detail. Section 4 gives results and analysis followed by Sect. 5 presents conclusion.

2 Literature Survey

A video authentication scheme using DCT and dither modulation (DM) for security and copyright protection is proposed in [7]. Dither modulation is a quantization index modulation (QIM) scheme. For authentication of digital video, a discrete wavelet transform based blind digital watermarking method is proposed in [8], based on the scene changes different parts of single watermark is embedded in different scenes of video. General attacks for content-based authentication schemes have been discussed in [9] and they showed that the existing content-based video stream authentication schemes cannot detect content-changing attacks. An algorithm for multicast video streaming is proposed in [10]. This technique at the transmitter side combines a digital signature with a hash-chain which is pre-computed and is embedded into the video stream. In this technique received blocks can be authenticated on-the-fly and the bandwidth overhead introduced is negligible.

A fragile digital watermarking scheme [11], where the watermark is made up of time information and camera ID, the secret key based on features of the video is embedded into the B frame chromatic components. This technique can detect frame cut, foreign frame insertion and frame swapping and the watermark information picked up from the carrier can be displayed on the video while the video is being played. A robust distance based watermarking for digital video is proposed in [12], the distances are calculated from the addresses, from the values of the watermark and these distances are used during embedding. Each watermark will have its own pattern of distances at different possible lengths of distance bits and 1-level DWT has been used for watermarking. A still image watermarking technique is extended to raw video by considering the video as a set of still images is proposed [13] in DFT domain.

Algorithm 1 WMG(*Original_video, Watermarked_video*)

1: **Input:** *Original_video*
2: **Output:** *Watermarked_video*
3: For the *Original_video* perform scene detection using equation (1).
4: Apply 5-level CDF9/7 wavelet on all detected scenes and decompose into sub bands.
5: Select LH5 and HL5 sub bands.
6: $T = 0.5 \times N$, where N is total number of coefficients in LH5 and HL5 subbands
7: From both the sub bands select T highest coefficients for watermarking, say $C_i = (c1_i, c2_i, ..., cT_i)$, where i denotes the number of detected scenes
8: [W, P] = GEN(pwd) //call algorithm2
9: Now watermark $C_i = (c_{ij})$ using $W = (w_1, w_2, .., w_T)$ and $P = (p_1, p_2, ..., p_T)$ as shown.
10: **if** $w_i' = 0$ **then**
11: $\quad c_{ij}' = c_{ij}$
12: **else**
13: $\quad c_{ij}' = c_{ij} + \alpha \times$ pi,
$\quad\quad$ where i denotes the ith coefficient in C and j denotes the scene that is being watermarked and α denotes user defined embedding strength.
14: **end if**
15: Now replace c_{ij} by c_{ij}' in the bands modified.
16: Now apply Inverse 5-level CDF9/7 wavelet on wavelet of the scene.
17: Resultant is watermarked video.

---

3   Proposed Work

In this section, Watermarking algorithm is explained in detail. This section is divided into two sections as watermark generation and watermark embedding.

3.1   Watermark Generation

A secured password is generated using key generator. This password is known to the sender and the receiver. From this password, generate two random keys K1 and K2, using Gollman Cascade F-FCSR. Now using K1 generate a sequence of random binary watermark W, and using K2 generate another sequence of random numbers P.

3.2   Watermark Embedding

In this work, the original video has to undergo scene change detection. After detecting the scene changes, these scenes are used for the watermarking. Scene detection can be performed by finding the histogram difference between the consecutive frames as shown in Eq. (1).

$$D_i = \sum_{k=1}^{N} (|H_i - H_{i-1}|) \tag{1}$$

where i denotes frame number, N denotes number of bins in the histogram, H_i and H_{i-1} correspond to the histograms of ith and $i - 1$th frames respectively, and D_i denotes difference between the histograms. Now select the frames that are having the difference $D_i > TH$ as the scene for the watermarking. *TH* denotes user defined threshold value.

Algorithm 2 GEN(*pwd*)

1: *pwd* is given as a key to random number generator for generating K1 and K2.
2: K1 is used as a seed to Gollman Cascade F-FCSR for generating a binary watermark W of size T.
3: K2 is used as a seed to Gollman Cascade F-FCSR for generating sequence of random numbers P of size T.

After detecting the scenes, apply CDF 9/7 lifting wavelet on them. Select LH5 and HL5 subbands for watermarking to reduce visual quality degradation and data authentication. From these coefficients, select highest coefficients to embed the watermark. Small change to the frame will lead to the large change in highly significant portion of the frame. So, by using the significant coefficients we can verify any modification has been done to the frame or not for data authentication.

Algorithm 3 Extract($Watermarked\_Video$, pwd, C_i)

1: Perform scene detection for watermarked video, using equation (1).
2: Apply 5-level CDF9/7 wavelet on all detected scenes and decompose into sub bands.
3: Select $LH5$ and $HL5$ sub bands.
4: $T = 0.5 \times N$, where N is total number of coefficients in LH5 and HL5 subbands.
5: From both the sub bands select T highest coefficients for watermarking, say $C_i' = (c_{1i}', c_{2i}', \ldots, c_{Ti}')$, where i denotes the index of detected scenes
6: [W, P] = GEN(pwd)
7: Now extract watermark $pi' = \frac{c_{ij}' - c_{ij}}{\alpha}$
   where i denotes the ith coefficient in C and j denotes the scene that is being watermarked and α denotes user defined embedding strength.
8: Find normalized correlation between inserted and extracted watermarks.
9: If the correlation is greater than the user defined threshold then data will be authenticated; otherwise not authenticated.

Now embed the watermark generated into the selected coefficients. If the corresponding watermark bit W_i is zero, then embed the corresponding random number P_i with the embedding strength α (user defined) into the selected coefficients. Otherwise, no change is done to the selected coefficient. Apply inverse CDF 9/7 lifting wavelet to the subbands. The resultant is the watermarked video. The block diagram of the proposed method is given in Fig. 1. The algorithm for watermarking the video is explained in detail in Algorithm 1. The watermark generation algorithm is discussed in Algorithm 2. The extraction algorithm for video is explained in Algorithm 3.

Fig. 1 Proposed watermarking scheme

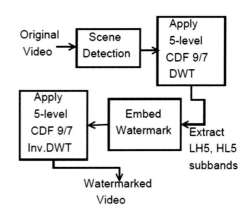

4   Results and Analysis

The proposed method has been tested using videos given in [14] (Container, Foreman, Sample and VIRAT data set videos). The proposed method watermarked video shows less bit error rate and plotted in Fig. 2. The PSNR values for the watermarked frames for the proposed technique and for the technique in [1, 15] is plotted in Fig. 3. It is evident that the proposed method outperforms the watermarking methods provided in [1, 15].

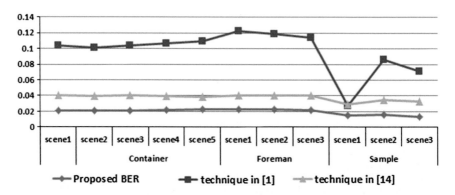

Fig. 2 BER for proposed, and for the technique given in [1, 15]

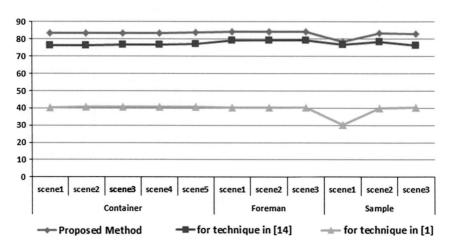

Fig. 3 PSNR for proposed, and for the technique given in [1, 15]

5 Conclusion

In this paper, a secure video watermarking using CDF 9/7 lifting wavelet and Gollman cascade F-FCSR is proposed. The CDF 9/7 lifting wavelet has a good compression efficiency compared to conventional convolution-based wavelets and DCT. The use of this wavelet will give most significant features of the frame. The watermarking has been performed only on the detected scenes and the watermark used is a random binary watermark. The watermark is generated using Gollman cascade F-FCSR, which is a secure random number generator to make our proposed watermarking scheme secure. The experimental results on video standard database shows that the proposed method outperforms.

References

1. Shoaib, S., Mahajan, R.C.: Authenticating using secret key in digital video watermarking using 3-level DWT. In: IEEE International Conference on Communication, Information and Computing Technology (ICCICT), pp. 1–5 (2015)
2. Doerr, G., Dugelay, J.-L.: A guide tour of video watermarking. Sig. Process. Image Commun. **18**(4), 263–282 (2003)
3. Zhao, H., Lei, F.: A novel video authentication scheme with secure cs-watermark in cloud. In: IEEE International Conference on Multimedia Big Data (BigMM), pp. 294–299 (2015)
4. Venugopala, P.S., Sarojadevi, H., Chiplunkar, N.N., Bhat, V.: Video watermarking by adjusting the pixel values and using scene change detection. In: IEEE 2014 Fifth International Conference on Signal and Image Processing (ICSIP), pp. 259–264 (2014)
5. Fallahpour, M., Shirmohammadi, S., Semsarzadeh, M., Zhao, J.: Tampering detection in compressed digital video using watermarking. IEEE Trans. Instrum. Meas. **63**(5), 1057–1072 (2014)
6. Voloshynovskiy, S., Pereira, S., Pun, T., Eggers, J.J., Su, J.K.: Attacks on digital watermarks: classification, estimation based attacks, and benchmarks. IEEE Commun. Mag. **39**(8), 118–126 (2001)
7. Kadam, B.D., Metkar, S.P.: Digital video watermarking based on dither modulation. In: Annual IEEE India Conference (INDICON), pp. 1–6 (2014)
8. Chetan, K.R., Raghavendra, K.: DWT based blind digital video watermarking scheme for video authentication. Int. J. Comput. Appl. **4**(10), pp. 19–26 (2010)
9. Lo, S.-W., Wei, Z., Ding, X., Deng, R.H.: Generic attacks on content-based video stream authentication. In: 2014 IEEE International Conference on Multimedia and Expo Workshops (ICMEW), pp. 1–6 (2014)
10. Chessa, S., Di Pietro, R., Ferro, E., Giunta, G., Oligeri, G.: Mobile application security for video streaming authentication and data integrity combining digital signature and watermarking techniques. In: IEEE 65th Vehicular Technology Conference, VTC2007-Spring, pp. 634–638 (2007)
11. Xu, D., Zhang, J., Pang, B.: A digital watermarking scheme used for authentication of surveillance video. In: IEEE International Conference on Computational Intelligence and Security (CIS), pp. 654–658 (2010)
12. Abdul-Ahad, A.S., Lindén, M., Larsson, T., Mahmoud, W.A.: Robust distance-based watermarking for digital video. SIGRAD **65** (2008)

13. Caldelli, R., Barni, M., Bartolini, F., Piva, A.: A robust frame-based technique for video watermarking. In: 10th IEEE European Signal Processing Conference, pp. 1–4 (2000)
14. Available: http://media.xiph.org/video/derf/
15. Majumder, S., Das, T.S., Sarkar, S., Sarkar, S.K.: SVD and lifting wavelet based fragile image watermarking. Int. J. Recent Trends Eng. **3**(1), 97–100 (2010)

Part II
Wireless Sensor Network

Quality Measurements of Fruits and Vegetables Using Sensor Network

Amol Bandal and Mythili Thirugnanam

Abstract In India most of the population survival is based on agricultural products. All the business and organizations that make, display, transport or prepare food for sale they will need to check food quality. A quality assessment system is required for farmers, the customer also in the fruit trading centers, to ensure quality of fruits and vegetables. If the fruits, vegetables are rotten, then a quality assessment system should intimate what are the items are not suitable to eat and what items we can keep in the storage so that other items should not get rotten. It will help farmers also, to classify fruits and vegetables as per quality and can easily do pricing of those products. This paper gives an overview about how wireless sensor network is designed and used to solve different problems related to agriculture, such as soil monitoring, Irrigation handling, etc. The same idea is extended to check the quality of agricultural products. Two methodologies used for assessing the quality of foods are destructive and non destructive. In case of destructive method we need to cut and then monitor internal section or take out the juice of fruit. A non destructive method just checks quality based on the external features like weight, color, appearance, etc. Some of the techniques like computer vision, image processing, hyper spectral imaging, etc. are used to assess food quality. As these methods are based on only external features of food item, there is a necessity to bring more accuracy in assessing the quality of food items. With this intention, this work aims to propose an automated framework using a multi sensor network for predicting quality of fruits and vegetables. To increase the availability of data related to research projects, results received from the described framework will be deployed on agriculture cloud.

Keywords Food safety · Multi-sensor network · Agri-cloud

A. Bandal (✉) · M. Thirugnanam
School of Computing Science and Engineering, VIT University, Vellore,
Tamilnadu, India
e-mail: amolbd1987@gmail.com

M. Thirugnanam
e-mail: tmythili@vit.ac.in

© Springer International Publishing Switzerland 2016   121
V. Vijayakumar and V. Neelanarayanan (eds.), *Proceedings of the 3rd International
Symposium on Big Data and Cloud Computing Challenges (ISBCC – 16'),*
Smart Innovation, Systems and Technologies 49, DOI 10.1007/978-3-319-30348-2_11

1 Introduction

In Indian scenario most of the population based on agriculture. Sometimes unknowingly spoiled or rotten food item is the part of storage then it will spoil other food items. Fruits and vegetables are marketable products for the increasing economic value of industries as well as satisfying customer demands. Farmers as well as a wholesaler or food organization must know the details of the maturity, evolution to decide how and when to sell each batch of product.

In recent years, almost in all places there is a deployment of sensor networks. Different applications like habitat monitoring, environmental monitoring, structural health monitoring, etc. are using wireless sensor network. Wireless sensor network it is nothing but a combination of 3 things: Sensing, CPU, i.e. processing unit and Radio for transmission of data. As soon as people start understanding the importance of sensor network hundreds of applications started to use it. Whenever it comes to the wireless device, we mainly think about cell phones, laptops, etc. Wireless sensor networks use small, low-cost embedded devices for a wide range of application. To implement it, we don't need any pre-existing infrastructure. In Agriculture scenario, we need sensors to check weather condition, soil information. In smart Irrigation system water level monitoring is done using sensor deployed in the field.

Literature survey is given in the following sections. First review gives idea about how smart agriculture is using wireless sensor network. To know environmental conditions different sensors like temperature, Humidity, pH, etc. are used. Second review gives idea about how the different non destructive methods such as machine vision, image processing, etc. assuring quality check. The last part of literature survey describes biological factors and attributes on which quality of food grains or vegetable is depending.

2 Literature Review

El-kader and El-Basioni [1] implemented wireless sensor network in cultivating the potato crop in Egypt. A literature review was carried out to understand the usage of different technology and sensors used in smart agriculture. Sensor nodes will contain sensors for temperature, humidity, light intensity, soil pH, and soil moisture. They have used sensor board such as MTS400 having sensors for temperature, relative humidity, light intensity, and also barometric pressure. Potato crop development phases can be monitored using above different factors. Potato crop modelling can be used as decision tool for farmers to do irrigation scheduling, fertilization scheduling and other plating practices scheduling which helps to improve potato crop and save of resources such as irrigation water and fertilizers, and this modelling can be efficiently and easily done by deploying the sensor nodes in the crop field which sense the required parameters and send it to the user on real-time where he can analyze this data, draw a complete accurate picture of the field characteristics, and take the suitable decision in the suitable time.

Majone et al. [2] describes how WSN deployment is done to monitor soil moisture dynamics. Using 135 soil moisture and 27 temperature sensor soil dynamics are tested. As per the soil dynamics, i.e. pH level and moisture, irrigation schedule is decided for apple trees. Test-bed implementation [3] of a wireless sensor network for automatic and real-time monitoring of soil and environmental parameters influencing crop yield is presented. In precision agriculture, continuous monitoring of sensor data at every minute may not be always needed. Instead, the data may be monitored on an hourly basis or at different times of the day, e.g., morning, noon, afternoon and evening. This, in turn, helps in conserving the battery power of sensor nodes. For this test bed implementation Mesh Network of sensors is used. These nodes have a bi-directional radio transceiver through which data and control signals are communicated wirelessly in the network and nodes are generally battery operated. Nandurkar et al. [4] designed and developed precision agriculture system. The main aim of this system is to give farmer information about temperature and moisture and as per farmer will regulate water supply. They used LM-35 temperature sensor, moisture sensor. Microcontroller ATMEGA 16 and using UART data of temperature and moisture is displayed on the LCD panel of the microcontroller.

Sakthipriya [5] designed and developed an effective method for crop monitoring. The main motive behind this work is to increase production of rice crop by automated control of water sprinkling. The decision is taken based on the information received from leaf wetness, soil moisture, soil pH sensor. Based on the value of soil pH action is taken to start or stop the water sprinkler. The system is designed with the help of MDA300 sensor board, micaZ mote, MIB510 Gateway.

Wireless sensor network for monitoring an agricultural environment is designed by Patil et al. [6]. Smart weather station is designed using a microcontroller to collect the value of the temperature, relative humidity, water level and soil moisture. ZigBee standard is used to send data over the wireless network. LM35 temperature sensor, SH220 humidity sensor, soil moisture sensor, WL400 water level sensor along with 89C52 microcontroller is used to monitor the environment. Chandraul and Singh [7] has simulated and experimented Agri-cloud in Indian scenario. They have stored data on the MAD-cloud, according to coordinate, then on physical and chemical requirement related to a particular crop.

3 Food Quality Check

This section gives overview about the different methods used to check the quality of food items or vegetables. Mainly nowadays image processing and computer vision widely used for food quality check. What are the advantages and disadvantages of these techniques are discussed here.

An app for safe food named as "Foodsniffer" is developed by the Greek research institute (NCSR-DEMOKRITOS 2014) with the help of 9 partners from 7 European countries. The smart phone app works in combination with a small electronic sensor. With this, the user can test a few simple handling products such as grapes, wine, beer and milk by placing a drop on it. The result appears than after 15–30 min on the screen. The Foodsniffer equipment uses potentially cheap disposable sensor chips and the detection is based on microscopic optical interferometry in a silicon chip suitable for mass production. AL-Marakeby and Aly [8] developed a framework for quality inspection of food products. Sorting tons of fruits and vegetable manually is time consuming. In this research a vision-based sorting system is developed to increase the quality of food products. The image of the fruit or vegetable is captured and analyzed. Four different systems for different food products are developed, namely apples, tomatoes, eggs and lemons. Acceptance of the fruit or vegetable it is based on different factors like color, crack edge detection in the case of eggs. The edge detection and segmentation methods are used to detect different color spots on the surface of fruits.

Parmar et al. [9] presented a unified approach for food quality evaluation using machine vision and image processing. Image of fruits or grains are captured and color value is converted into gray-scale. Object identification and measurement is done based on the external features, size, shape, color and texture. CCD camera and LED light panel is used to acquire the images. A frame grabber is used to do analog to digital conversion. Seng and Mirisaee [10] proposed a fruit recognition system based on the external features like color, shape and size. K-Nearest neighbor classification algorithm is used to classify captured images and clusters are formed. Al Ohali [11] designed and implemented date fruit grading system based on the computer vision. He collected 400 samples and extracted RGB images of the date fruit. Based on the external features, grading is defined as grade 1, grade 2 and grade 3. The System is tested using back propagation neural network. Kuswandi and Wicaksono [12] has given an overview of smart packaging with the help of chemical and biosensors. Smart packaging is done based on time temperature indicator, microbial spoilage sensor, leakage sensor, etc. A review is given on fruits and vegetable quality inspection with the help of hyperspectral imaging by Pu et al. [13]. Along with the external quality attributes like color, size, shape, etc. biochemical components can be measured with the help of hyperspectral imaging principle. High spatial and spectral resolutions equipment are required to do hyperspectral imaging.

As per the above mentioned papers there are some disadvantages of using these methodology's are:

1. Collected Attributes that are taken depends on the external features of food items, e.g., color, shape, etc. [8].
2. Object identification, being considerably more difficult in unstructured scenes, requires artificial lightning.
3. Most of the equipment used in image processing is too expensive, especially for those with high spatial and spectral resolutions used in hyperspectral imaging [13].

4. Spectral variations due to morphological changes of most fruits and vegetables (such as, round or cylinder-shaped objects) diminish the power of models; it might deteriorate the classification of food.
5. Some of the techniques are taking more time for predicting quality [14].
6. Some of the edge detection techniques that are only useful to predict only damaged surfaces.

4 Biological Food Quality Check

Every food Establishment uses, processes, and sells food in different ways. However, the general issues and key principles of food safety remains the same, whatever the style of the operation. Food poisoning is a serious health problem. It can cause severe illness and even death. Food poisoning can seriously damage the reputation of a business, damage the reputation of the food industry. To avoid mentioned issues in Sect. 3 related to image processing and computer vision we need to have some other way to predict quality with reference of proper attributes. Abbott [15] has summarized food and fruits related quality attributes are mentioned: Sensory attributes (e.g. Appearance, aroma, etc.), Nutritive values, Chemical constituents (e.g. Water level), Mechanical properties (e.g. Fruit texture, compression on surface), Optical properties (reflectance, transmittance, absorbance or scatter of light by the product).

Microbial contamination is the most important problem that needs to be solved. Food containing microbial contamination in terms of pathogenic microorganism can be extreme hazards to human health. Traditionally, quality control test includes odour sampling. Fungal spoilage is an important issue observed in case of fruits and bakery products. Most fruits and vegetables generate ethylene (C_2H_4) gas while they ripen, especially if they have been damaged [16]. If we mix fruits and veggies that either emit or are sensitive to ethylene gas, much of your fresh produce will age and decay faster than normal. Carbon dioxide (CO_2) is the likely output of some of the decompositions. Methane (CH_4) is generated from anaerobic decomposition [17]. Different Papers have used Cyranose-320 (carbon—black polymer sensor) to detect fungi and bacteria. Balasubramanian et al. [18] gives a solution to detect fungi and bacteria in food grains (Maize) using cyranose-320 black polymer sensor. The same sensor technique is used by Panigrahi et al. [19] identification of spoiled beef. Spoilage of Alaska pink salmon Fish is detected under different storage condition by Chantarachoti et al. [20] using same sensor technology. This cyranose polymer sensor it is also used in vegetables (Onion) by Li et al. [21] to check freshness. The cyranose polymer sensor is expensive to detect fungi. We can use MOSFET to detect fungi on fruits and vegetables, which is economical as compared to cyranose.

5 Objectives

Many of these techniques mentioned above are expensive or not just enough to solve the problems related to quality check. Some techniques need to be implemented to get a decision in less time. Cost effective solutions need to be implemented. The main objectives of this paper are to review the state of the art in fruit quality check and provide practical solutions with equipment as simple as possible. Some of the important objectives are summarized here:

(a) To introduce multi-sensor setup to extract the features, e.g. temperature, humidity, ripeness, Acidity, decomposition level. (b) Design an algorithm for identification of influence parameter to predict quality level. (c) To monitor and analyze decomposition stages of food items to predict quality and lifetime. (d) To use cost effective equipments with less processing time.

6 Proposed Framework

Based on the literature survey given in above sections, we propose a framework to check and predict the quality of fruits and vegetable. This framework will be designed based on multiple sensors. As per shown in Fig. 1 food quality check is divided into three phases.

Data acquisition system: In this phase we are going to use microcontroller and analog to digital converter. Different sensors will be attached to ADC to sense the different attributes of food items. Here we are planning to use temperature and humidity sensor along with CO_2 or CH_4 i.e. methane sensor and carbon black polymer sensor. The first phase will find out all values of these mentioned sensors and stored in the database.

Training Phase: In this second phase trained database will be generated. In this phase input database will be provided which is generated using data acquisition phase. In this phase training algorithm will be applied and classification will be done as per the quality. Classification done on the basis of quality attributes will be stored as trained database.

Testing Phase: Once the trained database is generated, we can go for food quality check. Whenever we have to test new food item, sensor values will be taken and will be compared to the trained database in the second phase. After comparison decision will be taken, i.e. food item is ok or not suitable to eat or spoiled. One more decision will be predicted, i.e. for another how many day food items can be kept in the current environment.

Algorithm

1. Initially on an hourly basis for a day, we will extract values from the fruits and vegetables using the mentioned sensors and stored in the database. Same

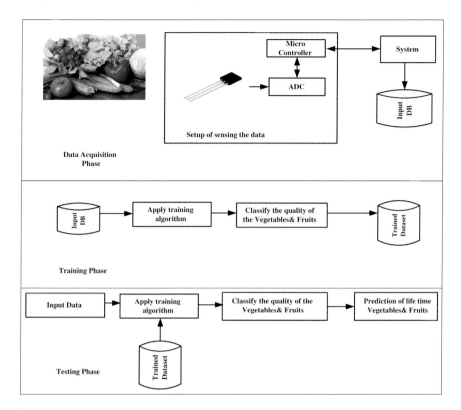

Fig. 1 Proposed framework

experiment will carry out after a few days. We will categorize these values into the two classes as 'Eatable' and 'spoiled'.

Estimate mean (μ) and variance (σ^2) for each sensor category as:

$$\mu = \frac{1}{N}\sum_{n=1}^{N} x_n \quad \text{and} \quad \sigma^2 = \frac{1}{N}\sum_{n=1}^{N}(x_n - \mu)^2$$

where N = Total readings of individual sensor.

Let, μ_{temp} & σ_{temp} are mean and variance of the Temperature sensor.

μ_{hmd} & σ_{hmd} are mean and variance of the Humidity sensor.

μ_{CO_2} & σ_{CO_2} are mean and variance of the CO_2 sensor.

μ_{CH_4} & σ_{CH_4} are mean and variance of the CH_4 (methane) sensor.

μ_{cbp} & σ_{cbp} are mean and variance of the Carbon Black Polymer sensor.

2. Let's say we have equiprobable classes, so P(Eatable) = P(Spoiled) = 0.5
3. When dealing with continuous data, a typical assumption is that the continuous values associated with each class are distributed according to a Gaussian distribution. Then the probability distribution of some value given a class can be

computed. For example, temperature value of a sample is given, and then the probability distribution can be computed as:

$$P(Sample_{temp}|Eatable) = \frac{1}{\sqrt{2\pi\sigma_{temp}^2}} \exp\left(\frac{-(Sample_{temp} - \mu_{temp})^2}{2\sigma_{temp}^2}\right)$$

where, $P(Sample_{temp}|Eatable)$ = Probability of the temperature value of a sample belongs to Eatable class.

4. Similarly, we can calculate the probability for the remaining features extracted from the sample as:

$$P(Sample_{hmd}|Eatable)P(Sample_{CO_2}|Eatable)P(Sample_{CH_4}|Eatable)P(Sample_{cbp}|Eatable)$$
$$P(Sample_{temp}|Spoiled)P(Sample_{hmd}|Spoiled)P(Sample_{CO_2}|Spoiled)P(Sample_{CH_4}|Spoiled)$$
$$P(Sample_{cbp}|Spoiled)$$

5. Using Decision making with the MAP (maximum a posteriori probability) rule, we can calculate

$$P(Eatable|Sample) = P(Sample_{temp}|Eatable) * P(Sample_{hmd}|Eatable) * P(Sample_{CO_2}|Eatable)$$
$$* P(Sample_{CH_4}|Eatable) * P(Sample_{cbp}|Eatable) * P(Eatable)$$

where $P(Eatable|Sample)$ = Probability of sample belongs to Eatable class

$$P(Spoiled|Sample) = P(Sample_{temp}|Spoiled) * P(Sample_{hmd}|Spoiled) * P(Sample_{CO_2}|Spoiled)$$
$$* P(Sample_{CH_4}|Spoiled) * P(Sample_{cbp}|Spoiled) * P(Spoiled)$$

where $P(Spoiled|Sample)$ = Probability of sample belongs to Spoiled class.

6. If $P(Eatable|Sample) > P(Spoiled|Sample)$ Then We label "Sample is Eatable" Otherwise "Sample is Spoiled"
7. End

7 Conclusion

Instead of going for a destructive method to test quality of food items in laboratories by checking their chemical compositions and then predicting quality, there is need of other non destructive methods that will help to predict quality without cutting or taking the juice out of fruit. Non destructive method, e.g., computer vision can detect injury of fruit surface, but not suitable to test internal quality. Here we are trying to propose a framework that will help to test internal quality attributes and microbial contamination with the help of sensors. Another advantage of this framework will be simple and cost-effective. New technique to predict food quality using sensor network will be enhanced. In the future, we are planning to deploy data generated for an individual food item as per the values received from the multi-sensor network on agriculture cloud. These values can be used anywhere through a cloud and the quality of food item can be predicted. It will help to enhance agricultural production and enhance the availability of data related to research projects in the field. Another advantage of using the cloud for agricultural research is to reduce the cost, time, and make the communication system much faster and easier.

References

1. El-kader, S.M.A., El-Basioni, B.M.M.: Precision farming solution in Egypt using the wireless sensor network technology. Egypt. Inf. J. **14**, 221–233 (2013)
2. Majone, B., Viani, F., Filippi, E., Bellin, A., Massa, A., Toller, G., Robol, F., Salucci, M.: Wireless sensor network deployment for monitoring soil moisture dynamics at the field scale. Proc. Environ. Sci. **19**, 426–435 (2013)
3. Roy, S., Bandyopadhyay, S.: A test-bed on real-time monitoring of agricultural parameters using wireless sensor networks for precision agriculture (2013)
4. Nandurkar, S.R., Thool, V.R., Thool, R.C.: Design and development of precision agriculture system using wireless sensor network. In: 2014 First International Conference on Automation, Control, Energy and Systems (ACES), pp. 1, 6, 1–2 Feb 2014
5. Sakthipriya, N.: An effective method for crop monitoring using wireless sensor network. Middle-East J. Sci. Res. **20**(9), 1127–1132 (2014)
6. Patil, S.S., Davande, V.M., Mulani, J.J.: Smart wireless sensor network for monitoring an agricultural environment. (IJCSIT) Int. J. Comput. Sci. Inf. Technol. **5**(3), 3487–3490 (2014)
7. Chandraul, K., Singh, A.: An agriculture application research on cloud computing. Int. J. Current Eng. Technol. **3**(5), 2084–2087 (2013)
8. AL-Marakeby, A., Aly, A.A.: Fast quality inspection of food products using computer vision. Int. J. Adv. Res. Comput. Commun. Eng. **2**(11), 4168–4171 (2013)
9. Parmar, R.R., Jain, K.R., Modi, C.K.: Unified approach in food quality evaluation using machine vision. In: ACC 2011, Part III, CCIS 192, pp. 239–248, 2011
10. Seng, W.C., Mirisaee, S.H.: A new method for fruits recognition system. In: 2009 International Conference on Electrical Engineering and Informatics, Selangor, Malaysia, 5–7 Aug 2009. 978-1-4244-4913-2/09, IEEE
11. Al Ohali, Y.: Computer vision based date fruit grading system: design and implementation. J. King Saud University—Comput. Inf. Sci. **23**, 29–36 (2011)
12. Kuswandi, B., Wicaksono, Y.: Smart packaging: sensors for monitoring of food quality and safety. Sens. Instr. Food Qual. **5**, 137–146 (2011)
13. Pu, Y.Y., Feng, Y.Z., Sun, D.W.: Recent progress of hyperspectral imaging on quality and safety inspection of fruits and vegetables: a review. doi: 10.1111/1541-4337.12123 (Institute of Food Technologists®)
14. An app for safe food Amsterdam. 24 Mar 2014, http://www.foodsniffer.eu/
15. Abbott, J.A.: Quality measurement of fruits and vegetables. Postharvest Biol. Technol. **15**, 207–225 (1999)

16. http://www.frugal-cafe.com/kitchen-pantry-food/articles/ethylene-gas1.html
17. http://www.madsci.org/posts/archives/2005-01/1106120458.Bc.r.html
18. Balasubramanian, S., Panigrahi, S., Kottapalli, B., Wolf-Hall, C.E.: Evaluation of an artificial olfactory system for grain quality discrimination. Food Sci. Technol. **40**(10), 1815–1825 (2007)
19. Panigrahi, S., Balasubramanian, S., Gu, H., Logue, C., Marchello, M.: Neural-network-integrated electronic nose system for identification of spoiled beef. Food Sci. Technol. **39**(2), 135–145 (2006)
20. Chantarachoti, J., Oliveira, A.C.M., Himelbloom, B.H., Crapo, C.A., McLachlan, D.G.: Portable electronic nose for detection of spoiling alaska pink salmon (*Oncorhynchus gorbuscha*). J. Food Sci. **71**(5), S414–S421 (2006)
21. Li, C., Schmidt, N.E., Gitaitis, R.: Detection of onion postharvest diseases by analyses of headspace volatiles using a gas sensor array and GC-MS. Food Sci. Technol. **44**(4), 1019–1025 (2011)

Part III
Cloud Computing

Associate Scheduling of Mixed Jobs in Cloud Computing

Dinesh Komarasamy and Vijayalakshmi Muthuswamy

Abstract In a cloud environment, the jobs are scheduled based on different constraints so as to complete the job within its deadline. However, the classical scheduling algorithms have focussed on processing the compute-intensive and data-intensive job independently. So, simultaneous processing of compute-intensive and data-intensive jobs is a challenging task in a cloud environment. Hence, this paper proposes a new technique called Associate Scheduling of Mixed Jobs (ASMJ) that will concurrently process compute-intensive and data-intensive jobs in a two-tier VM architecture using the sliding window technique to improve processor utilization and network bandwidth. The experimental results show that the proposed ASMJ improves the processor utilization, QoS, user satisfaction and network bandwidth compared with the existing techniques.

Keywords Cloud computing · Job scheduling · Compute-intensive · Data-intensive · Associate scheduling of mixed job

1 Introduction

In cloud computing, the networks of computers (i.e. servers) are connected together to process, store and manage the jobs using several remote servers instead of the local server or personal computer. Cloud computing delivers three major services such as Infrastructure as a Service (IaaS), Platform as a Service (PaaS) and Software as a Service (SaaS) [2]. The features of cloud computing are on-demand resource

D. Komarasamy (✉) · V. Muthuswamy
Department of Information Science and Technology, CEG Campus,
Anna University, Chennai, India
e-mail: dinesh@auist.com

V. Muthuswamy
e-mail: vijim@annauniv.edu

© Springer International Publishing Switzerland 2016
V. Vijayakumar and V. Neelanarayanan (eds.), *Proceedings of the 3rd International Symposium on Big Data and Cloud Computing Challenges (ISBCC – 16'),*
Smart Innovation, Systems and Technologies 49, DOI 10.1007/978-3-319-30348-2_12

provisioning, elasticity, measured service, broad network access and resource pooling [3]. The vast computing power of the cloud is partitioned and shared among the VMs using the virtualization technique [4]. A large number of users handed their jobs over to the cloud system and so scheduling plays a dramatic role in processing the jobs in a cost-effective manner. The job may be either compute-intensive or data-intensive job [5]. In compute-intensive, most of the computation time contributes for computing the job. Similarly, most of the processing time dedicates for data accesses in a data-intensive job [6].

Most of the existing scheduling algorithms only focused on executing either compute-intensive or data-intensive job [7, 8]. Thus, existing techniques optimally utilize either network bandwidth or processor utilization. So, this article proposes a new scheduling technique called Associate Scheduling of Mixed Jobs (ASMJ) that will simultaneous schedule both the compute-intensive jobs and the data-intensive jobs using the sliding window technique with the support of foreground VM and background VM in a two-tier VM architecture. The ASMJ embeds the sliding window technique and constructs the graph for scheduling the jobs between foreground and background VM efficiently. The rest of the paper is organized as follows. Section 2 discusses the literature review. Section 3 illustrates the proposed system (i.e. Associate Scheduling of Mixed Job) that composed of job model, system model and scheduling policies. Section 4 explains the experimental setup and simulation results. The conclusions and future work of this paper are given in Sect. 5.

2 Literature Review

This section briefly describes the existing scheduling technique to schedule the compute-intensive and data-intensive jobs independently [7, 8]. Here, the jobs scheduling algorithms are classified as static scheduling and dynamic scheduling. In a static scheduling, the resources are previously specified for the jobs. But in a dynamic scheduling, the jobs are assigned to the resources at the runtime [9]. In the existing scheduling algorithm, the jobs were prioritized based on various parameters such as length, deadline, cost, etc. [10, 11]. Here, the compute-intensive and data-intensive jobs scheduled separately in the existing scheduling algorithm that underutilized the processor of the VM and the bandwidth of the network. Here, the job does not fully utilize the processing speed of the VM while running the jobs. Hence, the computing power of the VM is dynamically partitioned between foreground and background VM [12]. Several algorithms were available to schedule the independent jobs as well as parallel jobs. The existing algorithms were AMSS (Adaptive Multilevel Scheduling algorithm), EASY backfilling, CMBF (Conservative Migration support Backfilling) and so on [13–15]. The Federated Job Scheduler (FJS) has been introduced to schedule compute-intensive and data-intensive jobs sequentially [6]. From the literature, several works were carried out to schedule the compute-intensive and data-intensive independently. Instead of that, this work proposes an Associate Scheduling of Mixed Jobs (ASMJ) to improve the processor utilization and network bandwidth by concurrently processing the compute-intensive and data-intensive job in a VM.

3   Proposed Model Design

The user submits an eternal number of jobs to run in the cloud in a cost-effective fashion. Hence, this article proposes a new algorithm called Associate Scheduling of Mixed Jobs (ASMJ) that composes of job model, system model and scheduling policies. Among these, the job model describes the characteristics of the job. The jobs are independent of each other (i.e. the job does not wait for the completion of another job to process in the VM). While submitting the jobs, the user must mention length and deadline along with its type. The job type mostly comes under either compute-intensive or data-intensive job. The job length and job deadline are represented as 'ls' and 'd' respectively. Similarly, the cloud system contains several Virtual Machine (VM). Here, the job does not utilize the entire processing speed of the VM and so computing capacity of the VM is dynamically shared between the foreground and background VM. The foreground and background VM information are maintained in the VM monitor. The VM monitor contains twin queue to store the information of the foreground and background VM as Q_{FVM} and Q_{BVM} respectively.

The jobs are scheduled similar to the centralized scheduling system. The ASMJ will schedule the jobs based on characteristics of the job and computing power of the VM. Among these jobs, the compute-intensive jobs run in the foreground VM and the data-intensive job process in the background VM to avoid the wastage of computational and bandwidth of the resource. In the proposed work, the jobs submitted by the users are initially stored in the job queue and later forwarded to the job-classifier. The job-classifier divides the job into compute-intensive and data-intensive job. Further, the compute-intensive jobs are stored in a computational queue (Q_c) and the data-intensive (i.e. I/O jobs) are stored in a data-intensive queue (Q_d).

The jobs in Q_c and Q_d move to the sliding window to maintain one to one relation between the jobs and resources. The size of the sliding window (S) dynamically varies depending on the available number of VMs in the VM monitor (η) as $S = \eta$. Here, the two-sliding windows are named as S_c and S_d as the size of η. The jobs in Q_c and Q_d move to the S_c and S_d respectively, based on an FCFS algorithm to avoid starvation. Suppose, the jobs are insufficient to fulfill S_c, at that time the jobs in Q_d will occupy the remaining space of the sliding window (i.e. S_c). Similarly, the jobs in Q_d is inadequate to accomplish the S_d and so the jobs in Q_c will engage the sliding window (i.e. S_c). The jobs persist in the sliding window are prioritized using the deadline along with the arrival time of the job. The prioritized compute-intensive job is represented as P_c. Similarly, prioritized data-intensive job is represented as P_d. The prioritized value is expressed in the equation below.

$$P_{cj} = d_j - a_j; \quad \forall_j \, in \, S_c \quad and \quad P_{dj} = d_j - a_j; \quad \forall_j \, in \, S_d$$

After prioritizing, the jobs need to map with the resource exist in the VM pool. The minimum required processing speed of the job is represented as PS and computed using the length and deadline of the job as given below.

$$PS_{cj} = \frac{ls_j}{d_j}; \quad \forall_j \in S_c \quad and \quad PS_{dj} = \frac{ls_j}{d_j}; \quad \forall_j \in S_d$$

After computing the PS, the job will map with the suitable VM based on the weightage value. Hence, the weightage value between the job and virtual machine is computed using the minimum required processing speed of the job with the processing speed of the VM. The weightage value is represented as W_t and computed as given below

$$W_t(jk) = \frac{PS_{cj}}{PVM_k}; \quad \forall_j \, in \, PS_c, \quad \forall_k \in (1, \eta)$$

$$W_t(jk) = \frac{PS_{dj}}{PVM_k}; \quad \forall_j \, in \, PS_d, \quad \forall_k \in (1, \eta)$$

where PVM represents the processing of the VM. Figure 1 represents the map between the job and the resource. The jobs and foreground VM along with background VM are represented as nodes. The nodes are independent of each other. Initially, the job connects with all the resource that looks like mesh topology (i.e. one-many relation). The jobs are depicted as a circle. Similarly, the foreground and background VM are represented as a square. The link established connecting the nodes is termed as an edge. The link established between the job in S_c with the foreground VM. Furthermore, the edge also established linking the job persist in S_d and background VM.

Fig. 1 Maps the job and resource

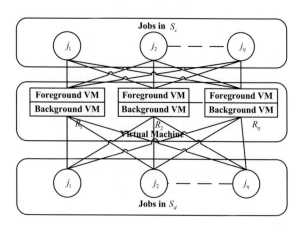

After establishing the edges, the job in the sliding window connects with either foreground VM or background VM. Among this several edges, the every job will choose only the edge that has the lowest weightage value among the existing edges and also the VM does not connect with another resource. The VM allocation algorithm shows effective scheduling of mixed workloads (i.e. compute-intensive and data-intensive) among the VMs (i.e. foreground VM and background VM).

Algorithm 1 VM allocation algorithm

1: Store a job J from Q_c and Q_d
2: Get a job j and assign to sliding window (S_c & S_d)
3: **if** $|Q_{pc}| \wedge |Q_{pd}| \geq |S_c|$ **then**
4:     j in $Q_{pc} \rightarrow S_c$ & j in $Q_{pd} \rightarrow S_d$
5: **else if** $|Q_{pc}| < |S_c| \& |Q_{pd}| \geq |S_d|$ **then**
6:     j in $Q_{pc} \rightarrow S_c$, & j in $Q_{pd} \rightarrow S_d \wedge idleS_c$
7: **else**
8:     j in $Q_{pd} \rightarrow S_d$ & j in $Q_{pc} \rightarrow S_c \wedge idleS_d$
9: **for** $\forall_j$ in S_c and S_d **do** $//ll^{ly}$ prioritize
10:     $P_{cj} = d_i - a_i : \forall_i in S_c$
11:     $P_{dj} = d_i - a_i : \forall_i in S_d$
12:     Compute the weightage value using
13:     $W_t(jk) = \frac{PS_{cj}}{PVM_k}$; $\forall_j in PS_c where in(1, \eta)$
14:     $W_t(jk) = \frac{PS_{cj}}{PVM_k}$; $\forall_j in PS_c where in(1, \eta)$
15: Design graph using job, fg,bg
16: Keep one edge between the job and VM have lowest value
17: Maintain one-one relation
18: Map the job and VM based on the weightage value
19: Goto (1) for scheduling of the remaining job untill $Q_{pc} \wedge Q_{pd} = \emptyset$

The VM allocation algorithm clearly explains the allocation of foreground VM and background VM for the submitted jobs.

4   Experimental Setup and Analysis

Cloudsim is one of the most important simulation tools to provide an extensible simulation framework for carrying out several experiments relevant to the cloud computing technologies [16]. Generally, the cloudsim supports job scheduling, load balancing, scalability, VM creation, etc. It provides the computing power of the VM [i.e. Infrastructure as a Service (IaaS)] for processing the jobs. It contains several data centers. Each of the data center may have one or more hosts that has several VMs to provide a computing power for processing the job simultaneously. The following classes of the Cloudsim need to modify for simultaneously scheduling of compute-intensive and data-intensive jobs effectively such as cloudlet scheduler, VM Scheduler, VM policy allocation and datacenter broker.

Table 1 describes the characteristic of job. The main constraint parameters of the job include length, deadline and type. The job length defines as Million Instruction (MI). The job deadline is millisecond (ms). Further, the job type generally belongs to either compute-intensive or data-intensive job. For the sake of simplicity test, the job spends more processing time for computation (i.e. compute-intensive jobs)

Table 1 Job description

Job_id	Job_type	Job_length	Job_deadline
0	0	15,000	10
1	1	20,000	15
2	1	14,500	14
3	0	17,400	28
4	1	15,300	20
5	0	8000	36
6	1	5500	25
7	1	5000	20

represents as type = 1. Similarly, the jobs dedicate more processing time for data access are considered as type = 0.

This paper proposes Associate Scheduling of Mixed Job (ASMJ) that will simultaneously schedule both compute-intensive and data-intensive jobs.

4.1 Waiting Time

The incoming jobs contain both compute-intensive and data-intensive jobs. The proposed work (i.e. ASMJ) focus on reducing the waiting time of the job compared with the existing scheduling algorithm such as FCFS, EDF by introducing the VM allocation algorithm in the two-tier VM architecture.

The VM allocation algorithm will simultaneously schedule both compute-intensive as well as data-intensive jobs among the foreground VM and background VM. Figure 2 describes the waiting time of the jobs. The proposed method (i.e. ASMJ) outperforms the classical scheduling algorithm by reducing the waiting time

Fig. 2 Waiting time

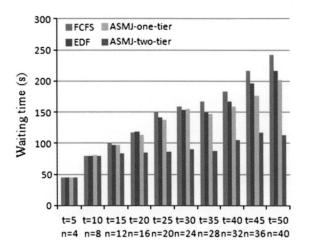

of the job using the two-tier VM architecture whenever $n \geq \eta$. The performance of the ASMJ degrades only when the number of incoming jobs is less than the total number of available VM $(n \leq \eta)$.

4.2 Throughput

The throughput of the VM is improved by partitioning the computing capacity of the VM with the aid of two-tier VM architecture. The existing scheduling algorithm run only one job at a particular time. Even though the compute-intensive jobs run in the VM, the computing power of the VM cannot be fully utilized because the compute-intensive job does not contain computing instruction alone. The performance measure is computed using the total number of jobs completed per unit time as given below.

$$Throughput = \frac{No.\ of\ job\ completed}{time\ duration}$$

Figure 3 describes throughput of the VM. The proposed ASMJ outperforms the existing scheduling algorithms by deploying the VM in two-tier VM architecture that will simultaneously process both the compute-intensive and data-intensive jobs. The proposed ASMJ algorithm underperforms only when the number of incoming job is less than the number of VM exists. The jobs are effectively mapped with the resource having minimum weightage value. Moreover, ASMJ effectively utilize the entire processing speed of the VM by deploying the VM allocation algorithm in the two-tier VM architecture.

Fig. 3 Throughput of the VM

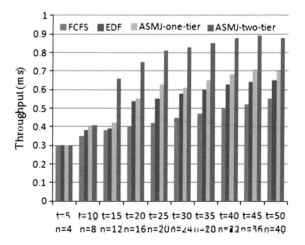

4.3   Processor Utilization

The ASMJ outperforms the other classical scheduling algorithm by partitioning the computing capacity of the VM as foreground VM and background VM. The processor cannot be fully utilized while running the compute-intensive jobs in a one-tier VM architecture.

The performance of the data center degrades while processing the compute-intensive jobs in the foreground and background VM. Further, the bandwidth cannot be effectively utilized while processing the compute-intensive jobs alone. Figure 4 denotes processor utilization. The utilization of the processor is increased by processing the compute-intensive jobs and data-intensive jobs simultaneously. Further, the job is mapped effectively with the VM by assigning the edge having the smallest weightage value.

4.4   Total Virtual Machine

The jobs which contains both data-intensive and compute-intensive jobs, are submitted to the foreground and background VM respectively. AMSJ needs only a less number of VM for processing the submitted jobs because the VMs utmost process two jobs concurrently in the single VM (i.e. foreground and background VM).

Figure 5 describes the number of required VM. It predicts the total number of VMs required for the processing the submitted jobs. AMSJ process the compute-intensive and data-intensive jobs in the foreground and background VM concurrently in order to avoid the degradation of the system performance. AMSJ algorithm surpasses the existing algorithm by reducing the number of VM required for processing the submitted jobs by simultaneously processing the jobs in the VM (i.e. foreground and background VM).

Fig. 4 Processor utilization

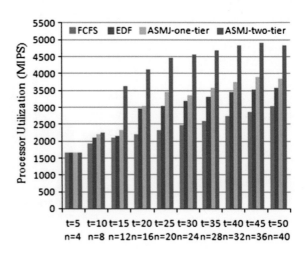

Fig. 5 Number of VM

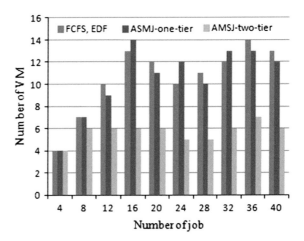

5 Conclusion and Future Work

Existing scheduling algorithms processed compute-intensive and data-intensive jobs separately. The ultimate aim of the proposed work is to simultaneously scheduling of both compute-intensive and data-intensive jobs. This proposed work (Associate Scheduling of Mixed jobs) come across various difficulties to concurrently schedule the jobs. In the proposed work, the jobs are simultaneously scheduled only with the deployment of the VM allocation algorithm in the two-tier VM architecture. Initially, the jobs were classified and then stored in a twin-queue. Further, the jobs move to the sliding window based on FCFS model to avoid starvation. The jobs establish a link with all the resources in the data center (i.e. mesh topology). Later, the jobs mapped with the VM having the smallest weightage value among all edges to improve processor utilization. The compute-intensive job process in the foreground VM and data-intensive job executes in the background VM to optimize the utilization of the processor as well as bandwidth. So, the ASMJ outperforms the classical scheduling algorithms by parallel running the compute-intensive jobs and data-intensive jobs in a particular VM. In future, the work can be extended for the dependent jobs.

References

1. Wang, Y., Ma, X.: A general scalable and elastic content-based publish/subscribe service. IEEE Trans. Parallel Distrib. Syst. **26**(8), 2100–2113 (2015)
2. Dinesh, K., Poornima, G., Kiruthika, K.: Efficient resources allocation for different jobs. Int. J. Comput. Appl. **56**(10), 30–35 (2012)
3. Mell, P., Grance, T.: The NIST Definition of Cloud Computing. NIST Special publication, pp 800–145 (2011)

4. Raghunathan, A.Chunxiao, Jha, K.Niraj: A trusted virtual machine in an untrusted management environment. IEEE Trans. Serv. Comput. **5**(4), 472–483 (2012)
5. Hovestadt, M., Kao, O., Kliem, A., Warneke, D.: Evaluating Adaptive Compression to Mitigate the Effects of Shared I/O in Clouds. In: IEEE International Symposium on Parallel and Distributed Processing Workshops and Phd Forum (IPDPSW), pp. 1042–1051 (2011)
6. Komarasamy, Dinesh, Muthuswamy, Vijayalakshmi: Efficient federated job scheduler for mixed workloads in cloud computing. Int. J. Appl. Eng. Res. (IJAER) **10**(69), 44–48 (2015)
7. Suresh, S., Huang, Hao, Kim, H.J.: Scheduling in compute cloud with multiple data banks using divisible load paradigm. IEEE Trans. Aerosp. Electron. Syst. **51**(2), 1288–1297 (2015)
8. Zhang, Qi, Zhani, M.F., Yang, Yuke, Boutaba, R., Wong, B.: PRISM: fine-grained resource-aware scheduling for MapReduce. IEEE Trans. Cloud Comput. **3**(2), 182–194 (2015)
9. Warneke, D., Kao, Odej: Exploiting Dynamic Resource Allocation for Efficient Parallel Data Processing in the Cloud. IEEE Trans. Parallel Distrib. Syst. **22**(6), 985–997 (2011)
10. Ding, Youwei, Qin, Xiaolin, Liu, Liang, Wang, Taochun: Energy efficient scheduling of virtual machines in cloud with deadline constraint. Future Gener Comput Syst **50**, 62–74 (2015)
11. Abdullaha, M., Othmanb, M.: Cost-based multi-QoS job scheduling using divisible load theory in cloud computing. In: International Conference on Computational Science, Procedia Computer Science, vol. 18, pp. 928–935 (2013)
12. Liu, X., Wang, C., Qiu, X., Zhou, B.B., Chen, B., Zomaya, A.Y.: Backfilling under two-tier virtual machines. In: IEEE International Conference on Cluster Computing (CLUSTER), pp. 514–522 (2012)
13. Komarasamy, Dinesh, Muthuswamy, Vijayalakshmi: Deadline constrained adaptive multilevel scheduling system in cloud environment. KSII Trans. Internet Inf. Syst. (TIIS) **9**(4), 1302–1320 (2015)
14. Liu, X., Wang, C., Zhou, B.B., Chen, J., Yang, T., Zomaya, AY.: Priority-based consolidation of parallel workloads in the cloud. IEEE Trans. Parallel Distrib. Syst. **24**(9):1874–1883 (2013)
15. Liu, X., Qiu, X., Chen, B., Huang, K.: Cloud-Based Simulation: The State-of-the-Art Computer Simulation Paradigm. In: ACM/IEEE/SCS 26th Workshop on Principles of Advanced and Distributed Simulation (PADS), pp. 71–74 (2012)
16. Calheiros, R.N., Ranjan, R., Beloglazov, A., De Rose, C.A.F., Rajkumar, B.: CloudSim: a toolkit for modeling and simulation of cloud computing environments and evaluation of resource provisioning algorithms. Softw. Pract. Experience. **41**(1):23–50 (2011)
17. Bin, Zhuge, Li, Deng, et al.: Resource scheduling algorithm and economic model in ForCES networks. Chine Commun. **11**(3), 91–103 (2014)

Towards an Analysis of Load Balancing Algorithms to Enhance Efficient Management of Cloud Data Centres

J. Prassanna, P. Ajit Jadhav and V. Neelanarayanan

Abstract Now a day's cloud computing breaks almost all the barriers of large scale computing and widens the scope of massive computational possibilities. Cloud computing provides various benefits to the whole computing societies such as on demand flexi pay access to techno business services and wide range of computing resources requires an exponential growth in its technology put forth serious challenges including VM load balancing especially in cloud data centers. It dynamically distributes the workload across multiple servers in the cloud data center so that not even a single server involved is underutilized or overutilized. If load balancing is not done properly in the cloud then it leads to the inefficiency in processor utilization that in turn risks the provider by creating a significant problem of increase in overall energy consumption and the world by increasing the carbon emissions. Lots of different techniques like Round Robin, Throttled and Equally Spread Current Execution are claimed to provide efficient mechanisms to resolve this problem. This paper compares and summaries the existing load balancing techniques which are used to solve the issues in cloud environment by considering the data center processing time and response time and propose an improved load balancing strategy believed to be a efficient solution for the cloud load balancing issues.

Keywords Cloud computing · Datacenters · Underutilization · Overutilization · Load balancing · Energy consumption

1 Introduction

Cloud computing is becoming a biggest buzz word today. Clients can access cloud as per their need for the services like software as a service, platform as a service or infrastructure as a service on the basis of pay as you go model.

J. Prassanna (✉) · P.A. Jadhav · V. Neelanarayanan
School of Computing Science and Engineering, VIT University,
Chennai 600 127, India

© Springer International Publishing Switzerland 2016   143
V. Vijayakumar and V. Neelanarayanan (eds.), *Proceedings of the 3rd International Symposium on Big Data and Cloud Computing Challenges (ISBCC – 16'),*
Smart Innovation, Systems and Technologies 49, DOI 10.1007/978-3-319-30348-2_13

Datacenter is the heart of cloud computing which are used to satisfy these requests. The main aim is to satisfy the customer request as well as the proper utilization of the resources. Load balancing plays key role in efficient utilization of the resources. If load balancing is not done properly then it will results in overutilization or underutilization of some resources [1]. This leads to the increase in energy consumption. Many algorithms like Throttled, Equally Spread Current Execution and Round robin are used to satisfy the client requests with minimum response time [2, 3]. Also many workload scheduling approaches like Green scheduler, the Energy aware network scheduler [4] deals optimizing and stabilizing the energy consumption of the data centres.

Load balancing algorithms are classified as static algorithms and dynamic algorithms. Static algorithm requires prior knowledge about the resources [5] but it does not consider the current state of the system [6]. Hence they are easy to implement. But they are not suitable for dynamic situations. Dynamic algorithm overcomes this problem. They do not require any prior knowledge [6] but they need current state of the system. Hence they are complex to implement but give better results to the dynamic situations [5].

This paper discusses some existing load balancing algorithm like Round Robin, Throttled and Equally Spread Current Execution and proposes a new improved approach for load balancing in cloud. The analysis result of these algorithms is done by using cloud Analyst tool.

2   Load Balancing in Cloud Computing

Load balancing is a computer networking method that effectively distribute the inbound web traffic or work load across network server computers to guarantee the availability of application and its performance. As a result of the distribution more work can be done in a same amount of time and in general all users can get server faster. The Load balancing algorithms will help to fine-tune the inbound traffic distribution across several connections. The huge volume of information generated by cloud applications and a bottleneck in the network through a cloud server would result an increase in the delay on data transactions in the cloud computing that leads towards the necessity of the load balancing for cloud inbound traffic demands [7] (Fig. 1).

2.1   Benefits of Load Balancing

Scalability
The servers can be added smoothly without any interruption and the application can be scaled without any barriers through load balancing the servers in the cloud.

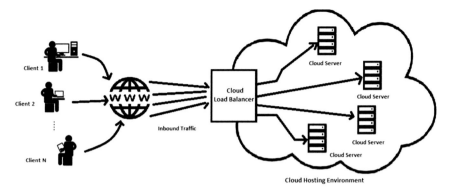

Fig. 1 Cloud load balancing

Availability
Even in the case of unavailability of few or more servers the load balancing can guarantees the services remains up and in full swing through distributing the load effectively, monitoring the routine health and maintaining the presentence of every session over multiple servers.

Performance
Load balancing make cloud applications and services to respond much faster than usual and the transaction time get reduced to a greater extends through efficient caching mechanism, compression techniques and speeding up of SSL for improvement of cloud application services.

Efficiency
The efficiency of the cloud services are increased in greater extend by which the demanded resources are ease to avail by the client, cloud serviced resources are utilized efficiently even under the circumstance of load imbalance, the clients value for money will be increased and the energy can be saved.

Reliability The reliability of the cloud services are ensured by server redundancy through load balancing which makes an application to be hosted at various cloud hub anywhere in the world. The cloud serving resource even in any worst case of failure will not let down from its hosted cloud by redirect those traffic to any other cloud location to host those service without any interruption.

3 Load Balancing Algorithm for Cloud

3.1 Round Robin Algorithm

One of the best examples of static load balancing algorithm is Round Robin algorithm [8]. The inbound requests are dispatched to the available servers in round robin fashion [9]. This algorithm uses the concept of time slicing or time quantum. In this approach, the virtual machine is randomly selected from the group of virtual machines to assign first request and then next requests are assigned in round robin manner using time slice concept. After assigning the request to the virtual machine that virtual machine is moved to the end of the list. In case if the VM is not free then that request should stand in the queue. This approach has one issue i.e. some nodes are heavily loaded or some are lightly loaded. This algorithm is easy to implement but main task is to decide the time quantum. If time quantum is small then it leads to increase in number of context switches. If time quantum is too large then this algorithm will work same as FFCS scheduling.

3.2 Equally Spread Current Execution Algorithm (ESCE)

This is the example of dynamic load balancing algorithm. This approach attempts to preserve the equal workload on all the available virtual machines. Initially an index table is maintained by ESCE load balancer for the virtual machines as well as the number of requests currently assigned to each virtual machine. All virtual machines are assigned with 0 allocation count. When request will come from the DataCenterController to allocate VM, ESCE load balancer scans the table to identify least loaded VM. Suppose it will give more than one VM then first identified VM is chosen for handling the request. This VM id is send to the DataCenterController by the ESCE load balancer. By considering this VM id, DataCenterController communicates the request to that VM. The allocation count of that VM is increased by one. This change is updated in index table. After completion of request assigned to that VM, DataCenterController will receive the response cloudlet and then notifies the ESCE load balancer for the VM de-allocation. The allocation count of that VM by is decreased by one. The ESCE load balancer again updates this change in index table. This leads to the additional overhead to scan the queue again and again [10].

3.3 Throttled Load Balancing

Throttled load balancing algorithm is another example of dynamic load balancing algorithm which is completely based on the virtual machines. Throttled load balancer maintains the index table of virtual machines along with their states i.e. available or busy. All the virtual machines are initialized as available. In this algorithm, initially client requests the throttled load balancer to find the right virtual machine which will perform the tasks given by the client. DataCenterController receives the request from client and queries the throttled load balancer for the allocation of virtual machine. Starting from the top, the index table is scanned by the throttled load balancer until the first available VM will found or table is scanned completely. If the available VM is found then immediately that VM id is send to the DataCenterController. By considering this VM id, DataCenterController communicates the request to that VM. Then throttled load balancer acknowledges the new allocation and index table is updated accordingly to it. In case if proper VM is not found then throttled load balancer returns −1 and the request is queued by the DataCenterController. When VM completes the assigned request, response cloudlet is received by the DataCenterController and it notifies the throttled load balancer for the VM de-allocation. Then DataCenterController will check if any requests are pending in the queue. If so then these requests are further processed by following the same procedure. If no request is pending then it continues to next incoming request [6, 10].

4 Results and Analysis

Simulation is carried out by using CloudAnalyst [11] Simulator. CloudAnalyst simulator gives scenario of six different geographical locations (Fig. 2).

User base configurations are described in Table 1. It describes the information like which region the User Base belongs and Requests per user per hour. Table 2 shows data center characteristics like its region, what kind of operating system it have, which architecture it uses etc.

Fig. 2 Snapshot of CloudAnalyst simulator

Table 1 User base configuration

Name	Region	Requests per user per hour	Data size per request (bytes)	Peak hours start (GMT)	Peak hours end (GMT)	Avg peak users	Avg off-peak users
U01	0	60	100	3	9	1000	100
UB2	1	60	100	3	9	1000	100
UB3	2	60	100	3	9	1000	100
UB4	3	60	100	3	9	1000	100
UB5	4	60	100	3	9	1000	100
UB6	5	60	100	3	9	1000	100
UB7	0	60	100	3	9	1000	100
UB8	1	60	100	3	9	1000	100
UBS	2	60	100	3	9	1000	100
UB10	3	60	100	3	9	1000	100

Table 2 Dataenter configuration

Name	Region	Arch	OS	VMM	Cost per VM $/Hr	Memory Cost $/s	Storage Cost $/s	Data Transfer Cost $/Gb	Physical HW Units
DC1	0	x86	Linux	Xen	0.1	0.05	0.1	0.1	2
DC2	2	x86	Linux	Xen	0.1	0.05	0.1	0.1	1
DC3	4	x86	Linux	Xen	0.1	0.05	0.1	0.1	1

Table 3 Response time by Round Robin

	AVG (ms)	MIN (ms)	MAX (ms)
Overall response time	145.41	35.32	401.09
Data center processing time	0.62	0.03	2.71

Table 4 Response time by ESCE

	AVG (ms)	MIN (ms)	MAX (ms)
Overall response time	145.43	35.32	401.09
Data center processing time	0.65	0.03	1.81

4.1 Overall Response Time Summary

Tables 3, 4 and 5 shows the overall response time of Round Robin, ESCE and
Throttled respectively. Figures 3 and 4 shows comparison of the overall response
time as well as Data Center Processing time of given algorithms. The comparison
done with Algorithms which are used and Time taken to respond as well as to

Table 5 Response time by Throttled

	AVG (ms)	MIN (ms)	MAX (ms)
Overall response time	145.40	35.32	401.09
Data center processing time	0.61	0.03	1.33

Fig. 3 Comparison of overall response time between RR, ESCE and Throttled

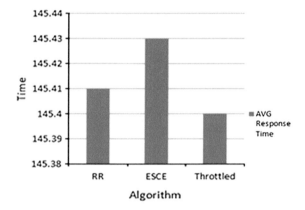

Fig. 4 Comparison of data center processing time between RR, ESCE and Throttled

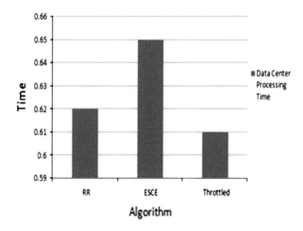

process the requests. By analysing this two figures we can say that the Throttled algorithm is little bit better than Round Robin and Equally Spread Current Execution Algorithm. Table 6 shows the response time by region using given algorithms and different user bases.

Table 6 Response time by region

User base	Round Robin	Equally spread current execution	Throttled
UB1	50.31	50.31	50.30
UB2	200.39	200.39	200.39
UB3	50.55	50.56	50.55
UB4	300.62	300.59	300.58
UB5	50.41	50.76	50.9
UB6	200.33	200.33	200.34
UB7	50.31	50.32	50.32
UB8	200.41	200.41	200.40
UB9	50.53	50.53	50.53
UB10	300.76	300.72	300.76

Table 7 VM allocation in Round Robin

Data center 1	Data center 2	Data center 3
0 → 19,153	0 → 15,400	0 → 15,489
1 → 19,153	1 → 15,400	1 → 15,489
2 → 19,153	2 → 15,400	2 → 15,488
3 → 19,153	3 → 15,400	3 → 15,488
4 → 19,153	4 → 15,400	4 → 15,488
5 → 19,153	5 → 15,400	
6 → 19,153	6 → 15,400	
7 → 19,153	7 → 15,399	
8 → 19,153	8 → 15,399	
9 → 19,152	9 → 15,399	
10 → 19,152	10 → 15,399	
11 → 19,152	11 → 15,399	
12 → 19,152	12 → 15,399	
13 → 19,152	13 → 15,399	
14 → 19,152	14 → 15,399	
15 → 19,152	15 → 15,399	
17 → 19,152	17 → 15,399	
16 → 19,152	16 → 15,399	
19 → 19,152	19 → 15,399	
18 → 19,152	18 → 15,399	

Tables 7, 8, and 9 depicts the VM allocation in each datacenter using three algorithms. In Throttled algorithm, 19 and 1066 requests are queued that is no VM is allocated to that requests. Figure 5 shows the allocation of virtual machines in datacenter 1, datacenter 2 and datacenter 3 using Round Robin, Equally Spread Current Execution and Throttled algorithm.

Table 8 VM allocation in ESCE

Data center 1	Data center 2	Data center 3
0 → 234,960	0 → 158,125	0 → 23,518
1 → 60,323	1 → 47,911	1 → 14,383
2 → 39,020	2 → 32,442	2 → 10,641
3 → 22,075	3 → 22,416	3 → 7855
4 → 13,161	4 → 15,261	4 → 5900
5 → 6973	5 → 10,596	5 → 4330
6 → 3601	6 → 7032	6 → 3307
7 → 1673	7 → 4910	7 → 2426
8 → 755	8 → 3307	8 → 1737
9 → 295	9 → 2252	9 → 1222
10 → 112	10 → 1471	10 → 799
11 → 45	11 → 968	11 → 552
12 → 26	12 → 591	12 → 340
13 → 9	13 → 343	13 → 203
14 → 6	14 → 172	14 → 118
15 → 5	15 → 88	15 → 60
17 → 3	17 → 46	17 → 29
16 → 2	16 → 28	16 → 20
19 → 3	19 → 13	19 → 9
18 → 2	18 → 15	18 → 8

Table 9 VM allocation in Throttled

Data center 1	Data center 2	Data center 3
0 → 235,029	0 → 158,135	0 → 47,408
1 → 60,329	1 → 47,873	1 → 16,457
2 → 38,963	2 → 32,492	2 → 7437
3 → 22,072	3 → 22,454	3 → 3528
4 → 13,133	4 → 15,221	4 → 1546
5 → 6993	5 → 10,562	−1 → 1066
6 → 3621	6 → 7064	
7 → 1679	7 → 4919	
8 → 745	8 → 3300	
9 → 289	9 → 2248	
10 → 105	10 → 1452	
11 → 46	11 → 957	
12 → 27	12 → 597	
13 → 9	13 → 341	
14 → 4	14 → 180	
15 → 3	15 → 83	
17 → 2	17 → 44	
	16 → 26	
	19 → 12	
	18 → 8	
	−1 → 19	

Fig. 5 VM allocation using
RR, ESCE and Throttled
algorithm

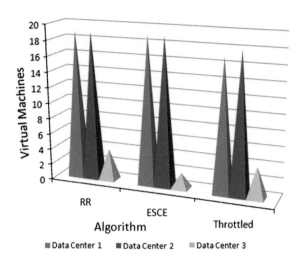

5 Proposed Work

Throttled load balancer checks the list from the top until first available VM will
found. But if numbers of VMs are large, then it takes time to parse the list. If we
will make two index tables one for available and another for busy VMs then this
will reduce the time taken by load balancer to parse the list of VMs. The proposed
algorithm is as follows (Figs. 6 and 7):

1. Initially maintain two index tables one for Available VMs and another for Busy
   VMs. Initially all the VMs are in Available table and the Busy table contains no VM.
2. New request is received by DataCenterController.
3. DataCenterController informs load balancer for next allocation of VM.
4. Load balancer will scan the Available table.

   If found:

   (i)   Load balancer will return the VM id to the DataCenterController.
   (ii) DataCenterController communicates with the VM identified by this VM id
   (iii) Load balancer is acknowledges the new allocation and updates the
         Available table as well as the Busy table. It removes VM from Available
         table and adds into Busy Table.

   If not found:

   (i) Load balancer returns −1.
   (ii) DataCenterController queues that request.

5. When processing request is finished by VM, DataCenterController will receive
   the response cloudlet. Load balancer receives notification for VM de-allocation.
   It removes Busy VM from table and adds into Available Table.

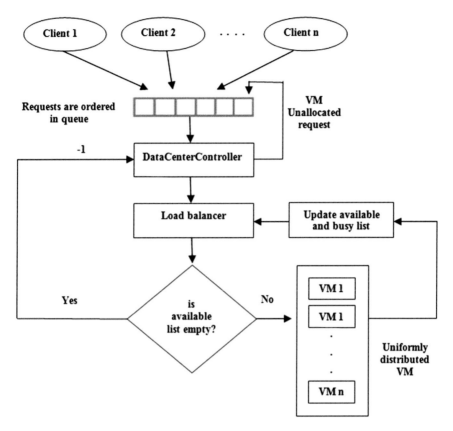

Fig. 6 Work flow of proposed cloud load balancing algorithm

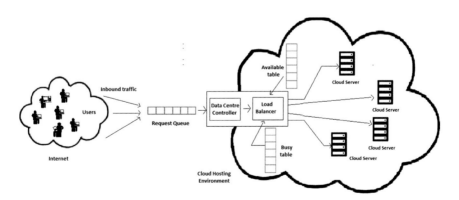

Fig. 7 Proposed cloud load balancing architecture

6. DataCenterController checks for the waiting requests in the queue. If queue is not empty then it continues from step 3.
7. Continue from step 2.

6 Conclusions

The greatest challenge of efficient load balancing is to minimize the response time to a greater extend. According to this, we have analyzed the three major load balancing algorithms by executing different user request in a simulated cloud environment. By analyzing these algorithms we can observe that the Throttled load balancing technique gives an optimal solution by reducing the response time than compared with Round Robin and Equally spread current execution algorithms. From this impact an improved and believed efficient alternative approach is proposed with two lists, availability and busy, used to efficiently harmonize the load to avoid the servers to be underutilized or over utilized.

7 Future Work

Future work will be focused on implementing proposed approach to get better result of minimum virtual machine allocation as well as response time. In future we can add priority to the VMs using different configuration of VMs.

References

1. Sreenivas, V., Prathap, M., Kemal, M.: Load balancing techniques: major challenge in cloud computing—a systematic review. In: International Conference on Electronics and Communication Systems (ICECS), vol. 1, no. 6, pp. 13–14, Feb 2014. doi:10.1109/ECS. 2014.6892523
2. Domanal, S.G., Reddy, G.R.M.: Load balancing in cloud computing using modified throttled algorithm. In: IEEE International Conference on Cloud Computing in Emerging Markets (CCEM), vol. 1, no. 5, pp. 16–18, Oct 2013. doi:10.1109/CCEM.2013.6684434
3. Domanal, S.G., Reddy, G.R.M.: Optimal load balancing in cloud computing by efficient utilization of virtual machines. In Sixth International Conference on Communication Systems and Networks (COMSNETS), vol. 1, no. 4, pp. 6–10 Jan 2014. doi:10.1109/COMSNETS. 2014.6734930
4. Karthikeyan, G.K., Jayachandran, P., Venkataraman, N.: Energy aware network scheduling for a data centre. Int. J. Big Data Intell. 2(1), 37–44 (2015). doi:10.1504/IJBDI.2015.067573
5. Nuaimi, K.A., Mohamed, N., Nuaimi, M.A., Al-Jaroodi, J.: A survey of load balancing in cloud computing: challenges and algorithms. In: Second Symposium on Network Cloud Computing and Applications (NCCA), pp. 137–142, 3–4 Dec 2012. doi:10.1109/NCCA.2012.29

6. Shoja, H., Nahid, H., Azizi, R.: A comparative survey on load balancing algorithms in cloud computing. In: International Conference on Computing, Communication and Networking Technologies (ICCCNT), vol. 1, no. 5, pp. 11–13, July 2014. doi:10.1109/ICCCNT.2014. 6963138

7. Wang, B., Qia, Z., Maa, R., Guana, H., Vasilakosb, A.V.: A survey on data center networking for cloud computing. Comput. Netw. **91**, 528–547 (14 Nov 2015)

8. Lee, R., Jeng, B.: Load-balancing tactics in cloud. In: International Conference on Cyber-Enabled Distributed Computing and Knowledge Discovery (CyberC), pp. 447–454, 10–12 Oct 2011

9. Teo, Y.M., Ayani, R.: Comparison of load balancing strategies on cluster-based web servers. Trans. Soc. Model. Simul. (2001)

10. Bagwaiya, V., Raghuwanshi, S.K.: Hybrid approach using throttled and ESCE load balancing algorithms in cloud computing. In: International Conference on Green Computing Communication and Electrical Engineering (ICGCCEE), pp. 1–6, 6–8 March 2014. doi:10. 1109/ICGCCEE.2014.6921418

11. Wickremaisinghe, B.: CloudAnalyst: a cloudsim-based tool for modelling and analysis of large scale cloud computing environments. MEDC Project, Cloud computing and Distributed Systems Laboratory, University of Melbourne, Australia, pp. 1–44, June 2009

Energy Efficient Cloud Computing

Subramaniyam Kannan and Suresh Rajendran

Abstract Cloud Computing is a developing technology which is revolutionizing the IT Infrastructure due to its high scalability and flexibility in providing the computing resources as services on demand. Cloud data centres are made up of large servers which consume a great amount of power which will in turn increase the operating costs and the environmental impact in terms of CO_2 emissions. Hence, Cloud Computing solutions are extremely effective in terms of processing power and storage but are often criticized for the high amount of energy they require. Therefore, this paper will look at how to model the energy usage of a simple cloud solution, taking into consideration primarily the running of the hardware servers, from the idle to full utilization, as well as the transport of data between servers. An energy saving mechanism named DVFS is used in this paper in order to reduce the energy consumption. Cloudsim toolkit is used to create the cloud data centre and analyze the efficiency of DVFS mechanism.

Keywords Cloud computing · Virtualization · Dynamic Voltage frequency scaling · Cloudsim

1 Introduction

Due to the fast growing internet in tandem with virtualization technology has led to a new form of utility computing which provides computing resources in a flexible and on-demand fashion. It significantly reduces the operational costs as the infrastructure is maintained by the cloud service providers. It is also characterized by pay per use Model in which customers pay only for what they use.

While extremely effective in terms of processing power and storage, cloud computing solutions are often criticized for the high amount of energy they require. Cloud infrastructure is equipped with large and power consuming data centres that

S. Kannan (✉) · S. Rajendran
School of Electronics Engineering, VIT University, Vellore 632014, Tamilnadu, India
e-mail: subramaniyam.k2011@vit.ac.in

© Springer International Publishing Switzerland 2016
V. Vijayakumar and V. Neelanarayanan (eds.), *Proceedings of the 3rd International Symposium on Big Data and Cloud Computing Challenges (ISBCC – 16')*,
Smart Innovation, Systems and Technologies 49, DOI 10.1007/978-3-319-30348-2_14

makes it challenging to model the energy consumption eco-efficiently. Data centres consume about 50–100 times more energy than typical office buildings. Data centres can even consume the same amount of electricity that is used by a city. It leads to extreme heat dissipation due to the high energy consumption of the data centres which would in turn increase the cooling costs and make the servers vulnerable to risks and failures. All these problems will increase the operational costs and environmental impact in terms of CO_2 emissions. Hence it is very important to model and reduce the energy usage of cloud data centres. An eco-efficient mechanism named DVFS is used to reduce the energy consumption which can help to increase the profitability of the Cloud data centres and in turn reduce the environmental impact.

2 Background

2.1 Cloud Computing

Cloud computing is one of the most significant shifts in information technology which has created a huge impact in everyone's life. Cloud Computing can be explained by separating the two terms "Cloud and Computing". Computing is defined as a model consisting of services that are rendered and delivered in a manner similar to utilities such as electricity, water, telephony and gas. In such a model, users can access services depending on their requirements, regardless of where the services are hosted. This model is called as Utility computing and recently (since 2007) being called as cloud computing. The latter term often denotes the infrastructure as a "cloud" from which users can access applications as services from anywhere in the world and on demand. Cloud Computing is basically an internet centric way of computing. The Services are delivered as Utilities using the Internet and hence the name Cloud Computing. According to Armbrust et al. [1] "Cloud computing refers to both the applications delivered as services over the Internet and the hardware and system software in the data centres that provide those services."

2.2 Virtualization

Virtualization is one of the fundamental components of cloud computing, especially in regard to infrastructure as a service. Virtualization is a term that refers to the creation of a virtual version of something, whether hardware, a software environment, storage, or a network [2]. A virtualized environment consists of three major components namely guest, host, and virtualization layer [3]. The guest represents the system component that interacts with the virtualization layer instead of the host. The host represents the actual hardware which manages the guest. The virtualization layer is responsible for creating the environment where the guest can operate.

Virtualization is used for creating a secure, customizable, and isolated execution environment for running applications, without affecting other user's applications [3]. The basis of this technology is the ability of a computer program or a combination of software and hardware to emulate an executing environment separate from the one that hosts such programs. For example, we can run Linux OS on top of a virtual machine, which itself is running on Windows OS. In virtualization, the guest is represented by a system image that consists of an operating system and installed applications. These are installed on top of virtual hardware that is controlled and managed by the virtualization layer, also called the virtual machine manager (VMM) or hypervisor. The host is represented by the physical hardware, and in some cases the operating system that defines the environment where the hypervisor is running.

2.3   CloudSim

In this paper, Cloudsim is used in order to create a cloud data centre and to analyze the Performance of DVFS. Cloudsim [4, 5] is a simulation toolkit which allows modelling and simulation of Cloud Computing systems. It is used for modelling Cloud system components such as data centres, virtual machines (VMs) and resource provisioning policies. Presently, it supports modelling and simulation of both single and inter-networked clouds (federation of clouds). It is widely used for investigation on Cloud resource provisioning and energy-efficient management of data centre resources.

Figure 1 shows the multi-layered architecture of the Cloudsim software framework. The Cloudsim simulation layer is responsible for allocating the hosts to the Virtual Machines and to manage the execution of Cloudlets or the tasks assigned to

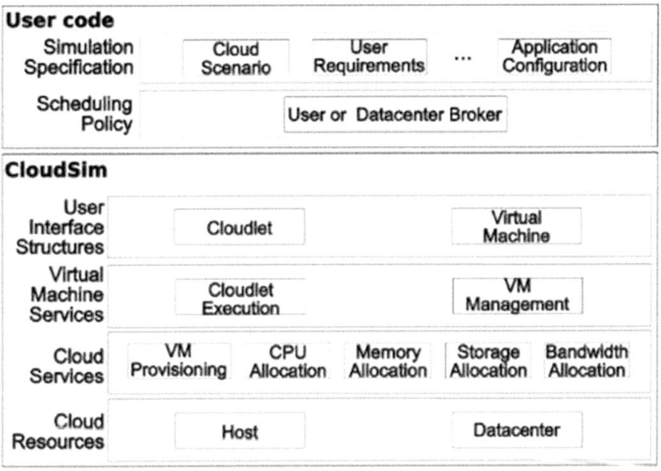

Fig. 1 Cloudsim architecture

the Virtual Machines. The top-most layer in the Cloudsim stack is the User Code which is used for specifying the Number of Hosts, Virtual Machines and Cloudlets along with their configuration.

The Hosts are assigned to one or more VMs using the VM allocation Policy. The Default VM allocation policy is used which allocates the VMs to the Hosts on First Come First Serve (FCFS) basis. At the host level, VM scheduling is used for allocating the processing power of host to VMs. At the VM level, cloudlet Scheduling is used in order to assign a fixed amount of available power to perform individual tasks. Cloudsim supports space-shared scheduling policy which as- signs specific CPU cores to specific VMs and time-shared scheduling policy which dynamically distributes the capacity of a core among VMs or assigns processing cores to VMs on demand. Thus, a data centre manages several hosts which will in turn manage Virtual Machines during their life cycle.

Cloudsim provides an abstract class called Power Model for the simulation of power consumption models and power management techniques. This class provides a function called getPower(), which returns the power consumed depending on the CPU Utilization of the host. The DVFS technique is implemented using the PowerModel class.

2.4 Management of Power Consumption

The energy which is supplied to the data centres is used for computational operation, cooling systems, networks, and other overheads. In case of computational opera- tions, there are some energy-saving techniques which can be deployed in order to monitor and control energy consumption. In terms of energy efficiency, it can be said that data centre X is better than data centre Y if X can consume less power and process the same workload as Y, or X can consume the same power but with more workload compared with Y. The following are the existing techniques used for reducing power consumption without degrading the performance in the data centres:

1. *Virtual Machine/Server Consolidation*: Firstly, a study by Corradi et al. indi- cates that Virtual Machine (VM)/Server consolidations is used for reducing the power consumption of cloud data centres [6]. This technique allocates more number of virtual machines on less number of host machines so that the host machines are fully utilized. For example, when there are two VMs, both the VMs are allocated to the same physical server instead of allocating each VM to a separate physical server. This technique is called as VM/Server consolidation which reduces the operational costs and increases the efficiency of energy usage by fully utilizing the host machines. However, the number of VMs in one host machine cannot be increased beyond a threshold as it will degrade the perfor- mance of VMs.

2. *Power and Thermal Management*: The study conducted by Pakbaznia et al. states that power and thermal management (PTM) technique can be used in order to

improve the energy efficiency of data centres [7]. In this approach, Server consolidation is used in accordance with efficient cooling in order to reduce the overall power consumption. The Server consolidation will reduce the number of physical servers that are kept ON and the supplied cold air temperature is also maintained in order to reduce the power consumption of the cloud data centre. But the number of cloudlets or the incoming workload in terms of requests per second should be known in prior in order to use this technique.

3. *Dynamic Voltage and Frequency Scaling (DVFS)*: Dynamic Voltage and Frequency Scaling (DVFS) is the most eco-efficient technique used in order to reduce the energy consumed in cloud data centre. This technique alters the CPU power consumption depending on the workload offered [8]. It is a hardware technology that dynamically changes the voltage and the frequency of the processor during execution depending on the CPU utilization. DVFS technology [9] can adjust the system voltage and frequency without having to restart the power supply. When the CPU working voltage is reduced depending on the CPU utilization, a large amount of energy is saved. The dynamic power consumption is defined by multiplying the voltage square with the system frequency.

$$P_d = a * c * v^2 * f \qquad (1)$$

where P_d is the dynamic power consumed, a is the switching activity, c is the physical capacitance, v is the supply voltage, and f is the clock frequency. The values of switching activity and capacitance are determined by the low-level system design. DVFS is the Dynamic Power Management (DPM) technique [8] which reduces the supply voltage and clock frequency in order to reduce the dynamic power consumed. The main idea of this approach is to intentionally scale down the CPU performance, when it is not fully utilized, by decreasing the voltage and frequency of the CPU. From Eq. 1, this should result in a cubic reduction of dynamic power consumption. DVFS is supported by most modern CPUs including mobile, desktop and server systems. The reduction of CPU frequency and the voltage will result in the degradation of the system performance and in turn increases the execution time. The DVFS decreases the execution speed of a task as the CPU frequency and voltage is decreased in order to achieve significant reduction in power consumption. The power consumption can be saved effectively only by reducing both the frequency and the voltage.

3   Design Approach

As discussed before, Cloudsim toolkit is used to implement Eco-efficient mechanisms in Cloud data centres. The main objective of the experiments is to calculate the energy consumed in data centres. The following steps are performed in order to model an eco-efficient cloud data centre.

- Create a Data Centre with hosts and then creating Virtual Machines over the hosts using Virtualization techniques and then assigning Cloudlets to the Virtual Machines. This is done using the Cloudsim tool kit by just specifying the technical parameters of the host and virtual machines.
- Calculate the energy Consumed by each host computer depending on its CPU utilization for performing the specific tasks assigned to them. The energy consumed is calculated by monitoring the CPU utilization and using the getPower() function present in the predefined class PowerModel [10, 11].
- Calculate the total energy consumed in order to perform all the tasks. This is the actual energy consumed by the data centre without implementing any Energy saving algorithm (Non Power Aware Mechanism).
- Implementing the energy saving algorithm called Dynamic Voltage and Frequency Scaling (DVFS) in the Cloudsim toolkit.
- Calculate the total energy consumed by the data centre with DVFS implementation.

3.1   Non Power Aware (NPA) Mechanism

NPA is the mechanism which calculates the energy consumed by the host computers without any power saving algorithm. In this mechanism the CPU utilization is calculated in order to determine the energy consumed by the host computers. In Cloudsim, the capacity of the host machine, the capacity of VM, and the cloudlet requested by the user is represented by using MIPS (million instructions per second). Each cloudlet will be distributed to VMs on different hosts. VM_{MIPS} is the amount of MIPS required for the VM to perform a particular task and $HOST_{MIPS}$ represents the amount of MIPS the host can support. The CPU utilization of a virtual machine V_{CPU} is calculated by Eq. 2 and the average CPU utilization of all the VMs (V_{avg}) is calculated using Eq. 3.

$$V_{CPU} = \frac{VM_{MIPS}}{HOST_{MIPS}} \tag{2}$$

$$V_{AVG} = \frac{\sum_{i=1}^{n} V_{CPU}(i)}{n} \tag{3}$$

where n is the number of virtual machines created on one host.

For Example, when there are 10 hosts and 20 VMs then each host supports 2 VMs. If the total capacities of hosts are 2660 MIPS, and that of the VM are 2500 MIPS then $V_{CPU} = \frac{2500}{2600} = 0.94$.

As there is no energy saving mechanism, the host computers consume maximum energy in NPA mechanism. In this mechanism, the total energy consumed is found by multiplying the total power required to switch on the host which is calculated using PowerModel class and the time period. Hence the total energy depends on the power consumed to switch on the host and is independent of the CPU utilization which implies the same amount of power is consumed for both the extremes of CPU utilization (too low and too high).

3.2 Dynamic Voltage and Frequency Scaling (DVFS) Mechanism

In DVFS, the measured CPU utilization is used for deter- mining power consumption. The MIPS needed to run a task in the VM and the available MIPS of the host is monitored at regular intervals and the CPU Utilization is calculated by using Eqs. 2 and 3. Depending on the CPU Utilization the DVFS mechanism will adjust the Supply voltage in order to reduce the consumed energy. When the utilization is too low, the voltage will be lowered to reduce the power consumption. In contrast, when the utilization is too high, the voltage will be increased in order to maintain users quality of service.

4 Technical Specifications

Cloudsim version 3.0.3 is used for the simulation as it provides power provisioning techniques which can be directly used. Cloud data centre is characterized by three classes namely Cloudlet, VM and Host.

- Cloudlet is the task or the application that the VM performs. It is characterized by Instruction length and data transfer overhead. The specifications for the cloudlets are given below in Table 1.
- Virtual Machine is the one which is virtually created over the host computer. In order to process the cloudlets efficiently without any overload, 4 types of VMs with different configurations are used in this project. The different types of VMs with the specifications are shown in Tables 2 and 3.

Table 1 Cloudlets

Cloudlets	
Cloudlet length	1,728,000 (MIPS)
No. of processingelements (Cloudlet PES)	1
Utilization seed	1
File size	300 (Bytes)
Output size	300 (Bytes)

Table 2 Four types of virtual machine

High-CPU medium instance	2.5 EC2 Compute Units, 0.85 GB
Extra large instance:	2 EC2 Compute Units, 3.75 GB
Small instance:	1 EC2 Compute Unit, 1.7 GB
Micro instance:	0.5 EC2 Compute Unit, 0.633 GB

Table 3 Virtual machines

Virtual machines	
No. of VM types	4
MIPS	2500, 2000, 1000, 500
RAM	870,1740, 1740, 613
No. of processing elements (VM PES)	1
Bandwidth	100 Mb/s
Storage	2.5 GB

- Host is the actual physical computer which is present in the data centre. In order to create VMs with different Configurations, 2 types of Hosts with different processors are used. The different types of Host processors with the specifications are given in Tables 4 and 5.

The Scheduling interval is taken as 300 s and the simulation limit is taken as 24 h. The values which are used have already been experimented and validated by Calheiros and Buyya [10, 11] to quantify Cloudsim efficiency in simulating Cloud Computing environments. Also, these technical specifications represent the typical specifications of the machines found in a data centre.

Table 4 Two types of Hosts

HP ProLiant ML110 G4	(1 × [Xeon 3040 1860 MHz, 2 cores], 4 GB)
HP ProLiant ML110 G5	(1 × [Xeon 3075 2660 MHz, 2 cores], 4 GB)

Table 5 Hosts

Hosts	
No. of hosts types	2
MIPS	1860, 2660
RAM	4096, 4096
No. of processing elements (HOST PES)	2
Bandwidth	1 Gb/s
Storage	1 TB

5 Results

Firstly, the actual energy consumed by the data centre is calculated by varying the three main parameters namely no. of hosts, no. of VMs and no. of Cloudlets. Secondly, the same set of scenarios are conducted with the deployment of a DVFS mechanism. Each scenario will have a number of experiments that will vary depending on the number of hosts, number of VMs and number of Cloudlets. The Scenarios are started with default values of 10 hosts, 20 VMs, and 20 Cloudlets and then all the parameters will be incremented to find variation of energy consumption with the key parameters and to determine the efficiency of DVFS Mechanism. The maximum values for each scenario that the Cloudsim can handle will also be determined.

In scenario-1, the No. of hosts are doubled whereas the No. of VMs and Cloudlets are kept constant at 20. The simulation is started with 10 hosts, 20 VMs, and 20 Cloudlets. The maximum values that the cloudsim can handle are 81,920 hosts, 20 VMs, and 20 Cloudlets.

Based on the results shown in Fig. 2 when the number of hosts are doubled, the energy consumption for the scenario without the deployment of DVFS also doubles because more host machines will be running for the same workload which leads to wastage of energy. But with the deployment of DVFS, the energy consumed is almost constant because only the number of host machines that are required to perform the workload will be running and the rest will not be running which will reflect in energy saving.

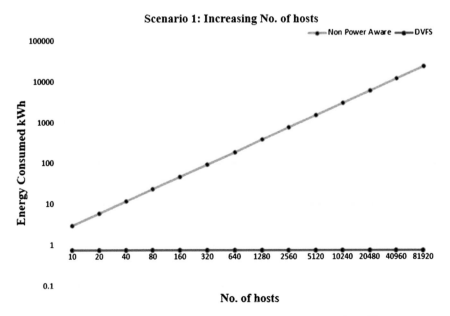

Fig. 2 Energy comparison between DVFS and NPA with increasing number of hosts

In scenario-2, the No. of hosts are doubled whereas the No. of VMs and Cloudlets are kept constant at 20. The simulation is started with 10 hosts, 20 VMs, and 20 Cloudlets. The maximum values that the cloudsim can handle are 81,920 hosts, 20 VMs, and 20 Cloudlets.

Based on the results shown in Fig. 3 when the number of hosts are doubled, the energy consumption for the scenario with and without the deployment of DVFS remains as a constant. Each VM has one processing element which is used to execute one cloudlet. As the number of VMs is 20 which implies only 20 cloudlets can be executed at one time and the remaining cloudlets are postponed. Hence the energy consumed is constant as only 20 cloudlets are executed in all the experiments. It is also clear that DVFS mechanism reduces the energy consumption by 75 % when compared to the scenario without deployment of DVFS mechanism.

In scenario-3, the No. of hosts and VMs are doubled whereas the No. of cloudlets are kept constant at 20. The simulation is started with 10 hosts, 20 VMs, and 20 Cloudlets. The maximum values that the cloudsim can handle are 81,920 hosts, 1,63,840 VMs, and 20 Cloudlets.

Based on the results shown in Fig. 4 when the number of hosts and VMs are doubled, the energy consumption for the scenario without deployment of DVFS doubles because more host machines will be running for the same workload which leads to wastage of energy. But with the deployment of DVFS, the energy consumed is almost constant because only the number of host machines that are required to perform the workload will be running and the rest will not be running which will reflect in energy saving (Same as Scenario 1).

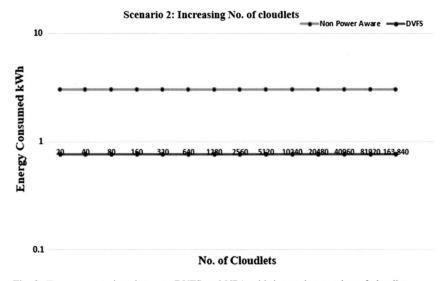

Fig. 3 Energy comparison between DVFS and NPA with increasing number of cloudlets

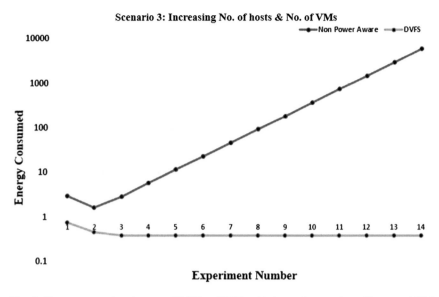

Fig. 4 Energy comparison between DVFS and NPA with increasing number of hosts and VMs

In scenario-4, the No. of hosts and cloudlets are doubled whereas the No. of VMs are kept constant at 20. The simulation is started with 10 hosts, 20 VMs, and 20 Cloudlets. The maximum values that the cloudsim can handle are 81,920 hosts, 20 VMs, and 1,63,840 Cloudlets.

Based on the results shown in Fig. 5 when the number of hosts and Cloudlets are doubled, the energy consumption for the scenario without the deployment of DVFS doubles but with the deployment of DVFS, the energy consumed is almost constant. The results obtained are same as the results of scenario 1. From scenario 4 and scenario 2 one can conclude that increasing the number of cloudlets will not have an impact on energy consumption.

In scenario-5, all the three parameters namely the No. of hosts, No. of VMs and No. of cloudlets are doubled. The simulation is started with 10 hosts, 20 VMs, and 20 Cloudlets. The maximum values that the cloudsim can handle are 2560 hosts, 5120 VMs, and 5120 Cloudlets. When the parameters are further doubled then the variation of energy consumed is no more linear which clearly indicates the Cloudsim's inability to handle it.

Based on the results shown in Fig. 6 when the number of hosts, VMs and Cloudlets are doubled, the energy consumed for both the scenarios with and without the deployment of DVFS doubles. But the energy consumed by the scenario without DVFS Mechanism consumes 4 times more energy than the energy consumed by DVFS Mechanism which implies 75 % of energy is saved by using DVFS.

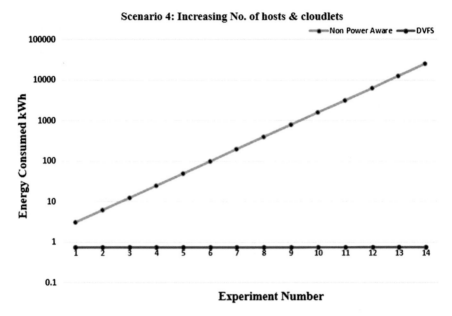

Fig. 5 Energy comparison between DVFS and NPA with increasing number of hosts and cloudlets

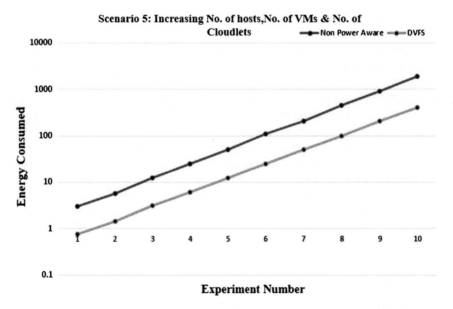

Fig. 6 Energy comparison between DVFS and NPA with increasing number of hosts, VMs and cloudlets

6   Design Constraints and Trade Offs

The Major realistic design constraints faced during the project design are:

- Cloudsim Toolkit is an Open Source software and it is in developing stage. It was very difficult to understand the toolkit as there were no user manual and tutorials available. Reverse engineering was used in order to understand the working of the tool kit.
- Cloudsim is not user friendly as it requires the users to have a prerequisite knowledge about Java Programming language
- Cloudsim is a raw java code without any graphical user interface which meant the entire simulation should be done by working with the libraries directly.
- The simulation of experiments was very long and because of the randomness of the toolkit each experiment was run multiple times. As the number of parameters increases, the simulation time also increases by great extent.
- The Cloudsim can handle only a certain amount of resources and it gives undesirable results when the resources are increased beyond the threshold limit.

The Significant trade-offs encountered during the project are:

- Cloud Sim Vs Green Cloud: Green Cloud is built on top of NS2 that can be used to determine the consumed energy of a data centre but its simulation takes long time and requires high memory usage. Hence its scalability is limited only to small data centres. As the Cloud Sim is scalable and have low simulation overheads, it is Preferred over Green Cloud.
- Energy Consumption VS Performance: DVFS dynamically reduces the voltage and Frequency of CPU in order to reduce the energy consumption of the entire data centre. The reduction of CPU frequency and the voltage will result in the degradation of the system performance and increases the execution time. Hence, DVFS compromises on the performance of the system in order to reduce the overall power consumption.

7   Conclusion

From Table 6, one can see that scenario 1 and scenario 4 are same. Considering scenario 1, 2 and 4, one can conclude that the number of cloudlets will not have an impact on the energy consumption of the data centres. One can also note that Scenario 1 and Scenario 3 are same which implies increasing the number of VMs alone will not have an impact on energy consumption. Hence the number of hosts is the significant parameter which will have an impact on the energy consumption of cloud data centres. Thus, from the results obtained, the deployment of DVFS in a data centre will reduce energy consumption significantly as it manipulates the voltage and clock frequency of a CPU depending on the CPU utilization.

Table 6 Comparison of different scenarios

Scenario	Variation of energy consumed with NPA mechanism	Variation of energy consumed with DVFS mechanism	Percentage of energy reduced
Scenario-I: Doubling the number of hosts	Doubles	Constant	Dynamically changes for each experiment but the Percentage of energy reduced increases as the number of hosts doubles
Scenario-II: Doubling the number of cloudlets	Constant	Constant	About 75 percent
Scenario-III: Doubling the num-ber of hosts and VMs	Almost double	Constant	Dynamically changes for each experiment but the Percentage of energy reduced increases as the number of hosts and VMs doubles
Scenario-IV: Doubling the number of hosts and cloudlets	Double	Constant	Dynamically changes for each experiment but the Percentage of energy reduced increases as the number of hosts and cloudlets doubles
Scenario-V: Doubling the number of hosts, VMs and cloudlets	Double	Double	About 75 %

References

1. Armbrust, M., Fox, A., Griffith, R., Joseph, A.D., Katz, R.H., Kon- winski, A., Lee, G., Patterson, D.A., Rabkin, A., Stoica, I., Zaharia, M.: Above the clouds: A Berkeley view of cloud computing. EECS Department, University of California, Berkeley, Tech. Rep. UCB/EECS- 2009-28, Feb 2009
2. Zhang, Q., Cheng, L., Boutaba, R.: Cloud computing: state-of-the-art and research challenges. J. Internet Serv Appl **1**(1), 7–18 (2010)
3. Buyya, R, Vecchiola, C., Selvi, S.T.: Mastering cloud computing: foundations and applications programming. Newnes, 2013
4. Calheiros, R.N., Ranjan, R., Beloglazov, A., De Rose, C.A., Buyya, R.: Cloudsim: a toolkit for modeling and simulation of cloud computing environments and evaluation of resource provisioning algorithms. Soft ware: Pract. Exp. **41**(1), 23–50 (2011)
5. Buyya, R., Ranjan, R., Calheiros, R.N.: Modeling and simulation of scalable cloud computing environments and the cloudsim toolkit: challenges and opportunities. In: International Conference on High Performance Computing & Simulation, 2009 (HPCS'09), pp. 1–11. IEEE (2009)
6. Corradi, M.F., Foschini, L.: Increasing cloud power efficiency through consolidation techniques. In: IEEE Symposium on Computers and Communications (ISCC), pp. 129–134, June 2011
7. Pakbaznia, E., Ghasemazar, M., Pedram, M.: Temperature-aware dynamic resource provisioning in a power-optimized datacenter. In: Design, Automation Test in Europe Conference Exhibition (DATE), pp. 124–129, Mar 2010

8. Kliazovich, D., Bouvry, P., Audzevich, Y., Khan, S: Greencloud: a packet-level simulator of energy-aware cloud computing data centers. In: Global Telecommunications Conference (GLOBECOM 2010), pp. 1–5. IEEE, Dec 2010
9. Lee, L.-T., Liu, K.-Y., Huang, H.-Y., Tseng, C.-Y.: A dynamic re- source management with energy saving mechanism for supporting cloud computing. Int. J. Grid Distrib. Comput. **6**(1), 67–76 (2013)
10. Calheiros, R.N., Buyya, R.: Energy-efficient scheduling of urgent bag-of-tasks applications in clouds through DVFS. In: IEEE 6th International Conference on Cloud Computing Technology and Science (CloudCom), pp. 342–349. IEEE (2014)
11. Gurout, T., Monteil, T., Costa, G.D., Calheiros, R.N., Buyya, R., Alexandru, M.: Energy-aware simulation with DVFS. Simul. Model. Pract. Theory **39**, 76–91 (2013)

Power Optimization System for a Small PC Using Docker

G. Anusooya and V. Vijayakumar

Abstract In this paper, we have proposed a technique using Raspberry Pi board as a small PC with Docker to implement the concept of virtualization. Docker which is used for running distributed applications in an open platform. This implementation shows how efficient it will be to implement server virtualization technique on simple computers like raspberry pi. The outcome will be a virtualized Raspberry Pi which will run various applications on the hypervisor. This method involves booting Arch Linux into the Raspberry Pi, and then installing a hypervisor into it. Virtual machines will be installed and keep running in the background. If the power consumed with virtualization is less than the power consumed when these processes are running in different hardware without any virtualization then it is concluded that virtualization is saving power and also reducing the hardware involved. Power monitoring software is used to measure the power consumed by the raspberry pi.

Keywords Power optimization · Raspberry pi · Docker · Virtualization

1 Introduction

1.1 Green Computing

Green computing the current trend which is responsible for eco-friendly environment [1]. The main aim of green computing in terms of computer technology is improve energy efficiency of central processing units (CPUs), peripherals and

G. Anusooya (✉) · V. Vijayakumar
VIT University, Chennai Campus, Chennai, Tamil Nadu, India
e-mail: anusooya.g@vit.ac.in

V. Vijayakumar
e-mail: vijayakumar.v@vit.ac.in

© Springer International Publishing Switzerland 2016   173
V. Vijayakumar and V. Neelanarayanan (eds.), *Proceedings of the 3rd International
Symposium on Big Data and Cloud Computing Challenges (ISBCC – 16')*,
Smart Innovation, Systems and Technologies 49, DOI 10.1007/978-3-319-30348-2_15

servers. The main aim of green computing is to reduce the carbon emission, to maximize the energy efficiency during the product life cycle and to improve the disposal of the e-waste.

Reducing the carbon emissions and taming energy efficiency is the major task of Green Computing. So, reducing the carbon emission is the major task. This can be achieved in terms of virtualization.

1.2 Virtualization

Virtualization [2] which makes the physical devices virtual for the resource, such as a server, storage device, network. An operating system which act as multiple framework for the execution of many environments. Our single hard drive which is partitioned into multiple drives is also one way of virtualization. We use the drives as if we execute only one applications at a time but where multiple operation is taking place virtually. The term virtualization is becoming trending now, and as a result the term is now associated with a number of computing technologies like:

- Combination of multiple storage units in a network which act as a single storage unit is known as storage virtualization.
- Making a single physical server to work virtually with multiple virtual servers for various operations is called as server virtualization.
- Operating system-level virtualization is a type of server virtualization works in the kernel layer.
- A logical segmentation of a single physical network resource is said to be network virtualization.

1.3 Docker

To build, ship and run distributed applications the developers uses Docker [3] which is an open platform. It consists of Docker Engine, a portable, lightweight runtime and packaging tool, and Docker Hub, a cloud service for sharing applications and automating workflows. Docker enables apps to be quickly assembled from components and eliminates the friction between development, QA, and production environments. As a result, it can ship faster and run the same app, unchanged, on laptops, datacentre VMs, and any cloud.

Docker uses the containers, which is very fast and light weight process. It is similar to any real life container, all the required resources for that application to work is made available inside the container. The Docker engine communicates with the operating system, thus eliminating the need of a guest operating system making it a lot quicker by reducing the overhead.

Components of Docker

- Images
- Registries
- Containers

Images

A Docker image is a read-only template. For example, an image could contain an Ubuntu operating system with Apache and your web application installed. Images are used to create Docker containers. Docker helps to provide a simple way to build new images or update already existing images, or we can download Docker images that other developers have already created. Docker images are the build component of Docker.

Registries

Docker registries store the images. These are public or private stores from which you can upload or download the images. The public Docker registry is also called the Docker Hub. It provides the collection of existing images for the use of new application development. These can be images you create for yourself or you can use the same images that others have previously created. Docker registries are the distribution component of Docker.

Containers

Docker containers are same as the directories we use. It holds everything that is needed for an application to run. Each container is created with the Docker image. Docker containers can be run, started, stopped, moved, and deleted. Each container is an unique and secure platform for running applications. Docker containers are the main source for running the component of Docker.

1.4   Arch Linux

Arch Linux [4] is an independently developed, i686/x86-64 general purpose GNU/Linux to suit any role of versatile distribution. The development of Arch Linux mainly focuses on simplicity, minimalism, and code elegance. It is installed as a minimal base system, configured by the user upon which their own ideal environment is assembled by installing only what is required or desired for their unique purposes. GUI configuration utilities are not officially provided, and most system configuration is performed from the shell by editing simple text files. Arch endeavour to stay in the edge, and typically offers the latest stable versions of most software.

Docker is placed in the Arch Linux in an Raspberry Pi board for the implementation analysis of the power consumption.

2 Existing Work

1. *Power Consumption benchmarking for green computing* [5]: It explored the
   techniques for accurately measuring of power consumption of computers in
   screensavers sleep mode, hard disk sleep mode, system stand by etc.,
2. *Optimization of operating systems towards green computing* [6]: It focuses on
   green computing by optimizing operating systems and scheduling of hardware
   resources. The architecture for optimized operating systems towards green
   computing to enable computer's power management features in operating systems for various techniques like virtualization, Terminal servers, Shared memory and power management etc.
3. *Profile-based Optimization of Power Performance by using Dynamic Voltage
   Scaling on a PC cluster* [7]: They proposed an optimization algorithm to select a
   gear using the execution and power profile by taking the transition overhead into
   an account. They have built and designed a power-profiling system, named
   PowerWatch. With this system they examined the effectiveness of their optimization algorithm on two types of power-scalable clusters namely crusoe and
   turion. They have created a positive effect without the impact of performance.
4. *Power Analysis and Optimization Techniques for Energy Efficient Computer
   Systems* [8]:

   It describes different techniques addressing a major challenge of reducing power
consumption of today's computer systems.

3 Proposed Work

The main goal was to implement server virtualization technique in Raspberry pi2.
The project was intended at reducing the power and hardware required to do a
certain job. Virtualization is a recent trend being followed in data center [9] to cut
down costs. This also saves a lot of power because the work of two to three
machines can be done by one.

3.1 Raspberry Pi with Docker

Implementation required installation of a hypervisor to virtualize the Raspberry Pi.
Arch Linux was chosen as the supporting operating system as it is the best for
customization and also for embedded systems. Docker was chosen as the hypervisor as the technology being used docker is quite new in the virtualization sector
and is far more efficient than the conventional hypervisors available.

Fig. 1 Architecture diagram

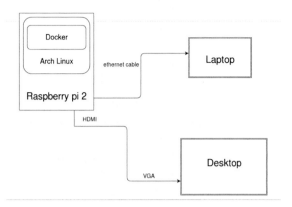

The raspberry pi2 board and virtualizing it using the appropriate hypervisor and there by analyzing the power to show that virtualization is a power efficient technique and implementing this would reduce the cost to a great extent (Fig. 1).

The Raspberry pi here is the computing device. The Raspberry pi is virtualized using the hypervisor called Docker. The base OS as represented in the diagram is the arch Linux and the hypervisor is installed over the base OS.

We are using raspberry pi2 as the hardware because it has the capacity to handle the virtualization and is also known as a mini pc. We can manage to run any OS on this. We have chosen Arch Linux as the OS to be used. We are using Docker software for the Virtualization of Raspberry Pi. To interface the same with the laptop we use an Ethernet Cable as shown in the figure. This is to create a Remote Desktop environment in the system.

3.2 Implemented Work

The Raspberry pi is connected to a laptop remotely using an Ethernet cable. This was done using the Vncserver and Vinagre remote desktop client. The raspberry pi was virtualized and multiple sessions were run to show that the using the same amount of power were able to perform the work of multiple hardware on a single hardware.

The laptop is used as an input output device. A remote server was installed inside the Raspberry Pi2 and a remote desktop client was installed inside the laptop. The Ethernet standard allows us to connect the laptop to the raspberry pi using the IP address of the pi board. The ssh protocol was used to connect with the raspberry pi using the internet protocol address of the pi. The raspberry pi was booted with Arch Linux (Fig. 2), so once you connect the raspberry pi the Arch Linux boots up.

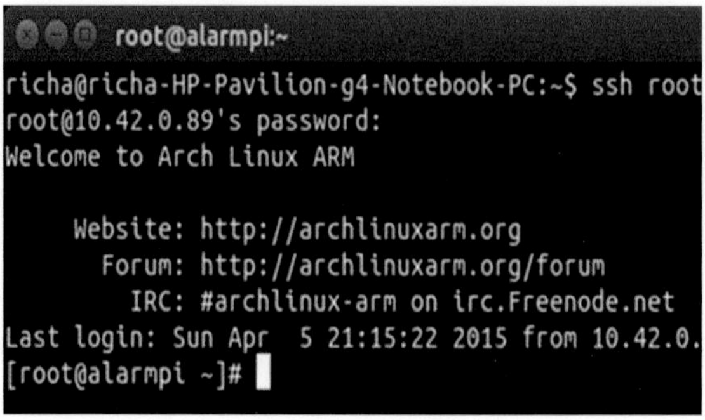

Fig. 2 Logging to Arch Linux

Arch Linux is a command line operating system so primarily we get to work on the command line interface only. We installed a desktop environment called Cinnamon to present Arch Linux with a desktop environment.

Once the Arch Linux is available to work on, we install Docker. Once the docker is installed the applications in the repository can be pulled to the system. An application called Own cloud was pulled down for the demonstration. Multiple sessions of docker containers with this application were run.

While this was being done a tool to measure the power consumption Fig. 3 (Powertop) was run in the background to measure the power consumed.

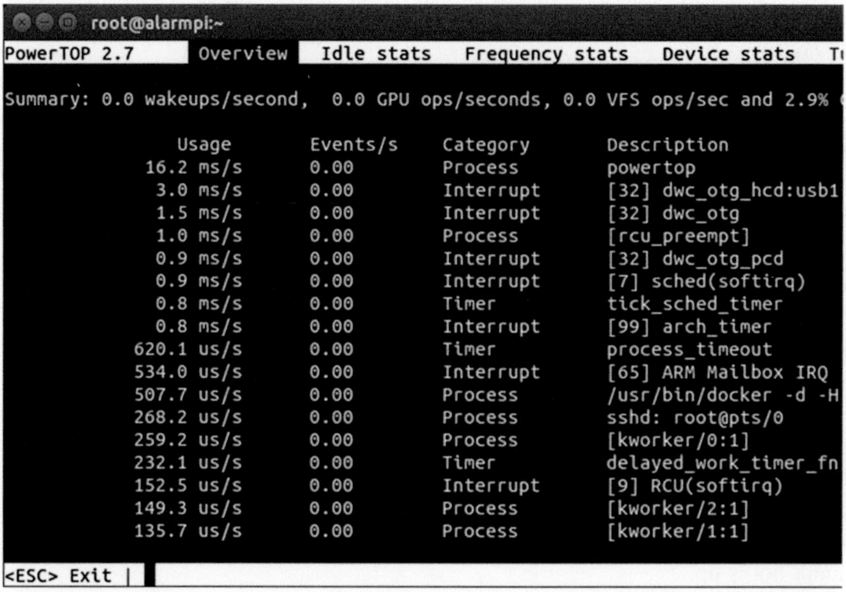

Fig. 3 PowerTop

The tool measures power (frequency measure) Fig. 4 in percentages and the power percentage remained almost constant while running multiple docker sessions.

This was again confirmed by using vcgencmd command. This command is used to check the voltages Fig. 5 and various other values like clock cycles of different components available on the pi board.

Analysis

The computation power of the raspberry pi might not be as great as a server but this can be compensated by using an array of raspberry pi boards which will still cost less than a server and with virtualization it can do the work more than one servers, thus cutting down cost by a huge amount along with power and the amount of hardware.

A typical server costs around 26,000. A server consumes around 18 kW per day, which converted to energy is 1800 kWh. Our project consumes around 100 Wh. If make the system efficient in the area of computing power, by taking an array of 5 raspberry pi boards it will still make the power consumed to 500 Wh. If the system takes the load this is a huge cut down. But analysing the load is yet to be done, so that is a disadvantage of our system compared to the existing system.

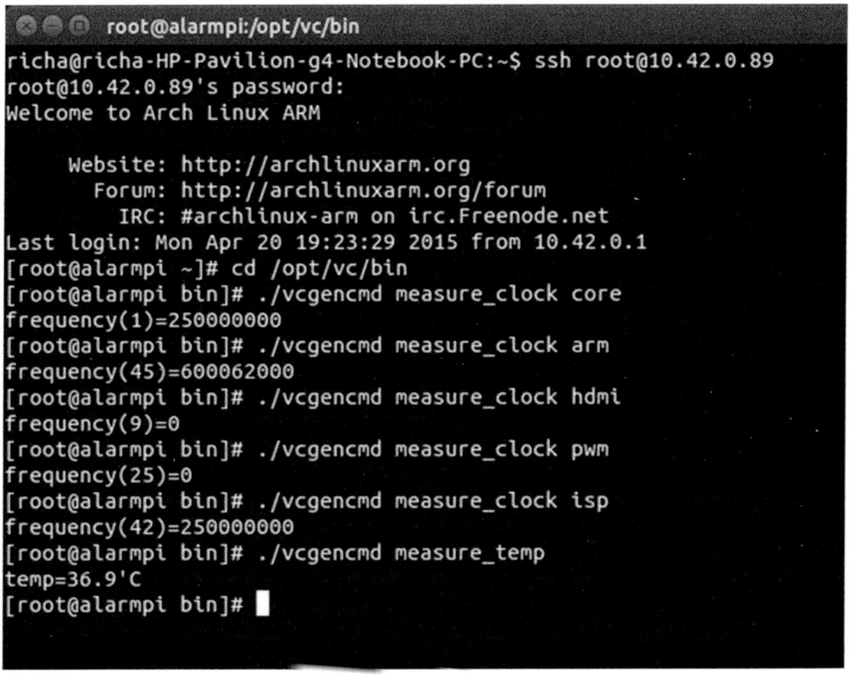

```
root@alarmpi:/opt/vc/bin
richa@richa-HP-Pavilion-g4-Notebook-PC:~$ ssh root@10.42.0.89
root@10.42.0.89's password:
Welcome to Arch Linux ARM

        Website: http://archlinuxarm.org
          Forum: http://archlinuxarm.org/forum
            IRC: #archlinux-arm on irc.Freenode.net
Last login: Mon Apr 20 19:23:29 2015 from 10.42.0.1
[root@alarmpi ~]# cd /opt/vc/bin
[root@alarmpi bin]# ./vcgencmd measure_clock core
frequency(1)=250000000
[root@alarmpi bin]# ./vcgencmd measure_clock arm
frequency(45)=600062000
[root@alarmpi bin]# ./vcgencmd measure_clock hdmi
frequency(9)=0
[root@alarmpi bin]# ./vcgencmd measure_clock pwm
frequency(25)=0
[root@alarmpi bin]# ./vcgencmd measure_clock isp
frequency(42)=250000000
[root@alarmpi bin]# ./vcgencmd measure_temp
temp=36.9'C
[root@alarmpi bin]# 
```

Fig. 4 Frequency measure

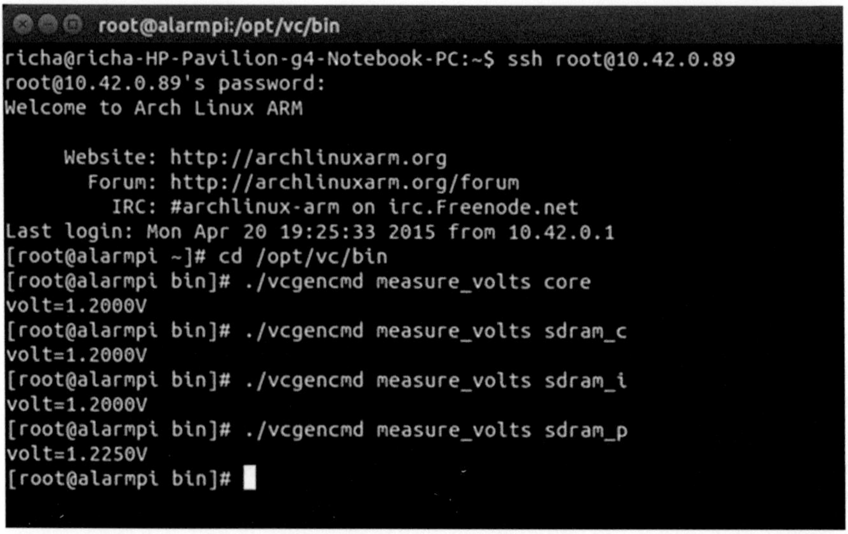

```
○ ○ ○   root@alarmpi:/opt/vc/bin
richa@richa-HP-Pavilion-g4-Notebook-PC:~$ ssh root@10.42.0.89
root@10.42.0.89's password:
Welcome to Arch Linux ARM

     Website: http://archlinuxarm.org
       Forum: http://archlinuxarm.org/forum
         IRC: #archlinux-arm on irc.Freenode.net
Last login: Mon Apr 20 19:25:33 2015 from 10.42.0.1
[root@alarmpi ~]# cd /opt/vc/bin
[root@alarmpi bin]# ./vcgencmd measure_volts core
volt=1.2000V
[root@alarmpi bin]# ./vcgencmd measure_volts sdram_c
volt=1.2000V
[root@alarmpi bin]# ./vcgencmd measure_volts sdram_i
volt=1.2000V
[root@alarmpi bin]# ./vcgencmd measure_volts sdram_p
volt=1.2250V
[root@alarmpi bin]# █
```

Fig. 5 Volts measure

All considered the proposed system will be cheaper. The power consumed by the monitor is not considered because it is the same in both the cases.

The powertop estimated 5.1 % of the total power consumption by the CPU when two applications were running on the pi.

4 Conclusion and Future Work

The server virtualization technique was implemented on Raspberry pi2, which enabled to run multiple Docker sessions at almost the same power requirement thus establishing that using the virtualization technique we can reduce the power and also hardware requirement. This technique when employed in large datacentres will reduce the cost significantly and also reduce the number of servers significantly making the data centre small and efficient.

The plan in the future is to implement the virtualized small computers in a data centre. The aim is to show that virtualizing the environment would lower the power consumed and also the hardware required. The research can also be extended into various power saving techniques like frequency scaling and also apply various intelligent algorithms to make an idol power consumption model which will give us the exact amount of power required to run any system.

References

1. Vandana, L., Sudhir, K.S., Nidhi G.: Going green:-computing for a sustainable future for economy, environment and eco-friendly technology. Int. J. Adv. Res. Comput. Sci. Software Eng. Christ University, India, IJARCSSE, 2014
2. https://en.wikipedia.org/wiki/Virtualization
3. https://www.docker.com/
4. https://www.archlinux.org/
5. Mujtaba, T.: Computer power consumption benchmarking for green computing, A Thesis Presented to the Faculty of the Department of Computing Sciences Villanova University, April 2008
6. Appasami, G.: Optimization of operating systems towards green computing. Int. J. Comb. Opt. Probl. Inf. 2(3), 39–51. ISSN: 2007-1558 (2011)
7. Hotta, Y., Sato, M., Kimura, H., Matsuko, H., Baku, T., Takahashi, D.: Profile-based optimization of power performance by using dynamic voltage scaling on a PC cluster. Graduate School of Systems & Information Engineering, University of Tsukuba, IEEE 2006
8. Chedid, W., Yu, C., Lee, B.: Power analysis and optimization techniques for energy efficient computer systems. Adv. Comput. 2005
9. Da Silva, R.A.: Green computing—power efficient management in data centers using resource utilization as a proxy for power. The Ohio State University (2009)

A Survey on Desktop Grid Systems-Research Gap

Neelanarayanan Venkataraman

Abstract To harvest idle, unused computational resources in networked environments, researchers have proposed different architectures for desktop grid infrastructure. The desktop grid system provide high computational capability at low cost and this motivates its use. However, there are several distinct differences between them in terms resource participation, resource sharing nature, application support, service quality, deployment effort required etc. Building a desktop system has to consider resource's heterogeneity, non-dedication, volatility, failures, security, etc. Therefore, it is important to comprehend current research approaches in desktop grid system development to identify research gaps. In this paper, we propose a taxonomy for studying various desktop grid systems. We also present the strength and weakness of existing desktop grid systems and identified the research gap.

1 Introduction

The concept of a "computing utility" providing "continuous operation analogous to power and telephone" can be traced back to the Multics Project in the 1960s [1]. The term the "Grid" has emerged in the mid 1990 s to denote a proposed computing infrastructure which focuses on large-scale resource sharing, innovative applications, and high-performance orientation [2]. The grid concept provides virtual organization environment for resource sharing and problem solving across multiple institution. The sharing involves a direct access to computing resources, storage devices, network resources, software, scientific instruments and other resources subjected to highly controlled sharing rules. The sharing rules define clearly what is shared, who is allowed to share, and the sharing conditions. A set of individuals and or/institutions defined by such sharing rules forms a virtual organization [2]. Researchers and corporations have developed different types of grid computing

N. Venkataraman (✉)
School of Computing Science and Engineering, VIT University,
Chennai 600127, India
e-mail: neelanarayanan.v@vit.ac.in

© Springer International Publishing Switzerland 2016
V. Vijayakumar and V. Neelanarayanan (eds.), *Proceedings of the 3rd International Symposium on Big Data and Cloud Computing Challenges (ISBCC – 16')*,
Smart Innovation, Systems and Technologies 49, DOI 10.1007/978-3-319-30348-2_16

systems to support resource pooling or sharing. Typically, such grid computing systems can be classified into computational and data grids. In [3], a taxonomy for grid systems is presented, which proposes three types of grid systems. As stated earlier, they are computational grids, data grids and service grids.

Now, let us focus on computational grids that optimizes execution time of applications that require greater number of processing cycles. The "Computational Grid" refers to the vision of a hardware and software infrastructure providing dependable, consistent, fast, and inexpensive access to highend computational capabilities [4]. Generally speaking, such grid computing platforms can be classi-fied into two main categories; classical high-end grids and volunteer or desktop grids [5]. Classical grids provide access to large-scale, intra and-inter institutional high capacity resources such as clusters or multiprocessors [4, 6]. For example, Globus [6] and Legion [7] provide a software infrastructure that enables applica-tions to handle distributed heterogeneous computing resources, normally dedicated, in multiple administrative domains. TeraGrid,[1] build using Globus Toolkit is one such example. However, installing, configuring, and customizing Globus middle-ware requires a highly skilled support team, such as the London e-Science Centre[2] or the Enabling Grids for E-science project.[3] Participating in such grid projects involves time consuming networking and training processes. The application developer must possess a knowledge of both the middleware being used and the underlying computational hardware. Using this information, task dependent libraries and binaries can be produced. These are typically managed by the user, who also has to possess some knowledge of the target architecture. This makes the process of application deployment both time consuming and error prone [8]. Globus based grid computing infrastructure requires third party resource brokers or meta-schedulers for task distribution. Meta-scheduler, a software scheduling sys-tem, allows a designated node to act as an active gateway for other passive nodes, for example GridWay.[4] Thus effort towards development, integration, testing and packaging of many components are substantial. For these reasons, the deployment and operational cost of such systems are substantial, which prevents its adoption and direct use by non-technical users. For example, NSF Middleware Initiative (NMI) program[5] has invested roughly $50M for development of various compo-nents. TeraGrid cyber infrastructure facility has allocated approximately 25 % of the staff to common integration functions and 75 % of the staff to resource provider facility functions [9].

Consider a situation where the participants are offering different resources to collaborate on a common objective. In this scenario, every participant wants to participate for a certain limited amount of time, normally till the participant has

---

[1]www.teragrid.org.
[2]www.lesc.imperial.ac.uk.
[3]www.eu-egee.org.
[4]www.gridway.org.
[5]www.nsf-middleware.org.

some utility interest in the participation. Administrative overheads erupting from classical grid participation make it impractical for such transient communities to undergo a formal grid establishment process.

Volunteer or Desktop Grids, in contrast, is designed to distribute computational tasks between desktop computers owned by individuals at the edge of Internet or idle desktop computers in an institution. Volunteer Grid computing systems allow formation of parallel computing networks by enabling ordinary Internet users to share their computer's idle processing power [10, 11]. Projects based on volunteer computing system provide computational infrastructures for complex computational intensive research problems that range from searching for extraterrestrial intelligence (SETI@home) to exploring new HIV/AIDS treatments (fightAIDS@home). Such systems require setting up a centralized control system responsible for managing the contributed clients, who in turn periodically request work from a central server. Volunteer computing is highly asymmetric; it is a 'master-slave' architecture in which volunteers supply computing resource but do not submit any work to be done. Public outreach and incentive structures (like high-score competitions) play a significant role in attracting volunteers. In volunteer computing, users require system administration and database expertise to create and operate projects [12].

Historically, the Condor project [10] pioneered using the idle time of organizational workstations to do parallel computing. Increasing computational power and communication bandwidth of desktop computers are helping to make distributed computing a more practical idea. By using existing desktop computers from a local network, the cost of such an approach is low compared with parallel supercomputers and dedicated clusters. The main difference in the usage of institutional desktop grids relatively to volunteer ones lies in the dimension of the application that can be tackled. In fact, while volunteer grid computing projects usually embrace large applications made up of a huge number of tasks, institutional desktop grids, which are much more limited in resources, are more suited for modestly-sized applications.

Peer-to-peer platforms provide an operating system independent middleware layer, which allows sharing of resources in a peer-to-peer fashion. Various protocols for supporting P2P service discovery (e.g. Gridnut [13] and GridSearch [14]) and P2P resource discovery [15] has been proposed. Grid computing focus on infrastructure. On the other hand, peer-to-peer computing focus mainly on scalability and not infrastructure. However, a convergence between peer-to-peer and Grid computing has been foreseen in literature [16, 17].

2 Overview of Desktop Grid Systems

In this paper we identify and comprehend concepts related to Desktop Grid Systems. We propose a new taxonomy considering resource provider, resource consumer and grid application perspectives. Then we map currently existing desktop grid systems using the above taxonomy to identify their strength and weakness. We also identify the research gap in existing research work related to desktop grid systems.

In this section we present some of the well known grid systems but with special focus to desktop grid systems. While this list is not exhaustive, but merely representative from this research point of view. The survey includes desktop grid systems that have been demonstrated with proof of concept application or working prototype or have been widely used or deployed in real environment. The survey excludes research work that reports theoretical model with no implementation and evaluation are excluded from the study.

2.1 BOINC

The Berkeley Open Infrastructure for Network Computing (BOINC) [11, 18] is a software platform for distributed computation using idle cycles from volunteered computing resources. BOINC is composed of a central scheduling server and a number of clients installed on the volunteers' machines. The client periodically contacts the server to report its availability and gets workload. BOINC is mainly based on voluntary participants connected through Internet.

Although projects using BOINC are diverse in their scientific nature, in general they are data analysis applications composed of independent tasks that can be executed in parallel. Each project must prepare its data and executable code to work with the BOINC libraries and client/server infrastructure. Also they need to set up and maintain their own individual servers and databases to manage the project's data distribution and result aggregation. Though each project requires individual server setup and maintenance, BOINC users can participate in multiple projects through single client interface.

The BOINC server consists of seven different daemon programs, some of which are provided by BOINC and others need to be implemented individually for each project [19]. The *feeder*, and *transitioner*, are components provided by BOINC. The BOINC server maintains a queue of work units that need to be sent to the clients. The feeder retrieves newly generated work units from the database to fill the queue. The transitioner controls the state transitions of work units and results throughout their lifecycles. The lifecycle of a work *unit* begins when it is generated by the *work generator* and is added to the database. The work generator daemon need to be developed by the application. The work units can then go through several state transitions as they are distributed to one or more clients for execution. If a client has received a work unit and has not returned the results within a predetermined amount of time, then the work unit is said to have *timedout* or *expired*. The transitioner detects work units that have timed out and redistributes them to different clients. The lifecycle of a work unit ends when enough valid results for that work unit have been collected and a single result called the canonical result is chosen for that work unit. Similar to work unit, the result can also undergo several state transitions. All results that are invalid or not selected to be the *canonical result* are deleted.

Database purge and file deleter daemons provided by BOINC. The file deleter and the database purge daemons remove files and work units that are no longer needed to keep the database size at constant size. The *validator* daemon attempts to determine which results are valid by comparing results from several different clients. The assimilator daemon processes valid results, which usually mean storing them to a separate database for later analysis.

The BOINC architecture is highly modular and scalable. If the project server becomes inundated with client requests, additional servers can be added to the project with daemons running on all the servers each handling only a fraction of the total incoming requests. With a sufficient number of project servers, the only bottleneck in the system is the MySQL[6] server [20].

BOINC, a powerful and robust system for public resource computing has significant limitations. The BOINC client has been ported to several platforms, but the BOIN server can only be executed on Linux-based operating systems. Hence, researchers or application developer need to have expertise in Linus system administration and database expertise [12]. Furthermore, project creators must have a large knowledge of C++ or FORTRAN programming [21]. Compared to the high complexity of BOINC system, there is very little documentation available about how to create a BOINC project. This lack of documentation is the largest barrier that researchers face when creating BOINC project [22].

Projects must have a large visibility in order to attract enough cycle donors (i.e., volunteers) [21]. Resource providers are concerned with the potential harm inflicted by Internet sharing systems, especially when the installation or use of the system requires administrator (or root) privileges [23].

2.2 DG-ADAJ

Desktop Grid—Adaptive Distributed Application in Java (DG-ADAJ) [24] is a middleware platform, facilitating Single System Image (SSI), to enable efficient execution of heterogeneous applications with irregular and unpredictable execution control in desktop grids.

DG-ADAJ is designed and implemented above the JavaParty and Java/RMI platforms. DG-ADAJ automatically derives graphs from the compiled bytecode of a multi-threaded Java application that account for data and control dependencies within the application. Then, a scheduling heuristic is applied to place mutually exclusive execution paths extracted from the graphs among the nodes of the desktop grid system.

---

[6]www.mysql.com.

DG-ADAJ does not include methods for resource brokering. Before scheduling any computational jobs resources need to be found/selected manually by the user. ADAJ applications had to be written utilizing the JavaParty, so the application had to be tightly coupled with the platform.

2.3   SZTAKI Local Desktop Grid

SZTAKI Local Desktop Grid [25] is built on BIONC technology but significantly extends the client concept of BOINC in order to enable the creation of hierarchical desktop grids within a large organization or community.

The hierarchical Desktop Grid allows a set of projects to be connected to form a directed acyclic graph. Task is distributed among the edges of the directed graph. By doing this, SZTAKI can reduce load on the primary BOINC server by using the second and third-level BOINC servers. But, each level of BOINC servers still has the same characteristics of the original BOINC, for which performance is not guaranteed. Communication and data transfer between the client and the desktop grid system is performed via HTTPS.

SZTAKI LDG focuses on making the installation and central administration of the local desktop infrastructure easier by providing tools to help the creation and administration of projects and the management of applications.

SZTAKI LDG supports DC-API for easy implementation and deployment of distributed applications on local desktop grid environment.

2.4   distributed.net

A very similar effort to BIONC is distributed.net [26]. However, the focus of the distributed.net project is on very few specialized computing challenges. Furthermore, the project releases only binary code of the clients and hence impossible to adapt in other projects. The project that attracted the most participant was an attempt to decipher encrypted messages. However, the volunteers were provided with cash prizes.

2.5   XtremWeb

XtremWeb [27], a research project from University of ParisSud, France aims to serve as a substrate for large scale distributed computing. Similar to BOINC, XtremWeb is based on the principle of cycle stealing.

XtremWeb supports the centralized setup of servers and PCs as workers. However, it can be used to build a peer-to-peer system with centralized control, where any worker node can become a client that submits jobs. It implements three distinct entities: the coordinator, the workers and the clients to create a XtremWeb network. Clients can be used by any user to submit tasks to the XtremWeb network. They submit the task to the coordinator and permits the end user to retrieve results. The workers installed on volunteer hosts execute the task.

Tasks are scheduled to workers according to their demand (i.e., pull model) in First In First Out (FIFO) order [28]. All actions and connections are initiated by workers. The coordinator registers every worker connection. The communication between coordinator and workers are encrypted for network security. The workers downloads the executable software and all other related components (e.g., the input files, the command line arguments for the executable binary file, etc). Workers sends the result (the output files) to the coordinator.

During registration workers provide configuration information such as XtremWeb worker version, operating system, CPU type, memory size, etc. XtremWeb coordinator performs matching about CPU type, OS version and Java version [29]. Coarse grained, massively parallel and applications that are not communication intensive are suited for deployment on XtremWeb [28]. XtremWeb uses replication and checkpointing for fault tolerance [30].

XtremWeb-CH [31] is an upgraded version of XtremWeb with major improvements in communication routines and improved support for parallel distributed application.

2.6 Alchemi

Alchemi [32] is an open source project from Melbourne University developed in C# for Microsoft .NET framework. It allows flexible application composition by supporting an object-oriented application programming model with a multithreading paradigm. Grid application consists of several grid threads that can be executed in parallel. Although tightly coupled with .NET platform, Alchemi can run on other platforms using Web services. Alchemi is based on the master-worker parallel programming paradigm.

Alchemi grid consists of three components: manager, executor, and owner. The manager manages grid application execution and thread execution. Executors sign up with manager. Owner submits grid threads to manager who adds them into to a thread pool. Then the manager schedules threads for execution on available executors. Executors after completion of execution of the threads submit the results to the manager. Later, the owner can retrieve the results from the manager.

Aneka [33], an improved version of Alchemi allows the creation of enterprise Grid environments. It provides facilities for advance reservation of computing nodes and supports flexible scheduling of applications.

2.7   *Bayanihan*

Bayanihan [34] is a web-based volunteer computing frame-work that allows users to volunteer their resources.

Bayanihan system consists of client and server. A client executes Java applet on a web browser. It has a worker engine that executes computation or a watcher engine that shows results and statistics. A server consists of HTTP server, work manager, watch manager and data pool. The HTTP server serves out Java class file. The work manager distributes tasks and collects result. Bayanihan basically uses eager scheduling, in which a volunteer asks its server for a new task as soon as it finishes execution of current task. Eager scheduling works by assigning distributed tasks to hosts until every task has been scheduled at least once. At this point, redundancy is introduced by rescheduling any task that has not yet completed. This process continues until every task completes. The client uses remote method invocation technique called HORB [35] on the server to get a task.

The applications are mainly compute-intensive and independent. In addition, Bayanihan supports applications running in Bulk Synchronous Parallel (BSP) mode, which provides message-passing and remote memory primitives [34].

2.8   *Condor*

Condor [36] developed at the University of Wisconsin—Madison scavenge and manage wasted CPU power from otherwise idle desktop workstation across an entire organization.

Workstations are dynamically placed in a resource pool whenever they become idle and get removed from the resource pool when they become busy. Condor can have multiple resource pool and each resource pool follows a flat resource organization. Condor collector listens for resource availability advertisements and acts as a resource information store. A Condor resource agent runs on each machine periodically advertising its availability and capability to the collector. Condor provides job management mechanism, scheduling policy, priority scheme, resource monitoring and resource management. Condor is comprised of a server and large number of volunteers. A central manager in the server is responsible for match-making, scheduling and information management about task and resources.

Condor provides ClassAd [37] in order to describe characteristics and requirements of both tasks and resources. It also provides a matchmaker for matching resource requests (tasks) with resource offers (i.e., available resources). Condor provides Directed Acyclic Graph Manager (DAGMan) [38] for executing dependable tasks. Condor enables preemptive resume scheduling on dedicated compute cluster resources. It can preempt a low-priority task in order to immediately start a high-priority task.

Condor-G [39] is the technological combination of the Globus and the Condor projects, which aims to enable the utilization of large collection of resources spanning across multiple administrative domains. Globus contribution composes of the use of protocols for secure inter-domain communication and standardized access to a variety of remote batch systems. Condor contributes with the user concerns of job submission, job allocation, error recovery and creation of user friendly environment.

2.9   Entropia

Entropia [40] facilitates a Windows desktop grid system by aggregating the desktop resources on the enterprise network.

The Entropia system architecture consists of three layers: physical node management, resource scheduling and job management layer. The physical node management layer provides basic communication, security, local resource management and application control. The resource scheduling layer provides resource matching, scheduling of work to client machines and fault tolerance. The job management layer provides management facilities for handling large number of computations and data files. A user can interact directly with resource scheduling layer by means of API or can use the management functionalities provided by job management layer. The applications are mainly compute intensive and independent.

Entropia virtual machine (EVM) runs on each desktop client and is responsible for starting the execution of the subjob, monitoring and enforcing the unobtrusive execution of the subjob, and mediating the subjobs interaction with the operating system to provide security for the desktop client and the subjob being run. The EVM communicates with the resource scheduler to get new task (subjobs), and communicates subjob files and their results via the physical mode management layer.

The resource scheduler assigns subjobs to available clients according to the client's attributes such as memory capacity, OS type, etc. The Entropia system uses a multi-level priority queue for task assignment.

The applications are mainly compute intensive, independent and involve less data transfer between server and clients [41].

2.10   QADPZ

Quite Advanced Distributed Parallel Zystem (QADPZ) [42] is an open source desktop grid computing system. The users of the system can transmit computing tasks to these computers for execution in the form of dynamic library, an executable program or any program that can be interpreted, for example Java, Perl, etc.

Messages between the components of the system are in XML format and can optionally be encrypted for security reasons.

A QADPZ consists of three types of one master, many slaves and multiple clients delegating job to the master.

- Master—A process running on the master computer responsible for jobs, tasks, and slaves accounting. The slaves talk to the master when they join or leave the system, or receive or finish tasks; the clients talk to the master when they start or control user jobs or tasks.
- Slave—A process running on the volunteer computer as a daemon or Win NT service. The slave communicates with the master and starts the slave user process. Without the slave running the slave computer cannot take part in collaborative computing.
- Client—A process running on a client computer. It communicates with the master to start and control jobs and tasks of a specific user, may also communicate directly with slaves and is responsible for scheduling the tasks of user jobs as required by particular user application.
- Slave computer—One of many computers where the distributed collaborative computation takes place, for example: a UNIX server, a workstation, a computer in a student PC Lab, etc.
- Client computer—Any computer that the user uses to start his application. A client computer may be a notebook connected to the network using a dial-up connection, a computer in the office, lab, etc.

All slaves participating in the system run a slave program that accepts the tasks to be computed. The master program keeps track of the status of the slave computers. The master registers the status of the slaves, i.e., idle or busy computing a QADPAZ task or disabled because a user has logged out. When a user wishes to use the system, he prepares a user application consisting of two parts: A *slave user program*, i.e., the code that will effect the desired computing after being distributed to the slaves, and the client that generates jobs to be completed.

The user can choose the QADPZ standard client, which allows him to set up and submit a job. The job description is saved into an XML-formatted project file and can be manually edited by more advanced users. Alternately, a user may want to write his or her own client application to have full control over the submission of tasks. It is possible to directly write a slave library to speed up the execution. In that case, the slave service or daemon will not start a new process with the downloaded executable but a dynamic shared library will be loaded by the slave process.

QADPZ is implemented in C++ and uses MPI as its communication protocol. The client and the slaves talk to each other by the use of a shared disk space, which is certainly a performance bottleneck and requires costly synchronization [43].

All the components need to be installed manually and have their own configuration files requiring manual configuration. The system has been used in the area of large-scale scientific visualization, evolutionary computation, and simulation of complex neural network models.

2.11   Javelin

Javelin, a Java-based infrastructure for global computing with a goal to harness the Internet's vast, growing, computational capacity for ultra-large, coarse-grained parallel applications. The work that started with SuperWeb has been continued with new versions named Javelin++, Javelin 2.0, Javelin 3.0, CX, JANET and currently JICOS, each with improvements in performance, scalability, computation and programming model. Javelin [44–46] is a Java based infrastructure. Applications run as Java applet in Javelin version or screen saver in Javelin++ version [47].

Javelin consists of three entities: broker, client and host. A broker is a system-wide Java application that functions as a repository for the Java applet programs, and matches the client tasks with hosts. A client submits a task to the broker through a Web browser by generating an HTTP POST request, and periodically polls the broker for the results. Hosts are Web browsers, generally running on idle machines, that repeatedly contact the broker for tasks to perform, download the applet code, execute it to completion, and return the results to the broker. The communication between any two applets is routed through the broker [44]. The Javelin is hindered by Java applet security model. Hence, Javelin 2.0 abandoned the applet-based programming framework and added support for problems that could be formulated as branch and bound computations. However, the essential architecture remained almost same.

In Javelin, the work stealing is performed by a host, and eager scheduling is performed by a client. Applications are mainly compute-intensive and independent. Currently, Javelin supports branch and bound computations. Javelin achieves scalability and fault-tolerance by its network of brokers architecture and by integrating distributed deterministic work stealing with a distributed deterministic eager scheduling algorithms. The main advantage of this architecture are:

- URL based computation model permits the worker hosts to perform the necessary computation off line and later reconnect to the Internet to return the results from the computation.
- Administrative overheads involved in preparing worker nodes (during deployment) is minimized by ubiquity of Web browsers.

However, centralized match making process and task repository by broker limits Javelin. Further, it assumes the user's a priori knowledge of the broker's URL address. Eager scheduling leads to excessive data traffic and presence of multiple copies of the same task.

2.12   Charlotte

Charlotte [48] developed at New York University is a web based volunteer computing framework and implemented using Java. Charlotte supports Java-based distributed shared memory over the Java Virtual Machine and parallelism. The key feature of the shared memory architecture is that it requires no support from the compiler or the operating system, as it is the case with most shared memory architectures. The implementation of distributed shared memory is at data level. Every basic data type in Java has a corresponding Charlotte data type. Charlotte maintains data consistency by using the atomic update protocol allowing concurrent reads but exclusive write/updates.

Charlotte, based on Calypso's programming model to provide parallel routines and shared memory abstraction on distributed machines. The main entities in Charlotte program are a manager (i.e., a master task) executing serial steps and one or more workers executing parallel steps called *routines*. A routine is analogous to a standard Java thread, except for its ability to execute remotely. A distributed shared namespace provides shared variables among routines. The manager process creates an entry in well-known Web page. Volunteer users load and execute the worker processes as Java applets embedded into web pages pointed by pointing their browsers to this page. A direct socket connection provides the manager worker communication.

Initially, it uses eager scheduling and then redundant assignment of tasks to multiple clients to handle client failures and slow clients. Multiple executions of a task can result in incorrect program state. Charlotte uses two-phase idempotent execution (TIES) to ensure idempotent memory semantics in the presence of multiple executions. TIES ensures guarantees correct execution of shared memory by reading data from the master and writing it locally in the worker's memory space. Upon completion of a worker the dirty data is written back to the master who invalidates all successive writes, thus maintaining only one copy of the resulting data [49]. Further, in order to mask latencies associated with the process of assigning tasks to machine by employing dynamic *granularity management*. Granularity management technique (bunching) assigns a set of task to a single machine at once. The size of bunch is computed dynamically based on the number of remaining task and number of available machines.

Charlotte makes use of existing Internet infrastructure such as HTTP servers and web browsers for running applets. As with Javelin project, Charlotte is also hindered by Java applet security model. Charlotte programs must conform to the parallel routine program structure and must be implemented as a Java applet. This approach is not very transparent or flexible because programmers must adhere to both the applet API and Charlotte's parallel routine structure [50]. Charlotte does not address the question of how a Web browser, i.e., the worker, finds the work. Earlier version requires manual input of the URL location. However, current version uses a registry and lookup service provided by KnittingFactory [50]. The primary function of the manager is not only scheduling and distributing works but also responsible for communication of workers as well as applet-to-applet communication. Hence, the manager could be a bottleneck. Further, Charlotte requires that a manager run on a host with HTTP server. If multiple applications are run by single machine, then multiple managers are needed to run on this machine. Hence, it would introduce additional overhead and leads to performance problem.

2.13 CPM

Compute Power Market (CPM) [51] is a market-based middleware system for global computing on personal computing devices connected to the Internet. It applies economic concepts to resource management and scheduling across Internet-wide volunteer computing.

It consists of a market, resource consumers and resource providers. The market consists of a market resource broker and market resource agent. The market resource broker is component of resource consumer and the market resource agent is a component of resource provider. It maintains information about resource providers and resource consumer—a kind of registry. The resource consumer buys computing power using the market resource broker. The market resource broker finds suitable resource based on the information provided by the market. It is responsible for negotiating cost with resource providers and task distribution to resource providers. The resource provider sells computational power using market resource agent. The market resource agents updates resource information and deploys and execute tasks. The resource broker selects the resource based on deadline or budget.

CPM takes the advantage of real markets in the human society. However, the centralized market servers introduce limitations, such as single point of failure and limited scalability. Furthermore, the centralized market server requires additional organizations for regular maintenance [52].

2.14 POPCORN

POPCORN [53] is a Web-based global computing system for scheduling Java-based parallel applications. POPCORN uses market model for matching sellers and buyers.

It consists of a market, buyers and sellers. The sellers provide their resources to a buyer by using Java enabled browser. The market has the popcorn, a currency used by buyer to buy resources. The seller can earn the popcorn for selling its resources. The market is responsible for performing matching between buyers and sellers, transferring task and result between them and for account management.

POPCORN uses several market models for scheduling. The Popcorn system implements three different kinds of auction mechanisms: Vickrey auctions, first-price double auctions, and k-price double auctions. Popcorn has a central repository by means of a Web platform for information aggregation. However, this platform may become a communication bottleneck. The applications are mainly compute-intensive and independent.

2.15   The Spawn System

Waldspurger et al. [54] proposes a distributed system that uses market mechanism for allocating computational resources called Spawn.

The Spawn system organizes participating resources as trees. A leaf resource can join randomly selected auction among the available auctions. The managers of each leaf serve as funding sponsors for their children. In case an agreement is reached after the auction process, a resource manager controls the communication with and monitoring of the supplied resources.

The Spawn system supports tree-based applications in which partial results are computed on different levels of the tree and are subsequently sent to a leaf on a higher level of the tree. On each leaf, a manager combines and aggregates the results received from its children. Subsequently, this manager reports the aggregated results to its higher leaf on the tree.

The Spawn system uses market models for scheduling. The managers of each leaf serve as funding sponsors for their children. This funding is used to purchase CPU resources for subtasks associated with each leaf. These resources are purchased by participating in Vickrey auctions which are instantiated by each node that offers idle CPU slice. In Vickrey auction uses sealed-bid that means bidding agents cannot access information about other agent's bid. Further, the winner pays the amount offered by the next highest competitive bidder. The auction instance and the pricing information are advertised to the neighbors of each node. Out of all available auctions, a requesting leaf randomly joins one particular auction. In case an agreement is reached after the auction process, a resource manager controls the communication with and monitoring of the supplied resources.

The authors through simulation of prototypical implementation show that Vickery auctions leads to economically efficient outcomes. In addition it has lesser communication costs compared to an iterative auction. However, as the information is only propagated to neighbors, new information is disseminated with delays. Besides, the seller-initiated auction is suitable only for heavily loaded systems. In lightly loaded system, the buyer-initiated auction is proved to outperform the seller initiated auction [55].

2.16   OurGrid

OurGrid system [56, 57] is an open, free-to-join, cooperative grid in which labs donate their idle computational resources in exchange for accessing other labs' idle resources when needed.

OurGrid is based on a peer-to-peer network, where each labs in the Grid correspond to a peer in the system. OurGrid has three main components: the OurGrid peer, the MyGrid broker and the SWAN security service. The MyGrid broker is responsible for providing the user with high-level abstractions for resource and computational task. The SWAN security service guarantees secure resource access. Each peer in represents a lab. Each peer has direct access to a set of resource. Each resource has an interface called GridMachine that provides access to the resource.

The OurGrid system mainly focuses on compute intensive and independent tasks. The OurGrid system provides resource matching and heuristic scheduling. The OurGrid architecture implements the idea of symmetric resource participation.

3 Classification of Desktop Grids

In this section, we present the various classification of desktop grids. There are several taxonomy of Desktop Grids. Baker et al. present the state of the art of Grid computing and emerging Grid computing projects. They hierarchically categorize Grid systems as integrated Grid systems, core middleware, user-level middleware, and applications/application driven efforts [58]. Krauter et al. proposes a taxonomy of Grid focusing on resource management. Their taxonomy classifies the architectural approaches from the design space [3]. Venugopal et al. proposes a taxonomy of Data Grids based on organization, data transport, data replication and scheduling. Sarmenta classifies volunteer computing into application-based and web-based [59]. Chien et al. [40] classifies Desktop Grid into Internet scalable Grid and Enterprise Grid. Fedak et al. [60] provide a taxonomy of Desktop and Service Grids based on distribution of computation units, scalability, type of resource contribution, and organization. Choi et al. [61] presents the taxonomy of Desktop Grid systems and mapping of taxonomy to the existing Desktop Grid systems. However, their taxonomy is focused on scheduling in Desktop Grid systems. We provide a taxonomy and mapping that is more generic and inclusive in nature. The objective of this taxonomy are to:

- Categorize based on attributes related to system architecture, resource provision, and grid deployment effort.
- Provide mapping of main desktop grid system according to the taxonomy.

3.1 Centralized Desktop Grids

Desktop grid system can be categorized into *centralized* and *decentralized* desktop grids. Centralized desktop grid system consists of three logical components: client, workers and server. Client allows platform users to interact with the platform by submitting jobs and retrieving results. Worker is a component running on the volunteer's desktop computer which is responsible for executing jobs. Server, a coordination service connects clients and workers. The server receives jobs submissions from clients and distributes them to workers according to the scheduling policy. Servers also manage fault tolerance by detecting worker crash or

Table 1 Centralized global desktop grid system

DG system	Platform	Scheduling policy	Deployment effort
BOINC	Middleware-based	Simple	Thick deployment
Entropia	Middleware-based	Simple	Thick deployment
distributed.net	Middleware-based	Not available	Not available
XtremWeb	Middleware-based	Simple	Thick deployment
DG-ADJ	Middleware-based	Deterministic	Thick deployment
POPCORN	Middleware-based	Economic	Thick deployment
Spawn	Middleware-based	Economic	Thick deployment
Bayanihan	Web-based	Simple	Thin deployment
QADPZ	Web-based	Simple	Thick deployment
Javelin	Web-based	Deterministic	Thin deployment
Charlotte	Web-based	Simple	Thin deployment

disconnection. If needed tasks are restarted on other available workers. Workers on completion of task execution return the results to the server. Finally, the server verifies the correctness of the results and stores them for the client to download them. Typical examples are BOINC, XtremWeb, Entropia etc. Properties of centralized desktop grid has been presented in Tables 1 and 2.

The desktop grid system can be classified based on deployment and resource location into *local* and *global* desktop grids. Local desktop grid, also known as Enterprise desktop grid, consists of desktop PC hosted within a corporation or University interconnected by Local Area Networks (LAN). Several companies such as Entropia and University projects such as Condor, SZTAKI local desktop grid, etc., have targeted these LANs as a platform for supporting desktop grid applications. Such grid environments have better connectivity and have relatively less volatility and heterogeneity than global desktop grids. Local desktop grids connect resources from the same administrative domains, normally by using local area networking technologies and poses low security risks. Internet desktop grids connects resources from different administrative domains, normally by using wide-area networking technologies. They aggregate resources provided by end-user Internet volunteer. Global grids are characterized by anonymous resource providers, poor quality of network connections, being behind firewalls, having dynamic addressing techniques (DHCP) and poses high security risks. Desktop grids based on BOINC such as SETI@HOME, Einstein@HOME, etc. falls under global desktop grids.

In order to maximize the potential work pool and minimize setup time, a volunteer computing system must be accessible to as many people as possible. Desktop

Table 2 Centralized local desktop grid system

DG system	Platform	Scheduling policy	Deployment effort
Condor	Middleware-based	Simple	Thick deployment
SZTAKI LDG	Middleware-based	Simple	Thick deployment

grid system can be categorized into Web-based and middleware-based desktop grid computing according to the accessibility platform running on volunteers. In the web-based desktop grid systems, clients write their parallel application in Java and post them as Applet on the Web. Then, volunteers can join by pointing to the web page using their browsers. The Applet gets downloaded and runs on the volunteer's desktops. Typical examples are Charlotte, Bayanihan, Javelin and so on. In the middlewarebased desktop grid computing systems, volunteers need to install and run a specific middleware that provides the services and functionalities to execute parallel applications on their machine. The middleware fetches tasks from a server and executed them on volunteer desktops, when the CPU is idle. Typical examples are BOINC, XtremWeb, Entropia, and so on.

In desktop grid environment, large number of people will want to share their resources. In addition, a higher number of application will be deployed. Therefore, special attention should be given to ease of deployment of new application and preparation of the execution environment. In [62], von Laszewski et. al. identify three deployment strategies: thick, thin and slender deployment based on the hosting environment. In thick deployment, an administrator uses a software enabling service, for example Grid Packaging Tool (GPT) [63] to install the Grid software on a resource participating in the Grid. In thin deployment, end user uses a Web browser to access Grid service in a transparent fashion, for example Charlotte, Bayanihan, Javelin, etc. In slender deployment, slender clients are developed in an advanced programming language such as Java and made available from a Web-based portal. Initially, end user can install these slender clients on their resources to participate in the Grid. Additionally, slender deployment provides an automatic framework for updating the component if a new one is placed on the Web server.

Scheduling policy matches tasks with resources by determining the appropriate tasks or resources. It can be classified into three categories: simple, model-based and heuristics based [64]. In simple approach, tasks or resources are selected by using First Come First Served (FCFS) or randomly. This scheduling policy is appropriate for high throughput computing requirement and commonly implemented by major desktop grid middleware like BOINC, Condor, XtremWeb, etc. The model-based approach is categorized into deterministic, economy, and probabilistic models. The deterministic model is based on structure or topology such as queue, stack, tree or ring. Tasks or resources are deterministically selected according to the properties of structure or topology. For example, in a tree topology, tasks are allocated from parent nodes to child nodes. In deterministic scheduling models, a set of jobs has to be processed by a set of machines and certain performance measures have to be optimized. In the economy model, scheduling decision is based on market economy. In the probabilistic model, resources are selected in probabilistic manners such as Markov, machine learning or genetic algorithms. In the heuristic-based approach, tasks or resources are selected by ranking, matching, and exclusion methods on the basis of performance, capability, weight, etc.

3.2   Distributed Desktop Grid System

A distributed desktop or peer-to-peer Grids constructs a computational overlay networks using tree, graph or distributed hash table. In the absence of a server, volunteers need to maintain partial information about other volunteers in the grid environment. Scheduling is performed at each volunteer depending on the computational overlay network. Volunteers exchange their information between other volunteers. The volunteers self-organize themselves into a computational overlay network based on a criteria such as resource capability or time zone. A client can submit a set of independent computational tasks to known volunteer. The known volunteer distributes tasks based on scheduling mechanism. Each volunteer executed the computational task and returns the result to parent volunteer (in the tree). Finally, the parent volunteer returns the result to the client. Distributed desktop grids have been presented in Table 3.

Typical examples are Compute Power Market (CPM), Our Grid, etc. Other P2P Desktop Grid platforms such as the Organic Grid [65], Messor [66], Paradropper [67], and the one developed by Kim et.al. [68]. These platforms are still under development and do not have any applications implemented on top of it. Further, they have not been deployed and evaluated. Hence they do not provide a solid substrate enabling the deployment, use, and management of a production level desktop grid environment [69]. As stated earlier, we have excluded such projects from the survey.

3.3   Computational Grids Based on JXTA Technology

The JXTA platform defines a set of protocols designed to address the common functionality required to allow any networked device to communicate and collaborate mutually as peers, independent of the operating system, development language, and network transport employed by each peer.

Several projects have focused on the use of JXTA as a substrate for grid services. The P2P-MPI project, the Jalapeno project, the JNGI project, the JXTA-Grid project, the Our Grid project, the P3-Personal Power Plant project, the Triana project, the Xeerkat project, the NaradaBrokering project, and Codefarm Galapagos are listed in the JXTA web-site (as on 2008). However, none of these projects are being actively developed [70].

Table 3 Distributed global desktop grid system

DG system	Platform	Scheduling Policy	Deployment Effort
CPM	Middleware-based	Economic	Thick deployment
OurGrid	Middleware-based	Deterministic	Thick deployment

Out of this 10 projects, 8 are FOSS projects. Currently, the P2P MPI project, the OurGrid project, the Triana project and the XeerKat project have switched to other platforms. The Jalapeno project, the JNGI project and the P3 project stopped their development around 2005–2006. The JXTA-Grid project is not yet in production [71].

JNGI uses software developed by Project JXTA to communicate in peer-to-peer fashion. Peer groups are used as fundamental building block of the system. The system consists of monitor groups, worker groups, and task dispatcher groups. The monitor group handles peers joining the system and redirects them to worker groups if they are to become workers. The system can have multiple monitor groups in a hierarchical manner. During task submission the submitter queries the root monitor group for an available worker group which will accept its tasks. The root monitor group decides the sub group to which the request should be forwarded to. The request would finally reach a worker group. The worker group consists of workers performing the computations. Once the request arrives at a worker group, a task dispatcher in the task dispatcher group of that worker group will send a reply to the submitter. The task dispatcher group distributes individual tasks to workers. The task dispatcher group consists of a number of task dispatchers, each serving a number of workers. If a task dispatcher disappears, the other task dispatchers will invite a worker to become a task dispatcher and join the task dispatcher group. Task dispatchers periodically exchange their latest results and thus results are not lost even if a task dispatcher becomes unavailable. The submitter polls the task dispatcher about the status and results of previously submitted tasks.

The Jalapeno system consists of manager peers, worker peers and task submitter peers. Each host can play one or more of these roles. Initially every host starts a worker peer that starts to search for available manager peers. When a manager is found, the worker tries to connect to it. Manager can have only a limited number of connected workers and rejects any worker when this limit has been reached. Accepted workers will join the worker group created by the manager and start executing tasks. If a worker is unable to connect to a manager within certain time, it will start a new manager peer on the local host and connect to it. The first host becomes a manager after some time and start to accept worker peers. When a manager becomes unavailable its workers will either find another manager or become managers themselves. Workers with a worker group can communicate with other workers or the manager. The task submitter submits a collection of tasks to a randomly chosen manager. The manager splits the bundle into a set of smaller bundles. The manager keeps a limited set of bundles from which tasks are extracted and handed to the connected workers. The rest of the bundles are forwarded to a number of other, randomly chosen, managers which repeat the process. Bundles which are not forwarded are returned to the task submitter. When a worker finishes a task it will return the result to its manager which in turn will forward the result to the task submitter.

In P3 system, peers form their own base peer group that is a subgroup of always existing JXTAs base group. A peer can run *host* daemon to share resources and/or *controller* tool to access resources shared by the others. To distribute a job to resource providers, a user first creates a distinct peer group for the job, job group, with the controller tool. After joining the group itself, controller publishes an advertisement of his job within the group. It includes a description of the job. Peers running a host daemon discover the advertisements of job groups made by controllers. According to their policy, they decide whether they want to join a job group and contribute to processing the job. If a host decide to join, it discover and obtain a Java Archive (JAR) from the controller. The archive contains compiled Java application code. P3 has an object passing library, by which an application can unicast or broadcast Java objects, and receive them synchronously or asynchronously. P3 provides two parallel programming libraries a master-worker library and a message passing library for application developers. The master-worker library supports master-worker style parallel processing and the message passing library supports MPI programming. If a worker leaves a job group, a work unit delivered to the worker but not completed is delivered to another worker.

3.4   Jini for Grid Computing

Jini [72], a Java-based infrastructure developed by Sun Microsystems provides a distributed programming model that includes dynamic lookup and discovery, support for distributed events, transactions, resource leasing and a security model. A number of research groups have considered Jini in the realization of their Grid middleware [73–79].

Jini is built on top of RMI, which introduces performance constraints [80, 81]. Further Jini is primarily concerned with communication between devices and hence file system access and processor scheduling need to be implemented [81]. Adaptive and scaLable internet-based Computing Engine (ALiCE) [77], a technology for developing and deploying general-purpose grid applications implemented in Java, Java Jini [72] and JavaSpaces[7]. The ALiCE system consists of a core middleware layer that manages the Grid fabric. It hosts compute, data and security, monitoring and account services. The core layer includes an Object Network Transport Architecture (ONTA) component that deals with the transportation of objects and associated code across the Grid fabric. The core layer uses Jini technology to support the dynamic discovery of resources within the Grid fabric. For object movement and communications within the Grid, a Jini based tuple space implementation called JavaSpaces is used.

JGrid [76], based on Jini aim at exploiting Jini's support for dynamic self healing systems. To virtualize a local computing resource, *Compute Services* are introduced that manage job execution on the local JVM. These Compute Services can be

---

[7]http://river.apache.org/doc/specs/html/js-spec.html.

hierarchically grouped using Compute Service Managers. Resource brokers schedule jobs onto these *Compute Service Managers* and Compute Services on a client's behalf. JGrid supports a number of application models including batch style execution using a Sun Grid Engine backend, Message Passing Interface (MPI). JGrid has extended Jini's discovery infrastructure to cope wide with wide area discovery.

4 Taxonomy Development

Taxonomy provides methods and principles for classification. Classification facilitates to summarize knowledge about objects under study. This paper aims to provide a means of classification through development of a taxonomy about desktop grid systems. In the previous section, we have presented the extensive literature review conducted to characterize the desktop grid system. In this section, we introduce the terms in the taxonomy to compare currently existing desktop grid systems and to identify the research gap in this domain.

4.1 Desktop Grid Architecture

Desktop grid system provide high computational power at low cost by reusing existing infrastructure of resources in an organization. However, most of the existing desktop grid systems are built around centralized client-server architecture [82]. This design could potentially face issues with scalability and single point of failure [61, 82]. In volunteer desktop Grid systems, central task scheduler would become a potential bottleneck when scheduling large number of tasks [65].

In contrast to existing Grid paradigms, models based on peer-to-peer architecture offers an appealing alternative [83]. Most of the current research on peer-to-peer architecture do adapt centralized approach in one form or other. For example, JXTA can be considered as hybrid peer-to-peer system as it has the concept of super peers. Jini uses lookup service to broker communication between the client and service and this approach appears to be a centralized model.

For the announcement of new auction instances, their states and current market prices, Popcorn installs a central repository by means of a Web platform. As a disadvantage of this central information aggregation the Web platform becomes a communication bottleneck [84]. In Spawn, market information is propagated only to neighbors and new information is disseminated with delays. This imperfect knowledge could result in performance trade offs [84].

4.2 Desktop Grid Adoption

Volunteer desktop Grid systems, in general, employ "many resource providers few users" model, meaning that any user can join and offer resources, but only a selected few users can make use of those resources. Hence, volunteer desktop computing is highly asymmetric [85] in which volunteers contribute computing resources but do not consume any. Thus the asymmetric nature of desktop Grid systems acts as one of the primary impediment to widespread adoption by users. From our survey, we could see that CPM and OurGrid alone support symmetric resource participation. CPM is based on market economy model and OurGird is based on P2P architecture model. Hence, we adopt distributed computational economy framework for quality-of-service (QoS) driven resource allocation. Such framework provides mechanisms and tools that realize the goal of both resource provider and resource consumer. Resource consumers need a utility model, representing their resource demand and preferences. And resource providers need an expressive resource description mechanism.

4.3 Grid Application Development

Currently, most of the grid applications are implemented with the help of computer scientists. Providing MPI-like grid application programming language is not sufficient, as many scientists are not familiar with parallel programming language [86].

One of the major resource for researchers is existing application toolkits. However, many existing applications are developed for desktop computers. Hence, grid enabling existing application with minimal efforts could be useful to the researchers.

Grid computing not only have technical challenges but also have deployment and management of grid environment challenge.

4.4 Quality of Service

Desktop computing systems, for example Entropia and P2P systems, for example Gnutella provide high levels of quality of service for specialized service. But are too specific for generalized use; e.g., it would be difficult to see how the Gnutella protocol[8] could be useful for anything but searching for files or data content. A Grid should be able to provide non-trivial quality of service, for example, this is measured by performance, service, or data availability or data transfer. QoS is not just about aggregation of capabilities of participating resources in the Grid. QoS is

---

[8]http://rfc-gnutella.sourceforge.net/.

application specific and it depends on the needs of the application. For example, in a physics experiment, the QoS may be specified in terms of throughput, but in other experiments, the QoS may be specified in terms of reliability of file transfers or data content. From our survey, we could see that only Compute Power Market (CPM) provides quality of service using differentiated price service.

4.5 Scheduling Policy

Popular desktop grid computing systems such as BOINC, Condor and XtremWeb employs the classical eager scheduling algorithm, First-Come-First-Served (FCFS). is for bag-of-tasks applications, where a task is simply delivered to the first worker that requests it. This scheduling policy is particularly appropriate when high throughput computing is sought and thus it is commonly implemented by major desktop grid middleware like BOINC, Condor and XtremWeb. However, FCFS is normally inefficient if applied unchanged in environments where fast turnaround times are sought, especially if the number of tasks and resources are in the same order of magnitude, as it is often the case for local user's bag-oftasks [87]. As an alternate distributed scheduling mechanism such as market-based scheduling can be considered. The evaluation results of computational and data grid environments demonstrate the effectiveness of economic models in meeting user's QoS requirements [88].

4.6 Limitations of Jini and JXTA

As both Jini and JXTA have Java based implementations, the features of the Java language for dealing with OS and hardware heterogeneity are inherited by both frameworks. Jini prescribes the use of leases when distributing resources across the network. The leasing model as a whole contributes to the development of autonomous self healing services that are able to recover from crashes and clear stale information without administrator intervention. The framework extends the Java event model in a natural way, incorporating measures that enable the developer to react to the intricacies of delivering events across a, possibly unreliable, network. Although Jini does not mandate the use of RMI for proxy to service communication, RMI is mandatory when interacting with the Jini event and transaction models. To interact with the reference implementation of the Jini lookup service and to renew leases, RMI is also required since the lookup service and lease objects are represented by a RMI proxy [89]. Java RMI is acceptable when transferring small or medium sized chunks of data over high-speed data links but being slower than TCP. Another issue is the performance degradation imposed by the HTTP tunneling technique [89]. JXTA project lacks in the documentation, hence design and implementation decisions are not documented. Often, it requires inspection of the

Table 4 Summary of market based systems

DG System	Model	Description
Spawn	Vickrey auction	The Spawn system provides a market mechanism for trading CPU times in a network of workstations. Tree-based concurrent programs are sub-divided into nodes. Each node holds vickrey auction independently to acquire resources
POPCORN	Auction	Each buyer (resource consumer) submits bid and the winner is determined through one of the three auction protocol implemented: Vickery, Double and Clearing House
CPM	Various models	CPM supports various market model such as commodity market, contract-net/tendering and auction

implementations source code to determine the different steps and policies used throughout the implementation of the JXTA protocols [89]. JXTA does have a good set of tools that allow using resources available in the network. However while working with this platform many mistakes can be noticed. The documentation for many important elements hardly exists. The tools for XML processing, used mainly with advertisements, are difficult to utilize. Despite of supposed popularity of JXTA there are still no professional products based on it. Implementation of JXTA protocols still has many undocumented elements [90] (Table 4).

5 Research Gap

Research in desktop grid system focus on two areas: desktop grid application development and desktop grid system development. Research on desktop grid application development focus on scheduling, resource management, communication, security, scalability, fault tolerance, trust, and architectural model.

The grid application development research focus on developing new applications suitable for desktop grid systems. However, writing and testing grid applications over highly heterogeneous and distributed resources are complex and challenging. Hence, the desktop grid system should support ease of application development and integration.

The strength and weakness of current approaches to desktop grid system development can be summarized as follows:

- Most of the current desktop grid system follow client server or centralized peer-to-peer architecture. These approaches are easy to implement and have high throughput. However, they are not scalable, have a single point of failure, and server component become a bottleneck.
- Distributed desktop grid system based on market approach has been proposed. However, use of multiple markets for resource sharing has not been explored.

- Most of the current desktop grid system focus on asymmetric resource utilization. However, support for symmetric resource utilization could motivate users to contribute their resources to the desktop grid initiative.
- Currently desktop grid system do not support quality of service based on non-trivial parameters.
- Current market-based resource management approaches in desktop grid computing focus on price-based mechanisms to allocate resources. In price-based mechanisms, the price represents supply/demand condition of resources in an economic market. Instead of price, a utility function can be used to represent these condition. However, this aspect has not been given adequate attention.
- Currently either thin or thick deployment technologies have been widely used. However, slender deployment fosters adoption of Gird by scientific community.
- Porting existing desktop application or development of new grid application should be possible with minimal efforts.

Apart from these issues, resource failure and lack of trust among participants are also important challenges to be addressed.

6   Conclusion

In this paper we have presented a taxonomy of desktop grid systems considering the perspectives of resource provider, resource consumer and grid application. The contribution of this paper are

- A taxonomy of desktop grid systems to characterize the demands of resource provider, resource consumer and grid applications.
- Mapping of existing Desktop Grid Computing System using the above taxonomy.
- Identifying the research gap in current research in the desktop grid domain.

References

1. Corbató, F.J., Vyssotsky, V.A.: Introduction and overview of the multics system. In: Proceedings of the November 30–December 1, 1965, fall joint computer conference, part I, AFIPS '65 (Fall, part I), New York, NY, USA, pp. 185–196, ACM (1965)
2. Foster, I., Kesselman, C., Tuecke, S.: The anatomy of the grid: enabling scalable virtual organizations. Int. J. High Perform. Comput. Appl. **15**, 200–222 (2001)
3. Krauter, K., Buyya, R., Maheswaran, M.: A taxonomy and survey of grid resource management systems for distributed computing. Softw. Pract. Exper. **32**, 135–164 (2002)
4. Foster, I., Kesselman, C.: Computational grids. pp. 15–51 (1999)
5. Kurdi, H., Li, M., Al-Raweshidy, H.: A classification of emerging and traditional grid systems. Distributed Systems Online, IEEE, vol. 9, pp. 1–1 (2008)

6. Foster, I.: Globus toolkit version 4: software for service-oriented systems. J. Comput. Sci. Technol. **21**(4), 513–520 (2006)
7. Chapin, S.J., Katramatos, D., Karpovich, J.F., Grimshaw, A.S.: The legion resource management system. In: Proceedings of the Job Scheduling Strategies for Parallel Processing, IPPS/SPDP '99/JSSPP '99, pp. 162–178, Springer, London, UK (1999)
8. Cafferkey, N., Healy, P.D., Power, D.A., Morrison, J.P.: Job management in webcom. In: ISPDC, pp. 25–30, IEEE Computer Society (2007)
9. Catlett, C., Beckman, P., Skow, D., Foster, I.: Creating and operating national-scale cyber infrastructure services (2006)
10. Litzkow, M., Livny, M., Mutka, M.: Condor—a hunter of idle workstations. In: Proceedings of the 8th International Conference of Distributed Computing Systems, pp. 104–111, IEEE Press (1988)
11. Anderson, D.P.: BOINC: a system for public-resource computing and storage. In GC '04: Proceedings of the Fifth IEEE/ACM International Workshop on Grid Computing, pp. 365–372, ACM Press, New York, NY, USA (2004)
12. Maurer, J.: A conversation with david anderson. Queue **3**, 18–25 (2005)
13. Talia, D., Trunfio, P.: A p2p grid services-based protocol: design and evaluation. In: Danelutto, M., Laforenza, D., Vanneschi, M. (eds.) Proceedings of Euro-Par 2004. Lecture Notes in Computer Science, vol. 3149, pp. 1022–1031, Springer, Berlin (2004)
14. Koh, M., Song, J., Peng, L., See, S.: Service registry discovery using GridSearch P2P framework. In: Proceeding of CCGrid, vol. 2, p. 11 (2006)
15. Pham, T.V., Lau, L.M., Dew, P.M.: An adaptive approach to p2p resource discovery in distributed scientific research communities. In: Proceeding of CCGrid, vol. 2, p. 12 (2006)
16. Foster, I., Iamnitchi, A.: On death, taxes, and the convergence of peer-to-peer and grid computing. In: 2nd International Workshop on Peer-to-Peer Systems, IPTPS03, pp. 118–128 (2003)
17. Trunfio, P., Talia, D., Papadakis, H., Fragopoulou, P., Mordacchini, M., Pennanen, M., Popov, K., Vlassov, V., Haridi, S.: Peer-to-peer resource discovery in grids: models and systems. Future Gener. Comput. Syst. **23**, 864–878 (2007)
18. Anderson, D.P., Christensen, C., Allen, B.: Grid resource management—designing a runtime system for volunteer computing. In: SC '06: Proceedings of the 2006 ACM/IEEE Conference on Supercomputing, p. 126, ACM Press, New York, NY, USA (2006)
19. Buck, P.D.: Unofficial boinc wiki: Overview of daemons (2005)
20. Anderson, D.P., Korpela, E., Walton, R.: High-performance task distribution for volunteer computing. In: Proceedings of the First International Conference on e-Science and Grid Computing, E-SCIENCE '05, pp. 196–203, IEEE Computer Society, Washington, DC, USA (2005)
21. Silva, J.N., Veiga, L., Ferreira, P.: nuboinc: Boinc extensions for community cycle sharing. In: Second IEEE International Conference on Self-Adaptive and Self-Organizing Systems Workshops, pp. 248–253 (2008)
22. Baldassari, J.D.: Design and evaluation of a public resource computing framework. Master's thesis, Worcester Polytechnic Institute, Worcester, MA, USA (2006)
23. Pan, Z., Ren, X., Eigenmann, R., Xu, D.: Executing MPI programs on virtual machines in an internet sharing system. In: Proceedings of the 20th International Conference on Parallel and Distributed Processing, IPDPS'06, pp. 101–101, IEEE Computer Society, Washington, DC, USA (2006)
24. Olejnik, R., Laskowski, E., Toursel, B., Tudruj, M., Alshabani, I.: DG-ADAJ: a java computing platform for desktop grid. In: Bubak, M., Turala, M., Wiatr, K. (eds.) Cracow Grid Workshop 2005 Proceedings. Academic Computer Centre CYFRONET AGH, Cracow, Poland (2006)
25. Balaton, Z., Gombas, G., Kacsuk, P., Kornafeld, A., Kovacs, J., Marosi, A., Vida, G., Podhorszki, N., Kiss, T.: Sztaki desktop grid: a modular and scalable way of building large computing grids. In: Parallel and Distributed Processing Symposium, 2007. IPDPS 2007. IEEE International, pp. 1–8 (2007)

26. distributed.net homepage. http://www.distributed.net/
27. Fedak, G., Germain, C., Neri, V., Cappello, F.: XtremWeb: a generic global computing system. In: Proceedings of First IEEE/ACM International Symposium on Cluster Computing and the Grid, pp. 582–587, IEEE Press (2001)
28. Petiton, S., Aouad, L., Choy, L.: Peer to peer large scale linear algebra programming and experimentations. In: Lirkov, I., Margenov, S., Wasniewski, J. (eds.) Large-Scale Scientific Computing. Lecture Notes in Computer Science, vol. 3743, pp. 430–437. Springer, Berlin (2006)
29. Lodygensky, O., Fedak, G., Cappello, F., Neri, V., Livny, M., Thain, D.: Xtremweb & condor sharing resources between internet connected condor pools. In: Proceedings of the 3rd International Symposium on Cluster Computing and the Grid, CCGRID '03, p. 382. IEEE Computer Society, Washington, DC, USA (2003)
30. Djilali, S., Herault, T., Lodygensky, O., Morlier, T., Fedak, G., Cappello, F.: Rpc-v: toward fault-tolerant rpc for internet connected desktop grids with volatile nodes. In: Proceedings of the 2004 ACM/IEEE Conference on Supercomputing, SC '04, p. 39. IEEE Computer Society, Washington, DC, USA (2004)
31. Abdennadher, N., Boesch, R.: Towards a peer-to-peer platform for high performance computing. In: Proceedings of the 2nd International Conference on Advances in Grid and Pervasive Computing, GPC'07, pp. 412–423, Springer, Berlin (2007)
32. Luther, A., Buyya, R., Ranjan, R., Venugopal, S.: Alchemi: a .netbased enterprise grid computing system. In: Arabnia, H.R., Joshua, R. (eds.) International Conference on Internet Computing. pp. 269–278, CSREA Press (2005)
33. Chu, X., Nadiminti, K., Jin, C., Venugopal, S., Buyya, R.: Aneka: next-generation enterprise grid platform for e-science and e-business applications. In: Proceedings of the Third IEEE International Conference on e-Science and Grid Computing, pp. 151–159, IEEE Computer Society, Washington, DC, USA (2007)
34. Sarmenta, L.F.G., Hirano, S.: Bayanihan: building and studying web-based volunteer computing systems using java. Future Gener. Comput. Syst. 15, 675–686 (1999)
35. Satoshi, H.: Horb: distributed execution of java programs. In: Masuda, T., Masunaga, Y., Tsukamoto, M. (eds.) World- Wide Computing and Its Applications. Lecture Notes in Computer Science, vol. 1274, pp. 29–42. Springer, Berlin (1997)
36. Tannenbaum, T., Wright, D., Miller, K., Livny, M.: Condor—a distributed job scheduler. In: Sterling, T. (ed.) Beowulf Cluster Computing with Linux. MIT Press, Cambridge (2001)
37. Raman, R., Livny, M., Solomon, M.: Matchmaking: distributed resource management for high throughput computing. In: Proceedings of the Seventh International Symposium on High Performance Distributed Computing, pp. 140–146 (1998)
38. Directed Acyclic Graph Manager Homepage. http://www.cs.wisc.edu/condor/dagman/
39. Frey, J., Tannenbaum, T., Livny, M., Foster, I., Tuecke, S.: Condor-g: a computation management agent for multi-institutional grids. Cluster Comput. 5, 237–246 (2002)
40. Chien, A.A., Marlin, S., Elbert, S.T.: Resource management in the entropia system. In: Nabrzyski, J., Schopf, J.M., Weglarz, J. (eds.) Grid resource management, pp. 431–450. Kluwer Academic Publishers, Norwell, MA, USA (2004)
41. Chien, A.A.: Architecture of a commercial enterprise desktop grid: the entropia system. In: Berman, F., Fox, G., Hey, T. (eds.) Making the Global Infrastructure a Reality, pp. 337–350. Wiley, Chichester, UK (2003)
42. Vladoiu, M., Constantinescu, Z.: An extended master worker model for a desktop grid computing platform (qadpz). In ICSOFT 2008—Proceedings of the Third International Conference on Software and Data Technologies, pp. 169–174, INSTICC Press (2008)
43. Fuad, M.M.: An autonomic software architecture for distributed applications. PhD thesis, Bozeman, MT, USA (2007)
44. Christiansen, B.O., Cappello, P.R., Ionescu, M.F., Neary, M.O., Schauser, K.E., Wu, D.: Javelin: internet-based parallel computing using java. Concurrency Pract. Experience 9(11), 1139–1160 (1997)

45. Neary, M.O., Christiansen, B.O., Cappello, P., Schauser, K.E.: Javelin: parallel computing on the internet. Future Gener. Comput. Syst. **15**, 659–674 (1999)
46. Neary, M., Phipps, A., Richman, S., Cappello, P.: Javelin 2.0: Javabased parallel computing on the internet. In: Bode, A., Ludwig, T., Karl, W., Wismller, R. (eds.) Euro-Par 2000 Parallel Processing. Lecture Notes in Computer Science, vol. 1900, pp. 1231–1238, Springer, Berlin (2000)
47. Neary, M.O., Brydon, S.P., Kmiec, P., Rollins, S., Cappello, P.: Javelin++: scalability issues in global computing. In: Proceedings of the ACM 1999 Conference on Java Grande, JAVA '99, pp. 171–180, ACM, New York, NY, USA (1999)
48. Baratloo, A., Karaul, M., Kedem, Z.M., Wijckoff, P.: Charlotte: metacomputing on the web. Future Gener. Comput. Syst. **15**, 559–570 (1999)
49. Batheja, J., Parashar, M.: A framework for opportunistic cluster computing using javaspaces. In: Proceedings of the 9th International Conference on High-Performance Computing and Networking, HPCN Europe 2001, pp. 647–656, Springer, London, UK (2001)
50. Baratloo, A., Karaul, M., Karl, H., Kedem, Z.M.: Knittingfactory: an infrastructure for distributed web applications. Technical Report TR 1997, 748, New York, NY, USA (1997)
51. Buyya, R., Vazhkudai, S.: Compute power market: towards a market-oriented grid. In: Proceedings of the 1st International Symposium on Cluster Computing and the Grid, CCGRID '01, pp. 574–, IEEE Computer Society, Washington, DC, USA (2001)
52. Xiao, L., Zhu, Y., Ni, L.M., Xu, Z.: Gridis: an incentive-based grid scheduling. In: IPDPS, IEEE Computer Society (2005)
53. Nisan, N., London, S., Regev, O., Camiel, N.: Globally distributed computation over the internet-the popcorn project. In: Proceedings of 18th International Conference on Distributed Computing Systems, 1998, pp. 592–601 (1998)
54. Waldspurger, C.A., Hogg, T., Huberman, B.A., Kephart, J.O., Stornetta, W.S.: Spawn: a distributed computational economy. IEEE Trans. Softw. Eng. **18**, 103–117 (1992)
55. Eager, D.L., Lazowska, E.D., Zahorjan, J.: A comparison of receiver-initiated and sender-initiated adaptive load sharing. Perform. Eval. **6**(1), 53–68 (1986)
56. Andrade, N., Costa, L., Germoglio, G., Cirne, W.: Peer-to-peer grid computing with the ourgrid community. In: Proceedings of the 23rd Brazilian Symposium on Computer Networks (2005)
57. Cirne, W., Brasileiro, F., Andrade, N., Costa, L., Andrade, A., Novaes, R., Mowbray, M.: Labs of the World, Unite!!! J. Grid Comput. **4**(3), 225–246 (2006)
58. Baker, M., Buyya, R., Laforenza, D.: Grids and grid technologies for wide-area distributed computing. Softw. Pract. Exper. **32**, 1437–1466 (2002)
59. Venugopal, S., Buyya, R., Ramamohanarao, K.: A taxonomy of data grids for distributed data sharing, management, and processing. ACM Comput. Surv. **38** (2006)
60. Fedak, G., He, H., Lodygensky, O., Balaton, Z., Farkas, Z., Gombas, G., Kacsuk, P., Lovas, R., Marosi, A.C., Kelley, I., Taylor, I., Terstyanszky, G., Kiss, T., Cardenas-Montes, M., Emmen, A., Araujo, F.: Edges: a bridge between desktop grids and service grids. In: Proceedings of the the the Third ChinaGrid Annual Conference (chinagrid 2008), CHINAGRID '08, pp. 3–9, IEEE Computer Society, Washington, DC, USA (2008)
61. Choi, S., Kim, H., Byun, E., Baik, M., Kim, S., Park, C., Hwang, C.: Characterizing and classifying desktop grid. In: Seventh IEEE International Symposium on Cluster Computing and the Grid. CCGRID 2007, pp. 743–748 (2007)
62. Von Laszewski, G., Blau, E., Bletzinger, M., Gawor, J., Lane, P., Martin, S., Russell, M.: Software, component, and service deployment in computational grids. In: Proceedings of the IFIP/ACM Working Conference on Component Deployment, pp. 244–256, Springer, London, UK (2002)
63. Grid Packaging Tools Homepage. http://grid.ncsa.illinois.edu/gpt/
64. Choi, S., Kim, H., Byun, E., Hwang, C.: A taxonomy of desktop grid systems focusing on scheduling. Technical Report KU-CSE-2006-1120-02. Department of Computer Science and Engineering, Korea University, Seong gbuk-gu, Seoul (2006)

65. Chakravarti, A., Baumgartner, G., Lauria, M.: The organic grid: self-organizing computation on a peer-to-peer network. IEEE Trans. Syst. Man Cybern. Part A Syst. Hum. **35**, 373–384 (2005)
66. Montresor, A., Meling, H., Babaoglu, Z.: Messor: load-balancing through a swarm of autonomous agents. In: Moro, G., Koubarakis, M. (eds.) Agents and Peer-to-Peer Computing. Lecture Notes in Computer Science, vol. 2530, pp. 125–137, Springer, Berlin (2003)
67. Zhong, L., Wen, D., Ming, Z.W., Peng, Z.: Paradropper: a general- purpose global computing environment built on peer-to-peer overlay network. In: Proceedings of the 23rd International Conference on Distributed Computing Systems, ICDCSW '03, p. 954, IEEE Computer Society, Washington, DC, USA (2003)
68. Kim, J.-S., Nam, B., Marsh, M.A., Keleher, P.J., Bhattacharjee, B., Richardson, D., Wellnitz, D., Sussman, A.: Creating a robust desktop grid using peer-to-peer services. In: IPDPS, pp. 1–7, IEEE (2007)
69. Anglano, C., Canonico, M., Guazzone, M.: The sharegrid peer-to-peer desktop grid: infrastructure, applications, and performance evaluation. J. Grid Comput. **8**, 543–570 (2010)
70. Antoniu, G., Jan, M., Noblet, D.A.: Enabling the p2p jxta platform for high-performance networking grid infrastructures. In: Yang, L.T., Rana, O.F., Di Martino, B., Dongarra, J. (eds.) High Performance Computing and Communications. Lecture Notes in Computer Science, vol. 3726, pp. 429–439. Springer, Berlin (2005)
71. Ferrante, M.: The jxta way to grid: a dead end? JXTA/Grid survey slide. Available at http://www.disi.unige.it/person/FerranteM/papers/JXTA-survey.pdf (2008)
72. Waldo, J.: The Jini specifications, 2nd edn. Addison-Wesley Longman Publishing Co., Inc., Boston, MA, USA (2000)
73. Juhasz, Z., Andics, A., Kuntner, K., Pota, S.: Towards a robust and fault-tolerant multicast discovery architecture for global computing grids. In: Proceedings of the 4th DAPSYS Workshop, pp. 74–81, Desprez (2002)
74. Juhasz, Z., Andics, A., Pota, S.: Jm: a jini framework for global computing. In: 2nd IEEE/ACM International Symposium on Cluster Computing and the Grid, p. 395 (2002)
75. Huang, Y.: Jisga: a jini-based service-oriented grid architecture. Int. J. High Perform. Comput. Appl. **17**, 317–327 (2003)
76. Hampshire, A., Blair, G.: Jgrid: exploiting jini for the development of grid applications. In: Guelfi, N., Astesiano, E., Reggio, G. (eds.) Scientific Engineering for Distributed Java Applications. Lecture Notes in Computer Science, vol. 2604, pp. 132–142, Springer, Berlin (2003)
77. Teo, Y.M., Wang X.: Alice: a scalable runtime infrastructure for high performance grid computing. In: Jin, H., Gao, G.R., Xu, Z., Chen, H. (eds.) NPC. Lecture Notes in Computer Science, vol. 3222, pp. 101–109, Springer, Berlin (2004)
78. Furmento, N., Hau, J., Lee, W., Newhouse, S., Darlington, J.: Implementations of a service-oriented architecture on top of jini, jxta and ogsi. In: Dikaiakos, M.D. (ed.) Grid Computing. Lecture Notes in Computer Science, vol. 3165, pp. 249–261, Springer, Berlin (2004)
79. Kent, S., Broadbent, P., Warren, N., Gulliver, S.: On-demand hd video using jini based grid. In: 2008 IEEE International Conference on Multimedia and Expo, pp. 1045–1048 (2008)
80. Hawick, K.A., James, H.A.: A java-based parallel programming support environment. In: Proceedings of the 8th International Conference on High-Performance Computing and Networking, HPCN Europe 2000, pp. 363–372, Springer, London, UK (2000)
81. de Roure, D., Baker, M., Jennings, N.R., Shadbolt, N.: The evolution of the grid (2003)
82. Fedak, G.: Recent advances and research challenges in desktop grid and volunteer computing. In: Desprez, F., Getov, V., Priol, T., Yahyapour, R. (eds.) Grids, P2P and Services Computing, pp. 171–185, Springer, US (2010)
83. Liu, L., Antonopoulos, N.: From client-server to p2p networking. In: Shen, X., Yu, H., Buford, J., Akon, M. (eds.) Handbook of Peer-to-Peer Networking, pp. 71–89, Springer, US (2010)

84. Schnizler, B.: Resource allocation in the grid: a market engineering approach. Dissertation, Universität Karlsruhe (TH). ISBN: 978-3-86644-165-1 (2007)
85. Rezmerita, A., Neri, V., Cappello, F.: Toward third generation internet desktop grids. Technical Report RT-0335, INRIA (2007)
86. Jin, H.: Challenges of grid computing. In: Fan, W., Wu, Z., Yang, J. (eds.) Advances in Web-Age Information Management. Lecture Notes in Computer Science, vol. 3739, pp. 25–31, Springer, Berlin (2005)
87. Kondo, D., Chien, A.A., Casanova, H.: Resource management for rapid application turnaround on enterprise desktop grids. In: SC, p. 17, IEEE Computer Society (2004)
88. Abdelkader, K., Broeckhove, J., Vanmechelen, K.: Economic-based resource management for dynamic computational grids: extension to substitutable cpu resources. In: Aboulhamid, E.M., Sevillano, J.L. (eds.) AICCSA, pp. 1–6, IEEE (2009)
89. Vanmechelen, K.: A performance and feature-driven comparison of jini and jxta frameworks. Master's thesis, University of Antwerp, Antwerpen, Belgium (2003)
90. mgr inz. Marcin Cie´slak: Boinc on jxta suggestions for improvements. Technical Report, Poland (2007)

Trust Based Resource Selection in Grids Using Immune System Inspired Model

V. Vijayakumar

Abstract Grid is emerging as a potential technology leading to the utilization of the computer resources that are underutilized. Resources scheduling is one of the core grid services and has a direct impact on the overall performance of the system and the service quality. Scheduling in grids helps to find an better resource which improves the performance and also helps in utilizing the available resources efficiently. Most scheduling and resource selection algorithms do not take the characteristics of the resource into account. This can ultimately lead to reliability and security issues which will affect the quality of service. In this work it is proposed to implement a trust module which predicts trust across the grid network drawing principles from the Immune System (IS). By integrating the trust module with dynamic scheduling, security is increased and ratio of job task failure decreases. The proposed immune inspired system improves the makespan by almost 35 % compared to traditional mechanism of manually selecting nodes based on trust and reputation.

Keywords Immune system (IS) · Trust and reputation systems (TRSs) · Trust management framework (TMF) · Artificial immune system (AIS) · Human immune system (HIS)

1 Introduction

Grid computing is the way of utilizing the resources in solving large scale problems and it supports wide range of applications, researches and for industries too. Security plays a very important role in grid environment and is involved in every stage of grid computing including resource selection, job submission, secure

V. Vijayakumar (✉)
SCSE, VIT University, Chennai Campus, Chennai, India
e-mail: vijayakumar.v@vit.ac.in

© Springer International Publishing Switzerland 2016
V. Vijayakumar and V. Neelanarayanan (eds.), *Proceedings of the 3rd International Symposium on Big Data and Cloud Computing Challenges (ISBCC – 16'),*
Smart Innovation, Systems and Technologies 49, DOI 10.1007/978-3-319-30348-2_17

communication, authentication and authorization [5, 14]. These security factor implementations can be termed as attribute based security mechanisms. Various mechanisms have been proposed in the related areas for grid security based on attributes. Another emerging area of research is the identification of trustworthiness of the resource based on its previous behavior and interaction with the grid [4, 6].

In grid computing, individuality is hard to be established because fresh accounts can be formed easily, agent may not employ in frequent communication together due to the grid size and diversity of required services (Malarvizhi and Rhymend [13]). The motivation for effort on Trust and Reputation Systems (TRSs) falls in understanding the dishhonest individuals. In this research we propose an immune system inspired collaborative filtering mechanism for recommendation of resources based on the user's requirement and risk appetite [7, 8].

2   Related Work

The barrier for the grid to be widespread is the using, configuring and maintaining difficulty, which wants extreme IT with human involvement. Also, interoperation amongst grids is on track. To be the core of grid systems, the resource management must be autonomic and inter-operational to be sustainable for future Grid computing. For this purpose, Liang and Shi [12] introduce HOURS, a reputation-driven framework for Grid resource management. HOURS is designed to tackle the difficulty of automatic rescheduling, self-protection, incentives, heterogeneous resource sharing, reservation and agreement in Grid computing. The work focused in designing a reputation-based resource scheduler and use emulation to test its performance with real job traces and node failure traces. To describe the HOURS framework completely, a preliminary multiple-currency-based economic model is also introduced, with which future extension and improvement can be easily integrated into the framework.

Recommendation-based reputation assessment in peer-to-peer systems relies on recommendations in predicting the reputation of peers [2, 15, 18]. In this they discuss the effectiveness and cost metrics in the recommendation retrieval and also they evaluate the following retrieval methods: flooding, recommendation tree and the storage peer. The simulation results show that overlay network construction significantly contributes to the performance of recommendation retrieval in terms of effectiveness and cost. Storage peer approach in structured network outperforms the other two approaches as long as the network is stable. Vijayakumar et al. [19] proposed a trust model by considering both trust and reputation with the user's feedback and also the other entities feedback in providing the secured resources for the users to submit their job in the grid environment.

3 Methodology

Artificial Immune System (AIS) algorithms are modeled on the Human Immune System (HIS) for solving complex real-world problems. The empirical knowledge of the HIS is used to create a process which is distributed, adaptive and self-organizing to tackle engineering problems. Depending on the pathogens HIS encounters, it triggers either innate or adaptive mechanism, to neutralize the pathogen [16]. The HIS has the ability to differentiate cells as self or non-self within the body. The HIS categorizes the self and non-self cells using distributed task force and its cleverness to act local and global perspective with the help of network messengers in communicating each other [1].

The immune system is the mechanism which never changes also it has the capability to neutralize formerly identified pathogens while the adaptive system responds to previously unknown foreign cells. The immune system acts both in parallel and sequential fashion. During an infection, the immune system responds by sending specific lymphocytes or antibodies to destroy the invading antigens [10]. Essentially, the immune system is matching of antigen and antibodies. Most of the AIS models are based on negative selection mechanism, clonal selection and the somatic hyper mutation theories [9].

In this paper it is proposed to implement a recommender system based on the Human Immune System (HIS). The hypothesis of Clonal selection theory specifies how B cell lymphocytes respond to an infection and how antibodies change and mutate for different type of infections. The theory specifies that when an organism is exposed to an antigen, immune system responds by producing Antibodies (Ab) from B lymphocytes which will neutralize the antigen. Each B lymphocytes produce specific type of antibodies, so while antigen matches with antibody, also binds to antigen and replicates in producing clones [3, 11]. With genetic mutation the binding strengthens with antigens exposure. The clone selection theory used in this work can be summarized as follows:

- Maintaining memory set of previous antigen attacks
- Selection and cloning of matched antibodies
- Mutating the cloned antibodies at a high rate
- Elimination of lymphocytes moving receptors which are self reactive
- propagation and separation happen while antigen contacts

The number of clone created is given by:

$$nClones = \left[\frac{a.M}{i} + 0.5 \right]$$

where, a is the clonal part, M is antibody pool size and i is rank assigned to the antibody.

The clones prepared for every antigen to the exposed system for m number of selected antibodies is given by:

$$N_c = \sum_{i=1}^{m} \left[\frac{a.M}{i} + 0.5 \right]$$

Appropriate encoding is key factor for the success of the algorithm. The antigens and antibodies in context of an application domain can be defined as, a target or solution is the antigen and rest of the data is antibodies. There may be one or more antigens and usually the antibodies are plentiful. The antigens and antibodies in the domain are similarly encoded. Usual forms of encodings are a string of numbers or features. Potential cell strength is calculated using:

$$E_g = \{ \ I,A,F,Au,S,In,SJE,Am\}$$

where,

Eg   Encoder
I   Intrusion Detection System capabilities
A   Anti virus capability
F   Firewall capability
Au   Authentication mechanism
S   Secure file storage capacity
In   Interoperability capacity of the resource
SJE   Secure Job Execution capability
Am   Authorization mechanism

The Self-Protection Capability requirement for the antibody is computed by:

$$SPC_i = \sum_{i=1}^{n} E_g(i) * x(i)$$

where, $x(i)$ is the current capability of the cell and $Eg(i)$ is the survival capacity. The encoded data can have values: $\{0,0.125, 0.250,0.375,0.5,0.625,0.75,1\}$, 0 represents no affinity and 1 represents required affinity for the cell, n = The total number of factors, x = The affinity of the cell.

Once the antibody has the desired strength measured by SPC, it is considered to be part of the memory pool for cloning and deployment. The best antibodies are selected by its previous reputation and is given by its Reputation vector R_V = {Consistency, Confidentiality, Truthfulness, Security, Privacy, Non-repudiation, Authentication, Authorization, Reliability, Robustness}.

The reputation vector of all the antibodies in the memory pool is given by:

$$R_M = \begin{bmatrix} SA_{11} & SA_{12} & SA_{13}......SA_{1j} \\ SA_{21} & SA_{22} & SA_{23}......SA_{2j} \\ SA_{i1} & SA_{i2} & SA_{i3}......SA_{ij} \end{bmatrix}$$

where, j is the number of chromosomes and i is the number of antibodies in the memory pool and S_{ij} can have values $\{0,0.125,0.25,0.375,0.5,0.625,0.75,1\}$.

Affinity measure is closely associated with encoding and is an important design criterion while developing AIS. In case of binary encoding, simple matching algorithms are used to obtain similarity score. These algorithms count the number of continuous bits that match or return the length of the longest matching bits as similarity measure. If the encoding is non-binary, then the 'distance' between the two strings is computed using the affinity relation:

$$r_m = \frac{\sum_i (x_i - \bar{x})(y_i - \bar{y})}{\sqrt{\sum_i (x_i - \bar{x})^2 (y_i - \bar{y})^2}}$$

with

$$r_{m(i)} = \begin{cases} 0 & when \quad r_{m(i)} \leq 0 \\ 1 & when \quad r_{m(i)} \geq 1 \end{cases}$$

In absence of ties differences between the ranks are calculated as r_m. The values equal to 1 show strong affinity and near 0 show weak affinity for the anticipated antigen.

The antibodies reputation with respect to other antibody is computed by:

$$R_i = \sum_{n=1}^{j} R_{v(i,j)}$$

The total capacity of an individual antibody by its self protecting capability and reputation is given by:

$$E_i = SPC_i + R_i$$

Given the antigen strength, the antibodies are segregated into three categories based on the distance from the protection capability required is given by:

$$\text{Antibody Clustering} = \sum_i (x_i - y_i)^2$$

For the trust values normalized in the range of −1 to +1 the search space of the proposed method is shown in Figs. 1 and 2.

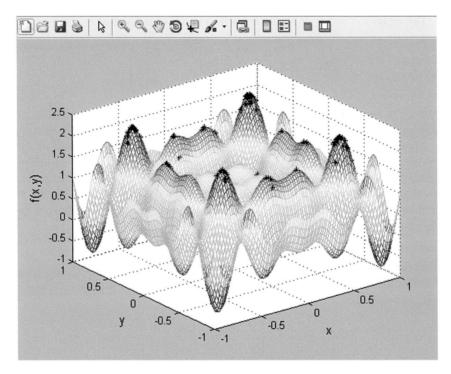

Fig. 1 For a given trust and reputation x and y, the search space as f(x, y)

Fig. 2 The antigen
generations

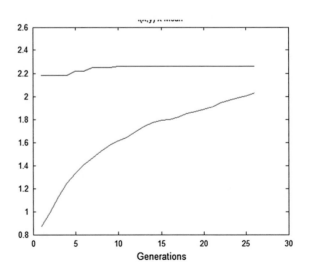

4 Results

Experiments were simulated in a grid environment containing 10 nodes are given by:

$$\{n_0, n_1, n_2, n_3, n_4, n_5, n_6, n_7, n_8, n_9\}$$

If the job completion time is less than the estimated time based on the scheduling strategy, only the first cluster is selected with high value of trust and reputation, else the first and the second cluster is selected with lowering of the trust. Let the required trust value to compute a given job be {0.6250, 0.1250, 0.8750, 0.8750, 0, 0.1250, 0.6250, 0.2500}. The Security parameters of each node during the first run is shown in Table 1. The SPC computed for each node is given in Table 2. The reputation snapshot for all the 10 nodes are given in Table 3. The trust value for each node computed from Tables 2 and 4 is shown in Fig. 3. The node centroids obtained for the above value is computed as 3.0184, 4.4282 and 3.9610. Based on these centroids the scheduling time for 30 jobs of random duration is shown in Fig. 4.

Table 1 The security parameters snapshot at first run

First run	I	A	F	AU	S	In	SJE	Am
n0	1	0.375	0.5	0.875	0.5	0.25	1	0.125
n1	0.875	0	0.625	0.375	1	0.25	0	0.75
n2	0.875	0.625	0.375	0	0.625	0.75	0.5	0.5
n3	0.25	0.5	0.875	0.25	1	0.875	0.375	0.125
n4	0.625	0.75	0.75	0.125	0.25	0.375	0.5	0.375
n5	0	0.75	1	0.25	0.75	0.875	0.75	0.625
n6	0.375	0.625	0.5	0.375	0.25	0.75	0.25	0.125
n7	0.25	0	0.25	0.5	0.75	0	0.875	0.75
n8	0.125	0	0	0.5	0.75	0.625	0.5	0.25
n9	0.125	0.25	0.625	0.875	0	0.375	0	1

Table 2 Computed SPC by the given method

First run	I	A	F	Au	S	In	SJE	Am	SPC
n0	0.625	0.0469	0.4375	0.7656	0	0.0313	0.625	0.0313	**2.5626**
n1	0.5469	0	0.5469	0.3281	0	0.0313	0	0.1875	**1.6407**
n2	0.5469	0.0781	0.3281	0	0	0.0938	0.3125	0.125	**1.4844**
n3	0.1563	0.0625	0.7656	0.2188	0	0.1094	0.2344	0.0313	**1.5783**
n4	0.3906	0.0938	0.6563	0.1094	0	0.0469	0.3125	0.0938	**1.7033**
n5	0	0.0938	0.875	0.2188	0	0.1094	0.4688	0.1563	**1.9221**
n6	0.2344	0.0781	0.4375	0.3281	0	0.0938	0.1563	0.0313	**1.3595**
n7	0.1563	0	0.2188	0.4375	0	0	0.5469	0.1875	**1.547**
n8	0.0781	0	0	0.4375	0	0.0781	0.3125	0.0625	**0.9687**
n9	0.0781	0.0313	0.5469	0.7656	0	0.0469	0	0.25	**1.7188**

Table 3 Normalized reputation snapshot

First Run	Consistency	Confidentiality	Truthfulness	Security	Privacy	Non repudiation	Authentication	Authorization	Reliability	Robustness
no	0.25	0.625	0.5	0.625	0.5	0.375	0.75	0.375	0.75	0.875
n1	0.75	0.75	0.625	0.5	0.625	0.5	0	0.125	0.375	0.375
n2	0.125	0.625	0.75	0.875	0.5	0.375	0.5	0.625	0.75	0.875
n3	0.25	1	0.375	0.25	0.75	0.75	0.5	0.25	0.75	0.375
n4	0	0.125	0.625	0.25	0.5	0.625	0.875	0	0.375	0.75
n5	0.625	0.75	0.375	0.125	1	0.75	0.5	0.75	0	0.375
n6	0.75	0.25	0.875	1	0.125	1	0.375	0.25	0.25	0.875
n7	0.5	0.125	0.875	0.625	0	1	0.75	0.375	0.375	0.75
n8	0.375	0.625	0.25	0.5	0	0.125	0.75	0.75	0.25	0.375
n9	0.625	0.5	0.625	0.625	0	0.125	0.5	0.375	0.125	0.125

Table 4 The correlation among the nodes to select the ideal node

	n0	n1	n2	n3	n4	n5	n6	n7	n8	n9
n0	1	0.1122	0.19	0.1682	0	0.2322	0	0	0	0
n1	0.1122	1	0	0	0.7152	0	0	0.2044	0.1587	0
n2	0.19	0	1	0.7075	0	0.425	0	0	0.1048	0.5148
n3	0.1682	0	0.7075	1	0	0	0	0.0377	0.1746	0.5698
n4	0	0.7152	0	0	1	0.2642	0	0	0.2236	0
n5	0.2322	0	0.425	0	0.2642	1	0	0	0	0.3024
n6	0	0	0	0	0	0	1	0.1291	0.1694	0.1661
n7	0	0.2044	0	0.0377	0	0	0.1291	1	0	0
n8	0	0.1587	0.1048	0.1746	0.2236	0	0.1694	0	1	0.5132
n9	0	0	0.5148	0.5698	0	0.3024	0.1661	0	0.5132	1

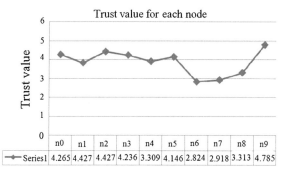

Trust value for each node

	n0	n1	n2	n3	n4	n5	n6	n7	n8	n9
Series1	4.265	4.427	4.427	4.236	3.309	4.146	2.824	2.918	3.313	4.785

Fig. 3 Trust value computed using the immune system based mechanism

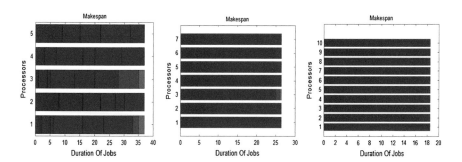

Fig. 4 The makespan for all the three scenarios

It can be seen from Fig. 4, the proposed immune inspired system improves the makespan by almost 35 % compared to traditional mechanism of manually selecting nodes based on trust and reputation. This clearly shows that quantity alone is not a criteria for providing good quality of service in grid environment.

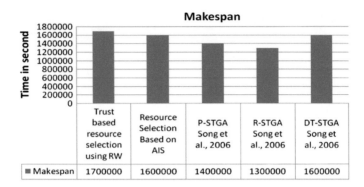

Fig. 5 The makespan obtained using proposed method

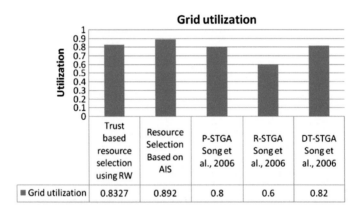

Fig. 6 The Grid utilization rate

In this proposed method the experimental setup is compared with Song et al. [17]. The makespan and grid utilization obtained is shown in Figs. 5 and 6.

The makespan of the proposed AIS method is 1.6×10^6 and the failure rate is 0.0937 compared to 0.19 obtained from DT-STGA method which was proposed by Song et al. [17]. Computing the percentage difference by the proposed method and DT-STGA method, the decrease in failure rate is 106.3 %. Similarly the grid utilization is also improved by 1.55 %.

5 Conclusion

In this paper it was proposed to implement an immune system inspired resource selection of grid resources based on trust and reputation. The paradigm of the immune system to identify antigens and clone antibodies is studied and applied with satisfying results. The proposed methodology was implemented using Matlab and makespan computed. The computed values look promising without compromising on the trust and security.

References

1. Aickelin, U., Bentley, P., Cayzer, S., et al.: Danger theory: the link between AIS and IDS? In: Timmis, J., Bentley, P.J., Hart, E. (eds.) ICARIS (2003)
2. Azzedin, F., Ridha, A.: Recommendation retrieval in reputation assessment for peer-to-peer systems. Electron. Notes Theor. Comput. Sci. **244**, 13–25 (2009)
3. Benhamini, E., Coico, R., Sunshine, G.: Immunology—A Short Course. Wiley-Liss Inc., USA (2000)
4. Cao, J., Yuand, W., Yuzhong, Q.: A new complex network model and convergence dynamics for reputation computation in virtual organizations. Phys. Lett. A **21**, 414–425 (2006)
5. Elmroth, E., Tordsson, J.: Grid resource brokering algorithms enabling advance reservations and resource selection based on performance predictions. Future Gener. Comp. Syst. **24**, 585–593 (2008)
6. Foster, I., Kesselman, C., Tsudik, G., Tuecke, S.: A security architecture for computational grids. In: ACM Conference Computers Security, pp. 83–91 (1998)
7. Foster, I., Kesselman, C., Tsudik, G., Tuecke, S.: The physiology of the grid: an open grid services architecture for distributed systems integration. Technical Report, Open Grid Service Infrastructure WG, Global Grid Forum (2002)
8. Gray, E., Jensen, C., O'Connell, P., Weber, S., Seigneur, J.-M., Chen, Y.: Trust evolution policies for security in collaborative ad hoc applications. Electron. Notes Theor. Comput. Sci. **157**(3), 95–111 (2006)
9. Hassan, M.W., McClatchey, R., Willers, I.: A scalable evidence based self-managing framework for trust management. Inf. Sci. **179**, 2618–2628 (2007)
10. Kubi, J.: Immunology, Fifth Edition by Richard A. Goldsby. Thomas J. Kindt, Barbara A. Osborne. W H Freeman, San Francisco (2002)
11. Leandro, N., Castro, D., Von Zuben, F.J.: The clonal selection algorithm with engineering applications. GECCO .Workshop on Artificial Immune Systems and Their Applications, Las Vegas, USA, pp. 36–37 (2000)
12. Liang, Z., Shi, W.: A reputation-driven scheduler for autonomic and sustainable resource sharing in grid computing. Grid J. Parallel Distrib. Comput. **70**, 111–125 (2005)
13. Malarvizhi, N., Rhymend, U.: Performance analysis of resource selection algorithms in grid computing environment. J. Comput. Sci. **7**, 493–498 (2011)
14. Muruganantham, S., Srivastha, P.K., Khanaa.: Object based middleware for grid computing. J. Comput. Sci. **6**, 336–340 (2010)
15. Papaioannou, T.G., Stamoulis, G.D.: A mechanism that provides incentives for truthful feedback in peer-to-peer systems. In: Trust and Privacy Aspects of Electronic Commerce (2010)
16. Rich, E., Knight, K.: Artificial intelligence. McGraw-Hill Chapter 20, Artificial Intelligence. McGraw-Hill, New York (1991)
17. Song, S., Hwang, K., Kwok, Y-K: Risk-resilient heuristics and genetic algorithms for security-assured grid job scheduling. IEEE Trans. Comput. **55**(6), 703–719 (2006)
18. Sumathi, P., Punithavalli, M.: Constructing a grid simulation for E-governance applications using GridSim. J. Comput. Sci. **4**, 674–679 (2008)
19. Vijayakumar, V., Wahida Banu, R.S.D., Abawajy, J.H.: An efficient approach based on trust and reputation for secured selection of grid resources. Int. J. Parallel Emergent Distrib. Syst. **27**(1), 1–17 (2012)

Part IV
Image Processing

Discrete Orthogonal Moments Based Framework for Assessing Blurriness of Camera Captured Document Images

K. De and V. Masilamani

Abstract One of the most widely used tasks in the area of image processing is automated processing of documents, which is done using Optical Character Readers (OCR) from document images. The most common form of distortion in document images is blur which can be caused by defocus, motion, camera shake etc. In this paper we propose a no reference image sharpness measure framework using discrete orthogonal moments and image gradients for assessing quality of document images and validated the results against state of the art image sharpness measures and accuracy of three well known Optical Character Readers.

Keywords Document image quality · Discrete orthogonal moments

1 Introduction and Literature Survey

With the invention of smart phone camera, a lot of documents are scanned as images from mobile phone and tablet cameras. The document images need to be processed automatically with the help of Optical Character Readers. For the accurate prediction of OCR the images need to be of very good quality. The most common distortion for document images is the introduction of blur in the image. The most common sources of blur are camera shake, motion, defocus. Blurred images are not accurately processed using OCRs.

Image Quality assessment algorithms are classified into three categories namely, (1) Full Reference Image Quality Assessment Algorithms (FR-IQA), (2) Reduced Reference Image Quality assessment algorithms (RR-IQA) and (3) No Reference Image quality assessment algorithms (NR-IQA) [1]. The first category of algorithms, FR-IQA need a reference image of the same scene and full reference image information is required for assessing quality of a target image. Peak Signal to Noise Ratio

K. De · V. Masilamani (✉)
Indian Institute of Information Technology Design & Manufacturing
Kancheepuram, Chennai 600127, India
e-mail: masila@iiitdm.ac.in

© Springer International Publishing Switzerland 2016
V. Vijayakumar and V. Neelanarayanan (eds.), *Proceedings of the 3rd International Symposium on Big Data and Cloud Computing Challenges (ISBCC – 16'),*
Smart Innovation, Systems and Technologies 49, DOI 10.1007/978-3-319-30348-2_18

(PSNR) and Structural Similarity Index Measure (SSIM) [2] are the most common examples. The next category of algorithms, RR-IQA also need a reference image of the same scene, but instead of full reference image data only certain features of the reference are used to perform the task of image quality assessment of the target image ex. [3]. Finally, the third category of image quality assessment algorithms, NR-IQA does not require any reference image information for the assessment of quality of target images. NR-IQA is an extremely difficult task to come up with an algorithm which will denote information about quality of image as different type of distortions have different statistical properties. For this reason distortion specific image quality assessment algorithms are developed for different type of distortions like noise [4], compression (JPEG [5], JPEG2000 [6]) and blur. The popular measures are based on Just noticeable blur (JNB) [7], Cumulative Probability of blur detection (CPBD) [8], S3 [9], Q-metric [10] etc.

There are different approaches for no-reference image blurriness/sharpness assessment like in spatial domain [11], wavelet domain [12], Phase coherence [13], using fuzzy logic [14], Auto regression space [15] and using deep learning approach to document image quality assessment, Kang et al. [16]. One of the recent work in this field using discrete orthogonal moments was proposed by Li et al. [17], where the authors use Tchebichef moments and image gradients with saliency detection to come up with a image sharpness measure. The proposed measure works excellent for natural images but does not give good performance for document images. In this paper we have adapted the concept of using discrete orthogonal moment energy and image gradients proposed by Li et al. [17], to develop a framework for image sharpness estimation specific for document images, where we can use any of the three discrete orthogonal moments Krawtchouk, Hahn and Tchebichef moments and image gradient calculation can be done by different methods as it will be explained later.

A brief survey of about the area of document image quality assessment is published in the article [18]. References [19–21] lists out some work related to OCR. Quality aspects for camera captured images for OCR other than blurring are discussed in [22]. The goal of document image sharpness/blurriness assessment is to develop a numerical measure which gives an estimate about the sharpness of the document images. To validate the proposed work we compare our proposed measures against accuracy of the state of the art Optical Character Readers (OCRs) and compare our results with state of the art image sharpness measures. The OCR considered for this study are ABBYY Finereader [23], Tesseract OCR [24] and Omnipage [25].

In Sect. 2 we will provide the theoretical concept on which our model is based. In Sect. 3, we will provide our proposed framework and in Sect. 4 we will present the results and finally we conclude with brief concluding remarks in Sect. 5.

2 Preliminaries

In this paper we propose a image sharpness/blurriness assessment framework based on discrete orthogonal moments and image gradients for document images. In this section we give a brief introduction of the theoretical concepts used.

2.1 Moments

Discrete Orthogonal moments like Hahn moments, Krawtchouk Moments and Tchebichef moments are used as features for different image processing and computer vision applications like object detection, structural analysis etc.

Krawtchouk Moments—Image analysis using a type of discrete orthogonal moments called Krawtchouk moments (K_{mn}) are discussed in detail [26] with applications and mathematical foundations.

Hahn Moments—Details of using Hahn moments in image analysis is mentioned in detail in [27]. The detailed equations involving generation of Hahn moments from Hahn polynomials with their mathematical background is explained in detail in [27] and we simply follow the procedure generate Hahn Moment matrix H_{mn}.

Tchebichef Moments—Li et al. [17] have explained the relationship between blurring in the image and Tchebichef moments in details. Similar relationships between blurring parameter of images and Krawtchouk and Hahn moments are found. The details process of generating the matrices of Tchebichef moments with mathematical proofs is available in [28]. In future if new moments with similar better properties are proposed by researchers we can use those kernels in our proposed framework.

2.2 Image Gradients

For calculating Image gradients we have considered two implementations. First one is the same one discussed by Li et al. using convolution operator. Refer [17] for the details of implementation. Another alternate implementation proposed in [29] is also used to test our proposed framework. The 5-tap derivative function is available in [30]. In future if more efficient and accurate ways to calculate image gradients are invented we can simply integrate in the proposed frame work.

3 Proposed Document Image Sharpness/Blurriness Framework

The block diagram of the proposed framework for document image sharpness assessment is shown in Fig. 1. Let the image be of size $M \times N$ be denoted by I. To extract text region from background classical Otsu's thresholding technique [31] is used and thus Binary image BW is generated from the original text image. Similarly gradient image G is generated by any one of the techniques discussed above. Now both the binary image BW and the gradient image G is split into $R \times R$ blocks. The set of blocks in the binary image BW are denoted as $\{B_{ij}^{BW}\}$ and set of blocks in the

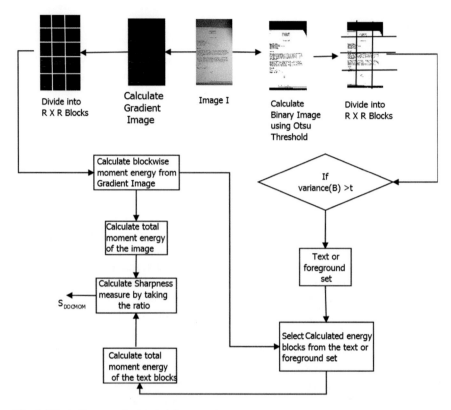

Fig. 1 Block diagram of proposed framework

gradient image G are denoted by $\{B_{ij}^G\}$, $i \in \{1, 2, 3, \ldots, X\}$, $j \in \{1, 2, 3, \ldots, Y\}$ where $X = \frac{\lfloor M \rfloor}{\lfloor R \rfloor}$, $Y = \frac{\lfloor N \rfloor}{\lfloor R \rfloor}$, $i\,j$ denotes the top most left most corner. Variance of each of the block in the binary image BW is computed. If variance of a particular block is below a low threshold t, then we classify the block as background block and if it is above the threshold t we classify the block as foreground or text block and put it sets $\{B^{BW Foreground}\}$ and $\{B^{G Foreground}\}$. Experimentally it has been observed that a low threshold $t = 0.005$ works very well and it performs well in cases where a small amount of noise is present in the image. Assumption is that only minute amount of additive noise must be present in the image, ideally it is preferred to remove noise before scanning a document image using OCR. Mathematical representation of both these sets

$$\{B^{G Foreground}\} \subset \{B_{ij}^G\} \text{ which means } \{B^{G Foreground}\} \text{ is a subset of } \{B_{ij}^G\}$$
$$\{B^{BW Foreground}\} \subset \{B_{ij}^{BW}\} \text{ which means } \{B^{BW Foreground}\} \text{ is a subset of } \{B_{ij}^{BW}\}$$

$$E_{i,j} = \sum_{p=0}^{m} \sum_{q=0}^{n} T_{pq}^2 - T_{00}^2 \qquad (1)$$

$$E_{i,j} = \sum_{p=0}^{m} \sum_{q=0}^{n} H_{pq}^2 - H_{00}^2 \qquad (2)$$

$$E_{i,j} = \sum_{p=0}^{m} \sum_{q=0}^{n} K_{pq}^2 - K_{00}^2 \qquad (3)$$

where T_{mn}, H_{mn} and K_{mn} are Tchebichef, Hahn and Krawtchouk moments and T_{00}, H_{00} and K_{00} are the dc components of Tchebichef, Hahn and Krawtchouk moments. The energy of only ac moment values is considered as only energies of edges and shapes are considered which is available in the ac component and thus dc component needs to be removed. Equations 1–3 give the energy terms for Tchebichef, Hahn and Krawtchouk cases respectively. Any of these energies can be used in the proposed framework. The first term of each $R \times R$ block is removed as the zeroth order term in the moment denotes the dc component and we are only interested in computing the ac components. The proposed measure S_{DOCMOM} will be the ratio of total sum of energy of components in set $\{B^{G_{Foreground}}\}$ divided by total sum of energy of components in the set $\{B_{ij}^G\}$ which means that the ratio of total energy of moments of the text region to the total energy of moments of the image. The energy of moments of text region can be represented mathematically $E' = E_{ij} || \mathrm{var}(E_{ij}) > t; \forall; 1 < i < X; 1 < j < Y;$ where var() denotes the variance of the block.

$$S_{DOCMOM} = \frac{\sum_{z \in E'} z}{\sum_{i=0}^{X} \sum_{j=0}^{Y} E_{ij}} \qquad (4)$$

4 Results

4.1 Analysis with DIQA Dataset

Document Image Quality Assessment Database (DIQA) [32] is a standard dataset available publicly for quality assessment of document images and it consists of 175 images of documents with various type of blurring distortions captured from different camera devices and also provides result data of three well known OCRs. To validate the proposed framework we use Spearman Rank Correlation Coeffcient (SRCC), Kendall Rank Correlation Coefficient (KRCC) to determine prediction monotonicity and Pearson Linear Correlation Coefficient for prediction accuracy. We find the correlation values between sharpness scores and the accuracy of state of

Table 1 Performance of proposed framewok on DIQA databse for differnt OCR

OCR		Krawtchouk			Hahn			Tchebichef		
		SRCC	KRCC	PLCC	SRCC	KRCC	PLCC	SRCC	KRCC	PLCC
Image gradient 5 tap	ABBYY	0.8184	0.6224	0.8121	0.8259	0.6301	0.8229	0.8259	0.6301	0.8229
Image gradient 5 tap	Omnipage	0.6516	0.4773	NA	0.6648	0.4895	NA	0.6648	0.4895	NA
Image gradient 5 tap	Tesseract	0.8045	0.6073	0.8063	0.8207	0.6248	0.8197	0.8207	0.6248	0.8197

Table 2 Standard image sharpness measures versus different OCR accuracy

	ABBYY finereader			Omnipage			Tesseract		
	SRCC	KRCC	PLCC	SRCC	KRCC	PLCC	SRCC	KRCC	PLCC
BIBLE [17]	0.4062	0.2851	0.2340	0.3377	0.2397	NA	0.4437	0.3104	0.4144
ARISM [15]	0.0051	0.0002	0.04	0.04	0.02	NA	0.04	0.02	0.12
JNB [7]	0.03	0.01	0.02	0.02	0.009	NA	0.083	0.06	0.03
LPC = SI [15]	0.7972	0.6040	0.8602	0.6901	0.5159	NA	0.8279	0.6379	0.8364
CPBD [8]	−0.376	−0.2607	−0.4384	−0.2991	−0.1936	NA	−0.2432	−0.1619	−0.2592
Q-metric [10]	0.7955	0.6067	0.6397	0.6293	0.4514	NA	0.7582	0.5629	0.7078
FISH [12]	0.7859	0.5951	0.6910	0.6427	0.4703	NA	0.7732	0.5797	0.7381
FISHBB [12]	0.7785	0.5902	0.7719	0.6414	0.4731	NA	0.7896	0.5981	0.7984

the art OCRs for the dataset. The OCRs considered for our work are ABBYY Finereader, Omnipage and Tesseract OCR. For testing purpose we generated gradient images using two techniques but the results are identical for both cases. The results obtained by using Hahn and Tchebichef moments were similar and found to be slightly better than Krawtchouk moments.

4.2 Comparison with State of the Art Image Sharpness Algorithms

In this section we compare our proposed image sharpness measures with the state of the art image sharpness measures (BIBLE [17], ARISM [15], JNB [7], LPC-SI [13], CPBD [8], Q-Metric [10], FISH [12] and block wise version of that FISHBB [12]) and we use SRCC, KRCC and PLCC for comparing the results. Table 2 gives the values for three well known OCRs namely, ABBYY Finereader [23], Omnipage [25] and Tesseract OCR [24]. The proposed measure work better than most of the standard image sharpness measures like ARISM [15], JNB [7], CPBD [8] and BIBLE [17] which uses discrete orthogonal moments and image gradient concepts works very badly with document images, but the improvement suggested in this paper makes it suitable for document image sharpness assessment. The measures Q-Metric [10], FISH [12], FISHBB [12] and LPC-SI [13] give decent performance for document images. Table 1 gives results for ABBYY Finreader [23], Omnipage [25] and Tesseract OCR [24] and we observe that all the measures of the proposed framework gives better performance than the state of the art measures.

5   Conclusion

No-reference image sharpness measure framework exclusively for document images based on discrete orthogonal moments and image gradients is proposed and validated using standard dataset DIQA which is publicly available and comparative study against latest state of the art image sharpness measures is performed and our proposed measure is found to beat the performance of all the measures.

References

1. Wang, Z., Bovik, A.C.: Modern image quality assessment. Synth. Lect. Image Video Multimedia Process. **2**(1), 1–15 (2006)
2. Wang, Z., Bovik, A.C., Sheikh, H.R., Simoncelli, E.P.: Image quality assessment: from error visibility to structural similarity. IEEE Trans. Image Process. **13**(4), 600–612 (2004)
3. Soundararajan, R., Bovik, A.C.: Rred indices: reduced reference entropic differencing for image quality assessment. IEEE Trans. Image Process. **21**(2), 517–526 (2012)

4. Pyatykh, S., Hesser, J., Zheng, L.: Image noise level estimation by principal component analysis. IEEE Trans. Image Process. **22**(2), 687–699 (2013)
5. Golestaneh, S.A., Chandler, D.M.: No-reference quality assessment of JPEG images via a quality relevance map. IEEE Signal Process. Lett. **21**(2), 155–158 (2014)
6. Sheikh, H.R., Bovik, A.C., Cormack, L.: No-reference quality assessment using natural scene statistics: JPEG2000. IEEE Trans. Image Process. **14**(11), 1918–1927 (2005)
7. Ferzli, R., Karam, L.J.: A no-reference objective image sharpness metric based on the notion of just noticeable blur (JNB). IEEE Trans. Image Process. **18**(4), 717–728 (2009)
8. Narvekar, N.D., Karam, L.J.: A no-reference image blur metric based on the cumulative probability of blur detection (CPBD). IEEE Trans. Image Process. **20**(9), 2678–2683 (2011)
9. Vu, C.T., Phan, T.D., Chandler, D.M.: S3: a spectral and spatial measure of local perceived sharpness in natural images. IEEE Trans. Image Process. **21**(3), 934–945 (2012)
10. Zhu, X., Milanfar, P.: Automatic parameter selection for denoising algorithms using a no-reference measure of image content. IEEE Trans. Image Process. **19**(12), 3116–3132 (2010)
11. Kumar, J., Chen, F., Doermann, D.: Sharpness estimation for document and scene images. In: 21st International Conference on Pattern Recognition (ICPR), pp. 3292–3295. IEEE (2012)
12. Vu, P.V., Chandler, D.M.: A fast wavelet-based algorithm for global and local image sharpness estimation. IEEE Signal Process. Lett. **19**(7), 423–426 (2012)
13. Hassen, R., Wang, Z., Salama, M.M., et al.: Image sharpness assessment based on local phase coherence. IEEE Trans. Image Process. **22**(7), 2798–2810 (2013)
14. Van Cuong Kieu, B., Cloppet, F., Vincent, N.: BNRFBE method for blur estimation in document images. In: Proceedings of Advanced Concepts for Intelligent Vision Systems: 16th International Conference, ACIVS, Catania, Italy, vol. 9386, p. 3. Springer, Berlin, 26–29 Oct 2015
15. Gu, K., Zhai, G., Lin, W., Yang, X., Zhang, W.: No-reference image sharpness assessment in autoregressive parameter space (2015)
16. Kang, L., Ye, P., Li, Y., Doermann, D.: A deep learning approach to document image quality assessment. In: IEEE International Conference on Image Processing (ICIP), pp. 2570–2574. IEEE (2014)
17. Li, L., Lin, W., Wang, X., Yang, G.M., Bahrami, K., Kot, A.C.: No-reference image blur assessment based on discrete orthogonal moments (2015)
18. Ye, P., Doermann, D.: Document image quality assessment: a brief survey. In: 12th International Conference on Document Analysis and Recognition (ICDAR), pp. 723–727. IEEE (2013)
19. Kumar, D., Ramakrishnan, A.: QUAD: quality assessment of documents. In: International Workshop on Camera based Document Analysis and Recognition, pp. 79–84 (2011)
20. Kumar, J., Bala, R., Ding, H., Emmett, P.: Mobile video capture of multi-page documents. In: IEEE Conference on Computer Vision and Pattern Recognition Workshops (CVPRW), pp. 35–40. IEEE (2013)
21. Obafemi-Ajayi, T., Agam, G.: Character-based automated human perception quality assessment in document images. IEEE Trans. Syst. Man Cybern. Part A Syst. Hum. **42**(3), 584–595 (2012)
22. Peng, X., Cao, H., Subramanian, K., Prasad, R., Natarajan, P.: Automated image quality assessment for camera-captured OCR. In: 18th IEEE International Conference on Image Processing (ICIP), pp. 2621–2624. IEEE (2011)
23. ABBYY Finereader 10 professional edition, build 10.0.102.74 (2009)
24. Tesseract-OCR: an OCR engine that was developed at hp labs between 1985 and 1995 and now at Google (2012). https://code.google.com/p/tesseract-ocr/
25. Omnipage professional version 18.0 (2011). http://www.nuance.com/for-business/by-product/omnipage/index.htm
26. Yap, P.T., Paramesran, R., Ong, S.H.: Image analysis by Krawtchouk moments. IEEE Trans. Image Process. **12**(11), 1367–1377 (2003)

27. Yap, P.T., Paramesran, R., Ong, S.H.: Image analysis using Hahn moments. IEEE Trans. Pattern Anal. Mach. Intell. **29**(11), 2057–2062 (2007)
28. Mukundan, R., Ong, S., Lee, P.A.: Image analysis by Tchebichef moments. IEEE Trans. Image Process. **10**(9), 1357–1364 (2001)
29. Farid, H., Simoncelli, E.P.: Differentiation of discrete multidimensional signals. IEEE Trans. Image Process. **13**(4), 496–508 (2004)
30. Kovesi, P.D.: Matlab and octave functions for computer vision and image processing (2000). Online: http://www.csse.uwa.edu.au/~pk/Research/MatlabFns/#match
31. Otsu, N.: A threshold selection method from gray-level histograms. Automatica **11**(285–296), 23–27 (1975)
32. Kumar, J., Ye, P., Doermann, D.: A dataset for quality assessment of camera captured document images. In: Camera-Based Document Analysis and Recognition, pp. 113–125. Springer, Berlin (2014)

An Analysis of BoVW and cBoVW Based Image Retrieval

P. Arulmozhi and S. Abirami

Abstract Bag-of-Visual Word (BoVW) is one of the popularly used Image Representation model. It is a sparse vector model based on the occurrence count of the visual words extracted from image features. One of the major limitations of BoVW model is the quantization error that is caused mainly because of false matches. In this paper, first the BoVW model based Image Retrieval is performed experimented and analyzed for various visual word size and vocabulary sizes. The novelty of this paper is based on the assumption that objects of our interest covers mostly the centre part of the images. In this paper, an algorithm where features belonging to the centre part of the images are given attention (cBoVW) and applied for online query Image Retrieval. Using CalTech 256 dataset, several evaluations are done and results seem to be promising in resolving quantization error when compared with the conventional BoVW model.

Keywords Bag of visual words · Vocabulary · Visual words · Quantization error · Image retrieval

1 Introduction

Earlier, half a century back, taking a photo copy of any kind of images, as and when an individual wish was almost not that easy as compared to today's scenario. This drastic change has been occurred as a result of revolution in Digital Technology. In addition to this is the growth of Internet and extensive usage of Social Media and Mobile Phones. As a result, accessing images is an inevitable part of today's human activity, which created a mandatory need for retrieving images both accurately and

P. Arulmozhi (✉) · S. Abirami
Department of Information Science and Technology, College of Engineering,
Anna University, Chennai, India
e-mail: arulmozhikec@gmail.com

S. Abirami
e-mail: abirami_mr@yahoo.com

© Springer International Publishing Switzerland 2016
V. Vijayakumar and V Neelanarayanan (eds.), *Proceedings of the 3rd International Symposium on Big Data and Cloud Computing Challenges (ISBCC – 16')*,
Smart Innovation, Systems and Technologies 49, DOI 10.1007/978-3-319-30348-2_19

efficiency. This paved the way for a new research area namely Content Based Image Retrieval (CBIR), where given an image as a query, the system should be able to return all the matched images from the stored image databases in ascending order based on their similarity score.

Image Representation, Image Indexing and Image Matching are the three stages to be performed for any kind of Image Retrieval. Initially images are represented by considering low-level features like lines, edges and corners etc. To further improve the quality of the image retrieval, next higher level of feature representation like part based, patches based representations are carried out. BoVW, is one such representation and it become popular now-a-days as it holds collection of codeword. It gets interest points initially using existing descriptors like SIFT [1], SURF, MSER, ASIFT, FAST and emerging binary descriptors like ORB, BRIEF etc. Then grouping of the descriptors using unsupervised method are performed and each image is represented as a histogram. Simplicity, efficiency, compact image representation, scalable retrieval in large databases at high precision make BoVW representation as the most common approach for image retrieval.

Apart from its popularity, it has limitations like low discrimination power, no spatial clues and quantization errors. As BoVW is the orderless collection of Visual Words based on the feature's frequency, it does not hold any spatial information and this ultimately lead to low discriminability. Feature quantization, one of a step in Image Representation using BoVW is responsible for creating Quantization error. In Hard Quantization, after finalizing the visual vocabulary, each feature of an image is quantized to a nearby vocabulary (visual word). This effect causes false matches by assigning to a visual word where originally it belongs to another visual word.

In this paper, an analysis of existing BoVW based Image Retrieval is experimented for different vocabulary sizes and visual word sizes. For centrally located images, the features covering the middle area is able to discriminate better than other part of the features as they may belong to background. Holding this idea, central part of the features are considered for testing the query images and performance is evaluated using a different image retrieval model called cBoVW.

This paper is organized in the following way. In Sect. 2, the papers related to BoVW Image Retrieval are discussed. The existing BoVW based Image Retrieval is analyzed and the usage of central feature (cBoVW) for Image retrieval is specified in Sect. 3. In Sect. 4, experimentation and result discussions are performed. Finally the paper is concluded with the information obtained from experimentation and the further work to be carried out is discussed in Sect. 5.

2 Related Works

Lot of works are performed in the literature to improve the retrieval accuracy and efficiency of image retrieval. In this paper, BoVW based improvements are listed based on category they belong to.

Zhang et al. [2] has taken the limited power discrimination of SIFT problem and proposed an edge based local descriptor for large-scale partial-duplicate image search on mobile platform. They preserved location and orientation of edges and created a compact representation. Liu et al. [3] proposed a new image representation method where the length of the signature is dependent on the image patch region. Weights are assigned by extracting region based patches using intensity and spatial information and developed a new texture based algorithm called Probabilistic Center-Symmetric Local Binary Pattern to achieve scale and orientation invariance. Liu et al. [4] proposed a method to do geometric verification by getting a binarized code using the spatial context information around a single feature. For each feature, a circle around it is formed and it is further divided into 3 regions based on the center feature's dominant orientation and binary code is generated. It is matched by computing hamming distance and achieved better geometric verification. But it is memory prohibitive to extend it to index billion scale of image database [5].

Zhou et al. [6] proposed a codefree feature quantization method, which efficiently quantize the SIFT descriptor to a discriminative bit-vector called binary SIFT (BSIFT) of size 128 or 256 bits and matched using hamming distance. As a result, no need of visual codebook training and can be applied in image search in some resource-limited scenarios. Yao et al. [7] proposed a descriptor that encodes the spatial relations of the context by order relation and adopted the dominant orientation of local features to represent the robustness. Zhou et al. [6] proposed a codebook free hashing method for doing mobile image search satisfying the memory constrains of it. Scalable Cascaded Hashing (SCH), another approximate nearest neighbor search method ensures recall rate by performing scalar quantization on the top k PCA features and verifies to remove the false positive feature matches.

Zhou et al. [8] created a new affine invariant descriptor, namely Sampling based Local Descriptor (SLD) to provide fast matching. For each image, elliptical sampling is done as per elliptical equation, and then orientation histogram is created over the gradient magnitude to achieve the orientation invariance. Tang et al. [9] found a remedy to reduce the semantic gap in visual word quantization by introducing contextual synonym dictionary by identifying the visual words with similar contextual distribution using the co-occurrence and the spatial information statistics and averaged over all the same image patches.

Yao et al. [7] proposed a new descriptor for image editing operations exploring the spatial relations of the context and considered the dominant orientation of local features for near duplicate images. With the calculation of appropriate weight, the features of the context are extracted and contextual descriptors are generated by encoding the contextual features to a single byte and the similarity between two images is measured. Chen [10] gave importance to codebook learning and estimated codebook weights using machine learning approaches based on the assumption that the weighted similarity between the same labelled images is larger than the differently labelled images with largest margin.

With respect to image indexing Zheng et al. [11] proposed IDF weighting scheme which provides a remedy for burstness problem. This lp norm pooling created a new IDF representation by involving term frequency, document frequency, and document length and codebook information and thereby optimizes the minimization of cost function. Feng et al. [12] proposed a multi-dimensional inverted index for generating feature point correspondences by ordering the visual words of feature descriptors called order quantization and it is able to distinguish the features belonging to the same visual words. When the network connection is slow or busy, Chen [10] suggested memory efficient image signature database of image signature stored entirely on mobile device is needed to have a fast access to local queries. They presented compact, efficient and highly discriminative representation to provide fast recognition.

Thus lots and lots of improvement using BoVW representation are happening repeatedly. Its simplicity, compact representation makes us to choose this BoVW representation for implementing image retrieval.

3 Proposed Methodology

As accessing the images using Internet increases day by-day, the retrieval accuracy and efficiency of images are becoming an essential need in today's environment. BoVW based Image Retrieval consists of the following phases: Image Representation, Image Indexing, Image Matching and Image Post Verification. In Image Representation, the features are extracted from the images that are invariant to translation, scale, and rotation and partially invariant to viewpoint. Image Indexing is subjected to provide fast access to the image features from the repos-itory of Image databases. In Image Matching, for the given query image, features are extracted, described and compared with the features of image databases using distance/similarity measures. New scoring is calculated for the matched features and used for displaying image similarity in descending order, i.e. the image having highest score is displayed first, second largest score as second image and so on.

In this paper, the existing BoVW Image representation based Image Retrieval is implemented and its performance is analysed for different vocabulary and visual word size. Later it has been extended to cBoVW to extract the central part of the image and obtain better result.

3.1 Image Retrieval Using BoVW

Among the various ways of performing Image Representation, Bag of Visual Word (BoVW) model is very popular due to its simplicity and ease of use.

BoVW Image representation needs to perform feature Extraction, Codeword Generation, feature Quantization steps. In the first step, it extracts features from the images by finding interest points using local descriptors like SIFT, SURF and the features are described in vector form so that they can be easily handled for matching and have geometric invariance. Next step is to generate a dictionary, where the feature descriptors are analysed, grouped based on their similarities and centroids are selected for each group. This centroid of each group is termed as Visual Word and collection of all the visual words are named as Dictionary, Vocabulary or Codeword. In Feature Quantization step, the features of the images are mapped to the nearby visual words. In Hard Quantization, a feature is assigned to the most nearest single visual word only.

Image Indexing provides a way to arrange the features of the dataset in such a way that accessing them for matching an image query is made efficient. Image matching is used to find the similarities among two images, one is a query image and another is a database image. A score is assigned to the images based on their similarity measures and displayed with their scoring order. Image Post Verification analyzes the ranked images in order to remove the false matched images and thus it helps further improve the retrieval accuracy. The algorithm for BoVW Image Retrieval is as follows

1. Extract features $\{f_j^i\}$, where f is the jth feature of image i obtained from SIFT descriptors and represented as 128 bit feature vectors.
2. For codeword generation, apply clustering over the features to get the visual words $\{vw_1, vw_2, \ldots, vw_c\}$ where c is number of visual words and vw_{ic} is the visual words and thereby constituting a visual dictionary $vd = \{vw_c\}$ where vd is collection of visual words.
3. For training Images in the database, hard quantize the features of each image generated using SIFT descriptor to the nearest visual words.
   $hq_i = \{hq_{i1}, hq_{i2}, \ldots, hq_{in}\}$, $hq_{in} = nearest\ vw_c$
4. For testing a query image, compare the database images using similarity mea-
   sure $Score_{qd} = \begin{cases} high\ score, & if \quad similar \\ low\ score, & if \quad dissimilar \end{cases}$ where q is the query image, d is
   the database image and evaluate them by finding the Mean Average Precision (MAP), Average Precision (AP).

3.2 Image Retrieval Using cBoVW

In most of the images, objects occupy the central area of the image. Having this assumption, features covering the central part is given more importance and the features of the central region are used to compare with the query image. This helps to reduce the false matches and thereby reduces the quantization error. So step 1, 2 and 3 remains the same and a small modification is done to the step 4 as shown below

4. (a) For the given query image take the central area of the image.
   (b) Fix the number of circles $\{c_1, c_2, \ldots, c_k\}$, k is the number of circles where the location and the radius of the circles are fixed.
   (c) Find the features that belongs to these fixed circles and use these features to compare with the database images.

4 Experimental Results and Discussion

Image Retrieval using BoVW is experimented on CalTech-256 dataset [13] using Matlab 2013a. This dataset consists of approximately 29,700 images with 256 object categories where each category has 85–800 images. Among 256 categories, five categories are considered for experimenting the performance namely Motorbikes, Airplanes, Butterflies, Mushrooms and Horses category.

Vocabulary sizes taken here are 150(30 for each category), 500(100), 1000(200) and a code book is created by taking the size of the visual words as 500, 1000, 2000. The training set consists of 1000 images and trained using different vocabulary and visual word sizes. The database created, consists of 1000 images which includes both the matching category images and other non-matching images. These database images are matched against the given query image using Bhattacharya similarity measure and displayed based on their ranking system.

To find the codebook, interest points are extracted from each raw image, and 128 bit feature vector is obtained by applying SIFT descriptor and the visual words are generated using K-means clustering. For training the datasets, using the generated codebook, hard vector quantization is performed. Then the query image is matched with the database images.

The metrics involved in evaluating the Image Retrieval are given in Eqs. (1) and (2).

1. **Average Precision (AP)**

   Precision is the fraction of retrieved images that are relevant

   $$\text{Precision} = \frac{\text{No. of relevant images retrieved}}{\text{No. of retrieved images}} = P(\text{relevant}/\text{retrieved}) \quad (1)$$

   Precision@k is precision list going down up to k. Average Precision is the average of the precision value obtained for the set of top k images existing after each relevant image is retrieved and this value is then averaged.

2. **Mean Average Precision (MAP)**

   Average of Average Precision

   $$\text{MAP value} = \frac{1}{N} * \sum_{k=1}^{k=N} \text{Precision@K} \quad (2)$$

Table 1 shows the Precision value for first 5, 10, 15 resultant image entries for vocabulary size and visual word size both taken as 1000. Among the considered five datasets, Mushroom dataset gives the lower precision value. Generally if the shape of a particular image is indefinite and irregular, then finding the exact match becomes very difficult and this could be the reason for getting lower precision value with respect to Mushroom dataset when compared with Horse and Butterfly datasets.

Table 1 Calculation of precision for first 5, 10 and 15 resultant images, AP@25 and MAP for vocabulary size as 1000 and visual word size as 1000

Query image	P for first 5	P for first 10	P for first 15	AP@25	MAP
Butterfly 1	5	8.66	12.14	79	86.5
Butterfly 2	5	10	14.49	94	
Horse 1	4.02	5.96	7.27	60	70
Horse 2	4.8	8.34	11.95	79	
Bike 1	5	9.69	13.39	87	86.5
Bike 2	5	9.36	13.22	86	
Mushroom 1	3.72	6.51	8.85	63	59
Mushroom 2	3.52	5.79	7.86	55	
Airplane 1	3.27	6	9.1	66	73.24
Airplane 2	2.72	5.96	9.2	71	
Airplane 3	4.35	8.06	12.06	83	

Bikes and Airplanes datasets obtained better precision. Among the Airplane dataset, Airplane 1 and 2 images has low and high precision values because the query Airplane 1 image is little different from the trained Airplane dataset. This requires the training set to get included of all possible modification of images of same category.

The average precision at 25 and their Mean Average Precision values for vocabulary size and visual word size as 1000 has been listed in Table 1. Again Mushroom dataset obtains the lower AP and MAP values. Here, Airplane 2 and 3 images consist of only five irrelevant pictures among the first 25 ranks. Even though, both the images have same irrelevant pictures, their P and AP exhibit large variations. This variation is due to the difference in their relevant picture's positions. Hence, it is observed that the precision depends on the position of the relevant images also.

Table 2 shows the Precision@10 and Average Precision@25 calculations for Vocabulary sizes 150, 500, 100 having visual words size as 1000. For certain datasets, as the vocabulary size increases, the AP value also increases. But this is not true in all the category datasets and this is visible from the Fig. 1. According to

Table 2 Comparison of precision for first 10 resultant images and average precision@25 for vocabulary sizes 150, 500, 100 having visual word size as 1000

Sl. No.	Query image	Vocabulary sizes					
		150		500		1000	
		P for first 10	AP@25	P for first 10	AP@25	P for first 10	AP@25
1	Butterfly	7.7	74	7.7	74	7.3	57
2	Horse	4	47	4.3	50	5.7	62
3	Bike	8.8	79	9.4	86	8.9	79
4	Mushroom	7.2	68	7.8	70	7	73
5	Airplane	6.5	71	5.9	67	5.7	67

Fig. 1 AP@25 for different
vocabulary sizes

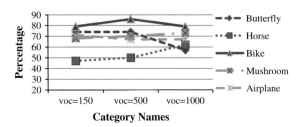

this graph, AP value for the Mushroom and the Horse data sets increases along with
vocabulary sizes. For other 3 categories, for vocabulary size 1000, it shows a
decline in AP performance. Even though performance increases along with the
vocabulary size, it is computationally expensive. Considering this point, for computational efficient task, low vocabulary size has been preferred and here taken as
500.

For the same category data sets, AP@25 is evaluated for visual word size as 500,
1000, 2000 having vocabulary size as 1000 and results are tabulated in Table 3.
A graph depicting these results has also been shown in Fig. 2. From the graph, it is
clear that, for visual word size 500, almost all the category images has nearly higher
value. Only for the butterfly category, visual word size of 2000 gives highest AP
value. Hence, 500 visual word size seems to be getting chosen in this case.

Table 4 presents the MAP value of conventional BoVW and cBoVW for the
vocabulary size of 1000. Here in cBoVW, for matching a given query image, only

Table 3 List of AP@25 for
different visual words (500,
1000, and 2000) and
vocabulary size 1000

Sl. No.	Query image	Visual words (P@10/AP@25)		
		500	1000	2000
1	Butterfly	7.9/72	7.3/57	9.22/79
2	Horse	5.1/60	5.7/62	6.4/63
3	Bike	9.9/90	8.9/79	8.36/72
4	Mushroom	7.3/69	7/73	6.7/69
5	Airplane	6.3/70	5.7/67	5.6/64

Fig. 2 AP@25 for different
visual word size having
vocabulary size as 1000

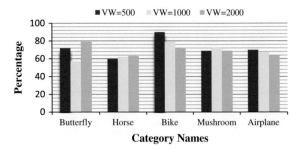

Table 4 Comparison of MAP between conventional BoVW and cBoVW for vocabulary size as 1000

Sl. No.	Query image	Conventional BoVW	cBoVW
1	Butterfly	45	52
2	Horse	62	50
3	Bike	79	84
4	Mushroom	53	51
5	Airplane	67	79

central area covering features are considered and matching algorithm is applied. In this table, for Butterfly, Bike and Airplanes images, cBoVW shows improvement in MAP value. For Mushroom category, cBoVW does not show improvements but the difference is not much deviated and this is charted in the Fig. 3. Since only fewer features are considered, their matching time is reduced when compared with the conventional BoVW.

In Table 5, in each row, the first column has query images belonging to the five category databases. The remaining columns of each row consist of the resultant images of retrieved images using BoVW representation along with their ranks. The different range of ranked images is presented in the above table.

Fig. 3 Comparison of BoVW with cBoVW

Table 5 Query images and their retrieved images along with their ranks

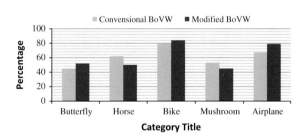

| Query images | Retrieved Images along with their ranks |

5   Conclusion

Image retrieval using BoVW representation is performed using CalTech 256 dataset and evaluated for various vocabulary size and visual word sizes. Then the features belonging to the central area of each image are considered for matching the query image along with the target images of the Image database and a small improvement in retrieval accuracy as well as computational efficiency for image matching has been obtained. Thereby this also lowers (scale down) the quantization error. But this algorithm could work fine only for centrally located pictures.

As an extension of this work, the central features can also be weighted more to obtain higher score compared with other features so that it may lead to better solution for quantization errors. As a future work, the central features should be learned and tested for various vocabulary sizes to get better retrieval accuracy.

References

1. Lowe, D.G.: Distinctive image features from scale-invariant keypoints. Int. J. Comput. Vis. **60**, 91–110 (2004)
2. Zhang, S., Tian, Q., Lu, K., Huang, Q., Gao, W.: Edge-SIFT: discriminative binary descriptor for scalable partial-duplicate mobile search. IEEE Trans. Image Process. **22**(7), 2889–2902 (2013)
3. Liu, L., Lu, Y., Suen, C.Y.: Variable-length signature for near-duplicate image matching. IEEE Trans. Image Process. **24**(4), 1282–1295 (2015)
4. Liu, Z., Li, H., Zhou, W., Zhao, R., Tian, Q.: Contextual hashing for the large scale image search. IEEE Trans. Image Process. **23**(4), 1606–1614 (2014)
5. Bhala, R.V.V., Abirami, S.: Trends in word sense disambiguation. Artif. Intell. Rev. **42**, 159–171 (2014). doi:10.1007/s10462-012-9331-5
6. Zhou, W., Yang, M., Li, H., Wang, X., Lin, Y., Tian, Qi: Towards codebook-free: scalable cascaded hashing for mobile image search. IEEE Trans. Multimedia **16**(3), 601–611 (2014)
7. Yao, J., Yang, B., Zhu, Q.: Near-duplicate image retrieval based on contextual descriptor. IEEE Signal Process. Lett. **22**(9), 1404–1408 (2015)
8. Zhou, W., Wang, C., Xiao, B., Zhang, Z.: SLD: a novel robust descriptor for image matching. IEEE Signal Process. Lett. **21**(3), 339–342 (2014)
9. Tang, W., Cai, R., Li, Z., Zhang, L.: Contextual synonym dictionary for visual object retrieval. In: Microsoft Research Asia, pp. 503–512. ACM, New York (2011)
10. Chen, D.M.: Memory-efficient image databases for mobile visual search multimedia data management in mobile computing. IEEE Comput. Soc. Multimedia Data **21**, 14–23 (2014)
11. Zheng, L., Wang, S., Liu, Z., Tian, Q.: Lp-norm IDF for large scale image search. In: IEEE Conference on Computer Vision and Pattern Recognition, pp. 1626–1633 (2013)
12. Feng, D., Yang, J., Liu, C.: An efficient indexing method for content-based image retrieval. Neurocomputing **106**, 103–114 (2013)
13. Griffin, G., Holub, A., Perona, P.: Caltech-256 Object Category Dataset. California Institute of Technology, California (2007)
14. Zhou, W., Li, H., Hong, R., Lu, Y., Tian, Q.: BSIFT: toward data-independent codebook for large scale image search. IEEE Trans. Image Process. **24**(3), 967–979 (2015)

15. Liu, Z., Li, H., Zhang, L., Zhou, W., Tian, Q.: Cross-indexing of binary sift codes for large-scale image search. IEEE Trans. Image Process. **23**(5), 2047–2057 (2014)
16. Sivic, J., Zisserman, A.: Video Google: a text retrieval approach to object matching in videos. In: Proceedings of the Ninth IEEE International Conference on Computer Vision (ICCV 2003) 2-Volume Set, pp. 1–8. IEEE Computer Society (2003)

Performance Analysis of Background Estimation Methods for Video Surveillance Applications

M. Sivabalakrishnan and K. Shanthi

Abstract Background subtraction is a very popular approach for foreground segmentation in a still scene image. A common approach is to perform background subtraction, which identifies moving objects from the portion of a video frame that differs significantly from a background model. Each video frame from the sequence is compared against a reference frame. There are several problems that a good background subtraction algorithm must resolve. To achieve robust background subtraction is required to have reliable and effective background estimation. In the existing system the background reference image is arbitrarily chosen. This paper presents a novel background estimation algorithm based on improved mode algorithm to obtain static background reference frame from input image. Evaluation of the background extraction is presented in this paper. The experimental results of using background image estimated by background estimation method are faster and accurate in background subtraction. The major goal is to obtain a clean static background reference image using a real-time video.

Keywords Background estimation · Background subtraction · Improved mode algorithm · Foreground image

1 Introduction

Computer-vision based people counting offers an alternate to other people counting methods such as photocell, thermal, 2D and 3D videos. The common problem of all computer-vision systems is to separate foreground from a background scene. Trouble-free motion detection methods compare a static background frame with the current frame of a video scene, pixel by pixel. This is the basic rule of background subtraction, which can be formulated as a method that builds a model of a

M. Sivabalakrishnan (✉)
School of Computing Science and Engineering, VIT University, Chennai, India

K. Shanthi
Department of ECE, MNM Jain Engineering College, Chennai, India

© Springer International Publishing Switzerland 2016   249
V. Vijayakumar and V. Neelanarayanan (eds.), *Proceedings of the 3rd International Symposium on Big Data and Cloud Computing Challenges (ISBCC – 16')*,
Smart Innovation, Systems and Technologies 49, DOI 10.1007/978-3-319-30348-2_20

background, and compares this model with the current frame, to detect zones where a major difference occurs. The use of a background subtraction algorithm is to differentiate the moving objects (hereafter referred to as the foreground) from the static or slow moving parts of the view (called background). While a static object starts moving, a background subtraction method detects the objects in motion as well as a gap left behind in the background (referred to as a ghost). A ghost is unrelated for motion analysis and has to be removed. Another description for the background is that it corresponds to a reference frame with a value, able to be seen most of the time that is with the maximum appearance probability, but this type of framework is not simple to apply in practice. Background estimation is related to, but distinct from, background modeling. Owing to the complex nature of the problem, our estimation strategy to static backgrounds (e.g. no waving trees), which is quite common in urban surveillance. The key technologies of background subtraction are the estimation and updating of background image quickly and correctly. Whereas a static background model may be suitable for analyzing small video sequences in a constrained indoor environment, the model is hopeless for most realistic situations. The background extracts from the sequence of images, which take care of variations in illumination, shadows, moving and non moving pixels. The efficiency and accuracy of continual image processing depends on the result of background extraction algorithm. Obviously, if there is a scene (frame) without any moving object in the image sequences, it can be used as a background frame. However, in real world, it is hard to get a pure background frame. Thus, in such systems, as in many other applications, a critical issue in this process is extracting the pure background image from the videos which included moving objects. Background extraction faces numerous challenges due to low resolution, more foreground color and scene which always contains some foreground elements and a complex background. Once a background model is established, the object in the video frames can be detected as the difference between the current video frame and the background model.

This paper presents work related to the next section. In Sect. 3, the proposed method for background extraction procedure is presented in detail. The application of the proposed technique is given in Sect. 4. In Sect. 5, the experimental results of background estimation algorithm under real-world conditions are shown and conclusions drawn in Sect. 6.

2 Related Works

Background estimation, as well as background modeling, is a common problem in many areas of computer vision. Background estimation from time image sequences or videos has been extensively explored in the context of background removal, specially in motion tracking applications. Recently, researchers of both home and abroad presented some classical algorithms of background extraction, including mean algorithm, median algorithm, determination of stable interval algorithm, the detection changes algorithm [1] and mode algorithm [2]. In addition, some high complex methods that have been used in background extraction algorithm too, such as textural and statistical features in [3] and genetic algorithm [4]. However, the first algorithm can be traced back 19 years ago to the work of [5] in estimating background for detecting moving objects. Xu and Huang [6] propose a simple, yet robust approach for background estimation based on loopy belief propagation.

A new method introduced for background subtraction by Agarwala et al. [7]. And also various background estimation methods are briefly described. More recently, Colombari et al. [8] has developed a technique for initialize background in cluttered sequence. Agarwala et al. [7] has developed a general and powerful framework for combining a set of images into a single composite image. The framework has been used for a wide variety of applications, which includes uncluttered background estimation, which is not always true in some applications. Granados et al. [9] propose a novel background estimation method for non-time sequence images.

The Temporal and Approximated Median Filtering (AMF) methods are based on the assumption that pixels related to the background scene would be present in more than half of the frames in the entire video sequence. In some cases the number of stored frames is not large enough (buffer limitations); therefore, the basic assumption will be violated and the median will estimate a false value, which has nothing to do with the real background model. The AMF [10] applies the filtering procedure, by simply incrementing the background model intensity by one, if the incoming intensity value (in the new input frame) is larger than the previous existing intensity in the background model. In fact this approach is most suitable for indoor applications.

A MOG model is designed such that the foreground segmentation [11] is done by modeling the background and subtracting it out of the current input frame, and not by any operations performed directly on the foreground objects (i.e. directly modeling the texture, colour or edges). The processing is done pixel by pixel rather than by region based computations, and finally the background modeling decisions are made based on each frame itself, instead of benefiting from tracking information or other feedbacks from the previous steps.

The Progressive Background Estimation Method [12] constructs the histograms from the preprocessed images known as the partial backgrounds. Each partial background is obtained using two consecutive input frames. This method is applicable to both gray scale and colour images, and is capable of generating the background in a rather short period of time, and does not need large space for storing the image sequences.

The Group-Based Histogram (GBH) algorithm constructs background models by using the histogram of the intensities of each pixel on the image. Since the maximum count (amplitude) of the histogram is much greater in comparison to the frequencies of the intensities related to the moving objects, there will not be any effects of slow moving objects or transient stops in the detected foreground.

From the above discussion, the characteristics of the various estimation methods are observed and summarized as follows:

- All the methods deal only with non moving pixels.
- It takes a longer time to create a background model.

In this work, it is decided to apply the improved mode method to have a unified and common framework for all kinds of images. To achieve this, moving pixels in the frame are to be handled.

3 Background Estimation Process

Estimation problem can be broadly classified into three categories:

(i)   Pixel-level processing,
(ii) Region-level processing,
(iii) A hybrid of the first two.

In the first category the simplest techniques are based on applying a median filter on pixels at each location across all the frames. The background is estimated correctly if it is exposed for more than 50 % of the time. In [5], the algorithm finds pixel intervals of stable intensity in the image sequence, then heuristically chooses the value of the longest stable interval to most likely represent the background. In [13], an algorithm based on Baye's theorem is proposed. For every new pixel it estimates the intensity value to which that pixel has the maximum posterior probability. In [8], the first stage is similar to that of [5], followed by choosing background pixel values whose interval maximises an objective function. All these pixel based techniques perform well when the foreground objects are moving but fail at regions where the time interval of exposure of background is less than that of foreground.

In the second category, the method proposed by Farin et al. [14] as rough segmentation of input frames into foreground and background regions by working on blocks. To achieve this, each frame is divided into blocks and temporal sum of absolute differences (SAD) of the blocks of successive frames is calculated and a block similarity matrix is formed. The matrix elements that correspond to small SAD values are considered as stationary elements and high SAD values correspond to non-stationary elements. The algorithm works well in most scenarios, however, the spatial correlation of a given block with its neighbouring blocks already filled by the background is not exploited, which can result in the blending of the foreground and background if the objects move slowly or are quasi-stationary for extended periods.

In [15], given an image sequence of T frames, each frame is divided into blocks of size N × N overlapping by half of their size in both dimensions. These blocks are clustered using single linkage agglomerative clustering along their time-line. In the following step the background is built iteratively by selecting the best continuation block for the current background using principles of visual grouping. The algorithm can have problems with blending of the foreground and background due to slow moving or quasi-stationary objects. Furthermore, the algorithm is not likely to achieve real-time performance due to its complexity.

In the third category, the algorithm presented [16] has two stages. The first stage is similar to that of [5], with the second stage estimating the likelihood of background visibility by computing the optical flow of blocks between successive frames. The motion information aids to classify an intensity transition as background to foreground or vice versa. The results are typically good but the usage of optical flow for each pixel makes it computationally intensive. In [17] the problem of estimating the background is viewed as an optimal labelling problem.

The method defines an energy function which is minimised to achieve an optimal solution at each pixel location. It consists of data and smoothness terms. The data term accounts for pixel stationarity and motion boundary consistency while the smoothness term looks for spatial consistency in the neighbourhood. The function is minimised using the expansion algorithm [18] with suitable modifications.

Another similar approach with a different energy function is proposed in [6]. The function is minimised using loopy belief propagation algorithm. Both solutions provide robust estimates; however, their main drawback is large computational complexity to process a small number of input frames. For instance, in [6] the authors report a prototype of the algorithm on Matlab takes about 2.5 min to estimate the background from a set of 10 images of QVGA (320×240) resolution.

When the background estimator is initiated, the pulse will take five frames from the observed video images. Each pixel at the same coordinate from the five frames will be used to find the median pixel to produce the background image. The steps are shown in Fig. 1.

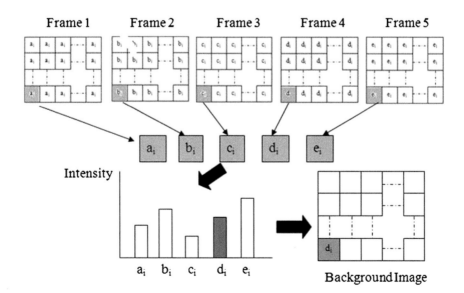

Fig. 1 Pixel level background estimation process

4 Proposed Method

4.1 Overall Architecture

In this section, the processing steps of the proposed background estimation approach are presented. The aim is to build automatic background estimation, which is able to accept different type of an image sequence.

Instead of using the trial-and-error procedure, a common preliminary step in any motion-based vision application is used to obtain image frames from a stationary or a moving camera, or from multiple cameras. After the acquisition of the image frames, image segmentation is performed using background subtraction, depending on the frame rate of the video sequences.

The input video is first converted to frames, followed by the implementation of background estimation, done using the improved mode algorithm. As a result, we get a reference frame which is taken as the refined background image.

Using this image as reference the other images are subjected to a basic background algorithm, where the current frame is compared to the reference frame (background image) and the accurate foreground image is extracted.

For high frame rate sequences, the adjacent frame subtraction is used since the change of motion between the consecutive frames is very small. This method eliminates the stationary background, leaving only the desired motion regions. In certain video systems, the acquisition process may be susceptible to erratic changes in illumination, reflection, and noise. The block diagram of the proposed methodology is shown in Fig. 2.

4.2 Proposed Algorithm

The proposed algorithm addresses the problems described in the previous section and has several further advantages. It falls into the region-level processing category. Specifically, image sequences are analysed on a block by block basis. For each block location a representative set is maintained which contains distinct blocks obtained along its temporal line.

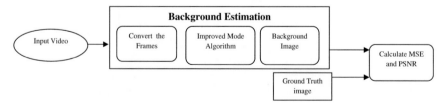

Fig. 2 Block diagram of proposed background estimation method

Background estimations are the process of distinguishing novel (foreground) from non-novel (background) elements, in a scene from a video sequence. The improved mode algorithm is applied to an input video. This method consists of two steps: First, the input video is converted into frames. Second, the background is estimated automatically from the successive frames. Background estimation is done using the improved mode algorithm.

4.2.1 Mode Algorithm

Mathematically mode refers to the most frequently occurring element in a series of data. This same concept is applied to the frames in video processing, here the pixel values are compared to each other to find the most frequently appeared pixel. This pixel value is considered to be the background image.

4.2.2 Median Algorithm

The term median refers to the element that is occurring in the middle of a series. Accordingly the pixel value which is occurring in the middle of the entire series of pixels is considered to be the background image.

4.2.3 Improved Mode Algorithm

Combining the concepts of mode and median algorithm a new concept called the improved mode is created.

The improved mode algorithm is implemented in the following manner.

Step 1. A movie is taken and converted into a number of successive frames (images).

Step 2. Two frames are obtained at fixed intervals from the video and saved in $In(x, y)$ and $In + 1(x, y)$. For the purpose of simple calculation and real-time speed, these two frames should be converted into grey images.

Step 3. Using $In(x, y)$ and $In + 1(x, y)$ the frame difference image can be obtained as $Dn(x, y)$. And then, the opening and closing operation of the mathematical morphology on $D(x, y)$ is applied and the computed result using Eq. (1) is saved in $BWn(x, y)$.

$$BWn(x, y) = \begin{cases} 0 = \text{unchanged background if } D(x, y) < T \\ 1 = \text{moving object otherwise} \end{cases} \quad (1)$$

Step 4. According to the value of pixels in $BWn(x, y)$, the pixels are classified into either background moving objects or unchanging objects. According to Eq. (2), make a flag as the pixel whether it is the moving objects or background. If the flag value is 1, it implies that the pixel belongs to a moving object and it's unavailable in mode calculation; otherwise its background is available. These values of the flag are saved in B_backn(x, y).

$$B\_backn(x, y) = \begin{cases} 1 & \text{if } BWn(x, y) = 0 \\ 0 & \text{otherwise} \end{cases} \quad (2)$$

Step 5. If n reaches the maximum set up, the procedure goes on to step 6; else the procedure should go to step 1.

Step 6. Bn(x, y) should be calculated including the video and saved as B(x, y, z), namely, the values of the background of all frames. Through the steps above, the background image can be estimated accurately. As a result of the pixels of moving objects being removed, even if the pixel background value just emerges once, it can be estimated accurately by the new method BG(x, y) (Fig. 3).

$$BG(x, y) = \text{mode}(Bt - N(x, y)X\alpha i - N, B\ t - N + 1(x, y)\alpha i - N + 1, Bt - N \\ + 2(x, y)X\alpha t - N + 2\ldots Bt - 2(x, y)x\alpha t - 2, Bt - 1(x, y)x\alpha\ t - 1) \quad (3)$$

$$\alpha n = \begin{cases} 1 & \text{if } B\_back\ n(x, y) = 1 \\ 0 & \text{otherwise} \end{cases}$$

αn determine the pixel (moving or background).

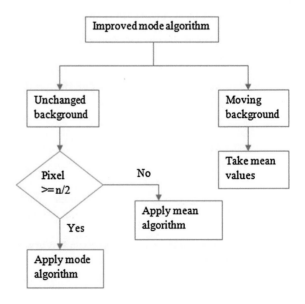

Fig. 3 Flow diagram for proposed algorithm

5   Experimental Results and Discussion

In our experiments, we apply the proposed method to many datasets to demonstrate that improved mode method is robust. The prototype of the algorithm using Matlab on desktop computer with 2 dual-core 2.2 GHz AMD Opteron processors with 4 GB memory. In improved mode experiments the testing was limited sequences. The size of each frame was set to 16 * 16. Images have been gathered from the dataset of several research groups such as Sequence of Gait dataset and Intelligent Room (IR) comes from Autonomous Agents for On-Scene Networked Incident Management (ATON) project.

The separate set of experiments to verify the performance of the proposed method was conducted. In the first case, the quality of the background was measured, The conventional method for measuring quality of image parameter Mean Square Error (MSE) and Peak Signal to Noise Ratio (PSNR) of each image. This technique gives better result than other techniques and their PSNR value is high and MSE is low.

5.1   Mean Square Error

Mean Square Error can be estimated in one of many ways to quantify the difference between values implied by an estimate and the true quality being certificated. MSE is a risk function corresponding to the expected value of squared error. The MSE is the second moment of error and thus incorporates both the variance of the estimate and its bias. Lower the value of MSE higher the quality of image. The MSE is defined as:

$$MSE = \frac{\sum_{M,N}[I_1(m,n) - I_2(m,n)]^2}{M * N} \tag{4}$$

$$RMSE = sqrt(MSE) \tag{5}$$

where M and N are the number of rows and columns in the input images.

5.2   Peak Signal to Noise Ratio

The Peak Signal to Noise Ratio (PSNR) is the ratio between maximum possible power and corrupting noise that affects representation of image. PSNR is usually expressed as decibel scale. The PSNR is commonly used as measure of quality reconstruction of image. The signal in this case is original data and the noise is the error introduced. High value of PSNR indicates the high quality of image. It is defined via the Mean Square Error (MSE).

$$PSNR = 10.\log_{10}[(255)_2/MSE] \qquad (6)$$

Here Max is maximum pixel value of image when pixel is represented by using 8 bits per sample. This is 255 bar color image with three RGB value per pixel. The MSE and PSNR values of each technique demonstrated below in via Bar chart and Table.

5.3 Result and Discussion

Background estimation results are shown below. Figure 4a show original image and Fig. 4b show the ground truth of estimated background. Figure 4c–e illustrate a mean, median and mode respectively. Figure 4e shows the result of the proposed background estimation algorithm. The MSE and PSNR values of each technique shown in Table 1 and Fig. 6 show the comparison of all techniques. Table 1 of MSE and PSNR of various background estimation techniques (Fig. 5).

In the experimental results, we mainly compare improved mode results with other methods. The comparison clearly shows that improved mode results are better than others. By contrast, improved mode method is much simpler. As a result, it is straightforward to re-implement improved mode method and make any possible improvement. Besides, the processing time is shorter when applying improved

Fig. 4 **a** Original frame. **b** Background model. **c** Mean algorithm. **d** Median algorithm. **e** Mode algorithm. **f** Proposed method

Table 1 MSE and PSNR of various background estimation techniques

	Mean		Median		Mode		Improved mode	
	RMSE quality value	PSNR	RMSE	PSNR	RMSE	PSNR	RMSE	PSNR
Seq1	78.6827	32.9389	99.1865	31.9332	70.2265	33.4327	65.7139	34.5255
Seq2	80.3755	32.8465	99.2343	31.9311	70.2265	33.4327	65.6997	34.5265

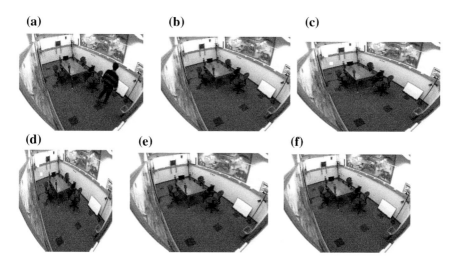

Fig. 5 **a** Original frame. **b** Background model. **c** Mean algorithm. **d** Median algorithm. **e** Mode algorithm. **f** Proposed method

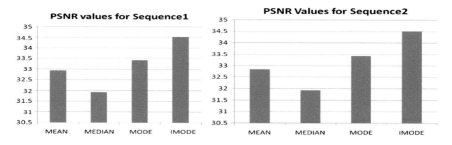

Fig. 6 Comparison of PSNR of proposed with other methods

mode method. Another strength of improved mode method is that it connects background estimation with other research areas. This method has many advantages compared to other background estimation methods, but it also has some drawbacks. The proposed method Root Mean Square Error (RMSE) value of improved mode is less, Peak Signal to Noise Ratio (PSNR) value of improved mode is high and improved mode is better than traditional methods (Mean, Median, Mode).

6 Conclusion

Image quality measurement plays an important role in various image processing applications. In this paper, different techniques like mode and median are applied to develop technique improved mode method. It is the best method for background estimation, because in this PSNR has maximum value and MSE has lower value. In the existing system, a stored video has been used for the experimental purpose but here it is implemented using a real-time video. Also the background image taken for subtraction is an arbitrary image, where as this experiment uses a refined image which is obtained after estimating using improved mode algorithm. In future, it is possible to find a refined image for background subtraction using the proposed method and implement it on a real-time video.

References

1. Bondzulic, B., Petrovic, V.: Multisensor background extraction and updating for moving target detection. In: 11th International Conference on Information Fusion, pp. 1–8, 30 June–3 July 2008
2. Yan, Z., Lian, Y.: A new background subtraction method for on-road traffic. China J. Image Graph. **13**(3), 593–599 (2008)
3. Jiang, Y., Qu, Z., Wand, C.: Video background extraction based on textural and statistical features. Optics Precis. Eng. **16**(1), 172–177
4. Yang, Y-Z., He, X-F., Chen, Y-Z., Bao, Z-K.: Traffic parameter detection based on video images. Control Eng. China **5**(7), 349–352
5. Long, W., Yang, Y.: Stationary background generation: an alternative to the difference of two images. Pattern Recogn. **23**(12), 1351–1359 (1990)
6. Xu, X., Huang, T.S.: A loopy belief propagation approach for robust background estimation. In: IEEE Computer Society Conference on Computer Vision and Pattern Recognition (CVPR), pp.1–7 (2008)
7. Agarwala, A., Dontcheva, M., Agrawala, M., Drucker, S.M., Colburn, A., Curless, B., Salesin, D., Cohen, M.F.: Interactive digital photomontage. In: Proceedings of ACM SIGGRAPH (2004)
8. Colombari, A., Cristani, M., Murino, V., Fusiello, A.: Exemplarbased background model initialization. In: Proceedings of the third ACM International Workshop on Video Surveillance and Sensor Networks, pp. 29–36 (2005)
9. Granados, M., Seidel, H., Lensch, H.: Background estimation from non-time sequence images. In: Graphics Interface, pp. 33–40 (2008)
10. McFarlane, N., Schofield, C.: Segmentation and tracking of piglets in images. Mach. Vision Appl. **83**, 187–193 (1995)
11. Stauffer, C., Grimson, E.: Adaptive background mixture models for real-time tracking. In: IEEE International Conference on Computer Vision and Pattern Recognition (CVPR), vol. 2, pp. 246–252 (1999)

12. Chung, Y., Wang, J., Chen, S.: Progressive background images generation. In: Proceedings 15th IPPR Conference on Computer Vision (2002)
13. Bevilacqua, A.: A novel background initialization method in visual surveillance. In: IAPR workshop on machine vision applications, Nara, Japan, pp. 614–617 (2002)
14. Farin, D., de With, P., Effelsberg, W.: Robust background estimation for complex video sequences. In: IEEE International Conference on Image Processing (ICIP), vol. 1, pp. 145–148 (2003)
15. Colombari, A., Fusiello, A., Murino, V.: Background initialization in cluttered sequences. In: Proceedings of the Conference on Computer Vision and Pattern Recognition Workshop (CVPRW'06), Washington DC, USA, pp. 197–202 (2006)
16. Gutchess, D., Trajkovic, M., Cohen-Solal, E., Lyons, D., Jain, A.: A background model initialization algorithm for video surveillance. In: International Conference on Computer Vision (ICCV), vol. 1, pp. 733–740 (2001)
17. Cohen, S.: Background estimation as a labeling problem. In: International Conference on Computer Vision (ICCV), vol. 2, pp. 1034–1041 (2005)
18. Boykov, Y., Veksler, O., Zabih, R.: Fast approximate energy minimization via graph cuts. In: International Conference on Computer Vision (ICCV), vol. 1, pp. 377–384 (1999)
19. Stauffer, C., Grimson, E.: Learning patterns of activity using real-time tracking. IEEE Trans. Pattern Anal. Mach. Intell. **22**(8), 747–757 (2000)
20. Cohen, S.: Background estimation as a labeling problem. In: Proceedings of the 10th IEEE International Conference on Computer Vision (ICCV), pp. 1034–1041 (2005)

Part V
Big Data

Interactive Big Data Management in Healthcare Using Spark

J. Archenaa and E.A. Mary Anita

Abstract This paper gives an insight on how to use apache spark for performing predictive analytics using the healthcare data. Large amount of data such as Physician notes, medical history, medical prescription, lab and scan reports generated by the healthcare industry is useless until there is a proper method to process this data interactively in real-time. Apache spark helps to perform complex healthcare analytics interactively through in-memory computations. In this world filled with the latest technology, healthcare professionals feel more comfortable to utilize the digital technology to treat their patients effectively. To achieve this we need an effective framework which is capable of handling large amount of structured, unstructured patient data and live streaming data about the patients from their social network activities. Apache Spark plays an effective role in making meaningful analysis on the large amount of healthcare data generated with the help of machine learning components supported by spark.

Keywords Healthcare · Big data analytics · Spark

1 Introduction

In today's digital world people are prone to many health issues due to the sedentary life-style. The cost of medical treatments keeps on increasing. It's the responsibility of the government to provide an effective health care system with minimized cost. This can be achieved by providing patient centric treatments. More cost spent on healthcare systems can be avoided by adopting big data analytics into practice [1].

J. Archenaa (✉)
AMET University, Chennai, India
e-mail: archulect@gmail.com

E.A.M. Anita
S.A. Engineering College, Chennai, India

© Springer International Publishing Switzerland 2016
V. Vijayakumar and V. Neelanarayanan (eds.), *Proceedings of the 3rd International Symposium on Big Data and Cloud Computing Challenges (ISBCC – 16')*,
Smart Innovation, Systems and Technologies 49, DOI 10.1007/978-3-319-30348-2_21

It helps to prevent lot of money spent on ineffective drugs and medical procedures by performing meaningful analysis on the large amount of complex data generated by the healthcare systems. There are also challenges imposed on the handling the healthcare data generated daily. It's important to figure out how the big data analytics can be used in handling the large amount of multi structured healthcare data.

1.1   What is the Need for Predictive Analytics in Healthcare?

To improve the quality of healthcare, it's essential to use big data analytics in healthcare.

Data generated by the healthcare industry increases day by day. Big data analytics system with spark helps to perform predictive analytics on the patient data [2]. This helps to alarm the patient about the health risks earlier. It also supports physicians to provide effective treatments to their patients by monitoring the patient's health condition in real-time. Patient centric treatment can be achieved with the help of big data analytics, which improves the quality of healthcare services. It also helps to predict the seasonal diseases that may occur. This plays an effective role in taking necessary precautions before the disease can spread to more people.

2   Spark Use Cases for Healthcare

Many organizations are figuring out how to harness big data and develop actionable insights for predicting health risks before it can occur. Spark is extremely fast in processing large amount of multi-structured healthcare data sets due to the ability to perform in-memory computations. This helps to process data 100 times faster than traditional map-reduce.

2.1   Data Integration from Multiple Sources

Spark supports fog computing which deals with Internet Of Things (IOT). It helps to collect data from different healthcare data sources such as Electronic Health Record (EHR), Wearable health devices such as fitbit, user's medical data search pattern in social networks and health data which is already stored in HDFS [3]. Data is collected from different sources and inadequate data can be removed by the filter transformation supported by spark.

2.2 High Performance Batch Processing Computation and Iterative Processing

Spark is really fast in performing computations on large amount of healthcare data set. It is possible by the distributed in-memory computations performed as different clusters. Genomics researchers are now able to align chemical compounds to 300 million DNA pairs within few hours using the Spark's Resilient Distributed Dataset (RDD) transformations [4]. It can be processed iteratively then for further analysis.

2.3 Predictive Analytics Using Spark Streaming

Spark streaming components such as MLib helps to perform predictive analytics on healthcare data using machine learning algorithm. It helps to perform real-time analytics on data generated by wearable health devices. It generates data such as weight, BP, respiratory rate ECG and blood glucose levels. Analysis can be performed on these data using k-clustering algorithms. It will intimate any critical health condition before it could happen.

3 Spark Healthcare Ecosystem

Spark architecture for healthcare system is shown in Fig. 1.

Today's world is connected through Internet of things (IOT). It is the source for the large amount of healthcare data generated. It fits into the category of big data as

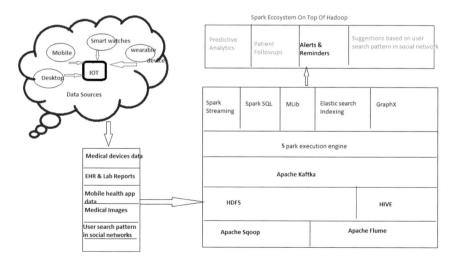

Fig. 1 Spark ecosystem for healthcare

it satisfies the three V's-Volume, Velocity and Veracity. Healthcare dataset contains structured data such as electronic health records, semi structured data such as case history, unstructured data such as X-ray images, DNA pairs and real time data generated by wearable health devices [2]. Different datasets are stored in hdfs and hive. Structured datasets are ingested through sqoop and unstructured data is handled by flume. Streaming data from twitter and facebook are handled through apache kaftka which is a messaging queue service. Spark streaming handles the user search patterns related to healthcare from social network such as google search engines. Spark SQL allows interactive querying on data generated by wearable medical devices. Prescriptive analytics can be made by analyzing the patient data and search pattern in social networks [5]. It can be achieved through the machine learning algorithms supported by spark. Elastic search indexing is used for faster data retrieval from large dataset.

3.1 How Does Spark Makes Healthcare Ecosystem as Interactive?

Spark is getting famous for the faster in-memory computations through the Resilient distributed objects (RDD) stored in distributed cache across different clusters.

It is a computational engine that is responsible for: scheduling, distributing and monitoring jobs across different clusters. The ability of iterative machine learning algorithm processing, high performance on batch processing computations and interactive query response supported by spark makes the healthcare system interactive.

3.2 Core Components of Spark

Spark Executor Engine:

It is the core component which contains the basic functionalities supported by spark API. Resilient distributed dataset (RDD) is the main element of spark. It is an abstraction of distributed collection of items operations and actions. Its in-built fault tolerance on node failures makes it as resilient. Fundamental functions supported by spark core includes: information sharing between nodes through broadcasting variables, data shuffling, scheduling and security.

Spark SQL:

It is an subset of HIVEQL and SQL which supports querying data in different formats. It allows querying data in structured, JSON and Parquet file format which is becoming popular for the columnar data storage where the data is stored along with the schema. BI tools can also query the data through Spark SQL by using classic JDBC and ODBC controls.

Spark Streaming:

It supports real-time processing from different sources such as flume, kaftka, twitter, facebook etc. Its in-built nature of recovering from failure automatically makes it more popular [6].

Spark Graphx:

It supports using graph data structures for implementing graph theory algorithms such as page rank, shortest path computations and others. Pergel message passing API supports large scale graph processing such as finding the nearest hospital based on patient location in the google map [7]. This feature really helps incase of patient's critical illness.

3.3   Spark Implementation in Scala for Finding People with Belly Fat

1. Loading an text file Patientdata.txt which contains data in the format of Patient id, name, age, BMI, Blood Glucose level, Exercising

```
Val conf = new SparkConf().setAppName("Belly Fat Symptoms");
Val sc = SparkContext(conf);
Val file = sc.textfile("hdfs://home/patientdata.txt");
```

RDD Transformations – Each transformation is stored in one individual partition

```
val counts = file.flatMap(line => line.split(" "))
.map(word => (word, 1)).countByKey()
/* Words are split into each line*/
/* To find out people with higher glucose level from 1 lakh records*/
Val hgl = counts.filter(1 =>1.contains("High")).cache
/*The above data in stored in cache for faster processing*/
/*To find out people with no exercising*/
Val exc = counts.filter(1 =>1.contains("No")).cache
```

RDD Action – Action is called upon on the transformed partition.

```
Hgl.union(exc).saveAsTextFile(BellyFat.txt)
```

Generated text file can be used for further analysis. This can be used as the training set for the Machine learning algorithm to predict the people who are at the risk of getting cardiac disease based on their age. Coding in spark is relatively easy when compared to map-reduce which involves java programming skills. To implement the above use case in map-reduce, developer needs to write 60 lines of code. Developers and data analyst prefer spark as it takes only less time to implement and execution time is faster.

3.4 Result of Healthcare Data Analysis Using Apache Spark

Figure 2 represents the spark workflow for healthcare data analysis. Input text file is split into RDD1 and RDD2 using filter transformations. RDD's are combined together using action—union. Result is written to the text file which consist of details about people who are in belly fat risk. Cached RDD1 and heart rate data generated by fitbit device are iteratively analyzed using machine learning algorithm. It performs predictive analysis about people who are in the risk of getting cardiac disease.

3.5 How Does Spark Outperform Map-Reduce in Health Care Analytics?

Spark is capable of handling large data sizes, real-time processing and iterative processing effectively when compared to map-reduce [8]. In-memory computations supported by spark allows to do the computations faster on large datasets by reducing disk input/output operations. In map-reduce more processing time is

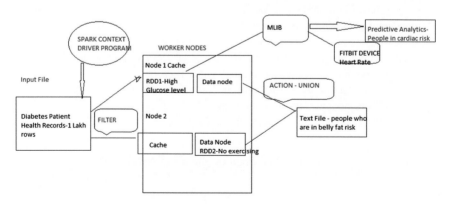

Fig. 2 Spark workflow for healthcare data analysis

consumed by incurring disk input-output operations for each mapper and reducer task. Still map-reduce is better in doing batch-processing jobs on large data set. Major drawback with map-reduce is that it does not support real-time processing. This feature is very important in analyzing health care applications. For an example: In-order to treat cancer effectively, 3 billion data pairs of DNA needs to be analyzed to identify the genetic pattern and how treatment techniques works on each gene pair. In the above case we can use map-reduce effectively for identifying the common type of cancer among gene pairs [9]. Map-Reduce will not be able to handle iterative processing—which involves processing different treatment methods for a gene pair to find out an effective cancer treatment. For the above use-case spark is more suitable as it also allows to run machine-learning algorithms iteratively. Since health-care analytics doesn't rely only on batch-mode processing, Spark will be the better option for the healthcare use case.

4   Conclusion

Real time healthcare data processing is now possible with the help of spark engine as it supports automated analytics through iterative processing on large data set. Map reduce is capable of performing the batch processing and after each operation the data will be stored in disk. For further processing data again needs to be read from the disk thus increasing the time in performing computations. It does not support the iterative processing. Spark's core concept in-memory computations overcomes this limitation imposed by map-reduce by caching data in distributed cache. This can speed up the execution time when compared to map-reduce. Our implementation in spark to find out the people who are at risk in getting belly fat from one lakh records takes 10 min execution time whereas in map-reduce it takes around 50 min to complete the same task. This result can be stored in cache and it can be used to predict people who are in risk of getting cardiac disease by analyzing heartbeat rate through the data generated by heart-rate monitoring devices. Combining Spark with Hadoop effectively unleashes the potential of predictive analytics in healthcare.

References

1. Morely, E.: Big data healthcare, IEEE explore discussion paper (2013)
2. Muni Kumar, N., Manjula, R.: Role of big data analytics in rural healthcare, IJCSIT (2014)
3. Feldman, K., Chawla N.V.: Scaling personalized healthcare with big data. In: International Big Data Analytics Conference in Singapore (2014)
4. Pinto, C.: A Spark based workflow for probabilistic linkage of healthcare data, Brazilian Research Council White Paper (2013)
5. Xin R., Crankshaw, D., Dave, A.: GraphX: unifying data-parallel and graph-parallel analytics, White Paper, UC Berkeley AMP Lab

6. Zaharia, M., Das, T., Li, H., Shenker, S., Stoica, I.: Discretized streams: an efficient and fault-tolerant model for stream processing on large clusters, White Paper, University of California
7. Bonaci, M.: Spark in action, Ebook
8. Armburst, M.: Advanced analytics with spark SQL and MLLib, White Paper, London Spark Meetup
9. Ahmed, E.: A Framework for secured healthcare systems based on big data analytics, IJASA (2014)
10. Hamilton, B.: Big data is the future of healthcare, cognizant white paper (2010)
11. Data Bricks: Apache spark primer, Ebook

Data Science as a Service on Cloud Platform

Aishwarya Srinivasan and V. Vijayakumar

Abstract Big firms generally have huge amount of data which needs to be analyzed and results have to be evaluated. When it comes to such a huge amount of data, we refer it with the term "big data", and its analysis is a tedious process. Companies employ people, who are trained data scientists and they are given the data sets along with the expected output. An integrated solution for data analytics comes as data science as a service (DSaaS), where the data scientists need not be employed by each company. DSaaS can be implemented on a worldwide basis with a global environment hosted on any platform. The proposal provides DSaaS on cloud platform with grid computing/multicore computing in cooperative technology for higher efficiency and reliability.

Keywords Big data · Data sets · Data science · Data scientists · Global environment · Cloud platform · Multicore computing · Cooperative technology

1 Introduction

In many companies, analysis of their data isn't done on a day-to-day fashion, rather it is done once or twice in a year. Employing data scientists for such analysis may be very expensive for the company.

Many start-ups are working on giving data science as a service where the companies can send their data, desired output or prediction models with some legal agreements, so that the data is protected, and these start-ups work on the data and revert back to these companies [1].

This is not very different from employing data scientists, as the same professionals are working in some firms and performing a similar task [2].

A. Srinivasan (✉) · V. Vijayakumar
Vellore Institute of Technology, Chennai Campus, Chennai, India
e-mail: aishwarya.srinivasan2013@vit.ac.in

V. Vijayakumar
e-mail: vijayakumar.v@vit.ac.in

© Springer International Publishing Switzerland 2016
V. Vijayakumar and V. Neelanarayanan (eds.), *Proceedings of the 3rd International Symposium on Big Data and Cloud Computing Challenges (ISBCC – 16')*,
Smart Innovation, Systems and Technologies 49, DOI 10.1007/978-3-319-30348-2_22

There comes two more challenges to overcome, one is that sometimes the data is of huge size, say in zeta bytes or hundreds of terabytes, which cannot be stored on normal computers and laptops and may require servers or multi core computers [3, 4].

Another problem faced is the environment which is needed to process on the data sets. For different data sets, depending on the requirement of the company and the nature of the output, the environment and the packages used may differ. Installing all such packages in a computer for processing different data sets becomes space consuming [5].

On the cloud platform, environments/platforms like Hadoop are available in Microsoft Azure. Lately, new softwares are developed like Spark using scala which is not wide spread in terms of usage [6]. Many other types of software do exist like R which is also used for analytics and need certain hardware, software and memory requirements before installation.

Keeping many platforms along with the softwares and data sets in a computer becomes tedious and with lesser reliability and accessibility.

2   Methodology

The cloud based web application can be built such that the access to the data server is classified differently. It can be moduled as (i) user module, (ii) cloud provider module and (iii) company module. It can be in the workflow as follows (Fig. 1).

The module 1 specifies the user, where the data scientist or the freelancer may access the cloud platform via the website. As per the interest of the user, the analysis project can be selected and applied for, after which the company (seen in later module) will shortlist the candidate. The user may require to sign the agreement as the data sets may be highly confidential and if leaked may cause severe financial damage to the company [7]. Once appointed, the user may access the private cloud specifically designed for each company and its freelancers, for the data. The users may be able to access the analytical tools and software that comes along with the compatible environments and IDEs. The user may get some term for which the cloud space would be free of cost, after which payment has to be done to the cloud provider. Once the output is produced as desired by the company within the specified span of time, the user can submit the work and get paid [7] (Fig. 2).

The most crucial module is the cloud provider module. Here the cloud provider has to simultaneously handle the users and the companies, with an interactive interface. The legal process, and agreements have to be judiciously designed. The software, the environment and IDE's provided must be accessible with good speed and should be dynamic. The data security must be given utmost importance. Hence, the cloud provider forms a bridge between the users and the company for the data analytics task (Fig. 3).

Fig. 1 Module 1—user end

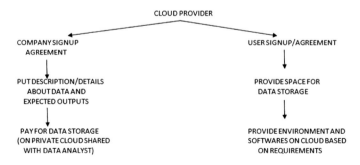

Fig. 2 Module 2—cloud provider

The company module is comparatively inactive from the other two modules. The company just needs to describe the input type of data, desired output, software specifications (if any), and other relevant information.

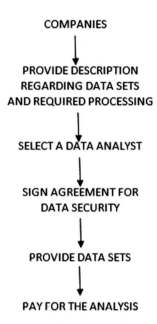

Fig. 3 Module 3—company's end

The company would have to pay some amount to the cloud provider for using the customized private cloud space shared between the company and the data scientist/user. In this scenario, the company acts as a superior module that assigns work for the shortlisted candidates or the data scientists.

3 Background and Related Work

Data Analytics field at its advent was more of a manual work and analysing upon the pattern and similarities within a data set and among few related data sets. As the data sizes increased, the analysis became more and more complex and hence were the development of Analytic Tools [8].

Around 10 years ago, a package named ARB was developed. It was a program comprising of a variety of directly interacting software tools or the maintenance of sequential database and its analysis controlled by graphical user interface. The ARB package was developed for analysis of bioscience data mainly [8].

The data analysis isn't dependent on the theme of the data, it may be of any branch. The major analysis consists of clustering, classification, merging etc. using map, reduce or other algorithms [9]. If the data is extracted continuously, i.e. dynamic data or streaming data, more sophisticated algorithms have to be observed for such analysis.

The existing programs for the analytics have a similar base architecture. It mainly consists of central database which contains processed data which is mostly primary.

Moreover, any additional data descriptions can be stored in the database in the fields assigned to the individual sequencing or the linkage through local or worldwide networks.

It also comprises of additional tools for the import, export, primary and secondary structuring, sequence alignment or editing, profile generation or filtering along with other components for the data analytics [10].

The analytics tools are generally developed in the UNIX system, similar to the ARB package on SuSE Linux [8]. It can be expressed as an improved and automated sequencing technique with certain interaction tools for better comprehensive packaging.

The general analytical tools are developed taking into account few objectives. It can be classified into two modules. Module 1 holds the maintenance of the structured integrative secondary database which combines the processed primary structure and also including any of the additional data assigned to certain individual sequencing entries. Module 2 handles the crucial selection of the software tools by interaction of two or more central databases which is controlled by a graphical interface.

Apart from business applications of data analytics, it also poses its application in the field of e-learning and management systems.

With the recent studies of the user driven data analytics, the framework is established with a sophisticated and complex architecture. This may reach up to ultimate optimized data analytic usage.

It can be classified as data analytics which works for the operational task and data infrastructure which controls the working and management of the distributed computation and storage resources.

The technological aspects currently existing contain intelligent crawler which collects relevant information and services which are already available in the cloud. The transfer learning concept is for adaptation of the modes of analysis among different domains.

Meta miner tool is used for recommendation module that uses optimum data analytics workflow. Visual analytics includes the representation of the output data for a better interactive module, which is a feature for the users in the data analytics for the workflow [11].

The usage miner focuses on adaptation with the collective patterns, its reusability and collaborative analysis.

The data infrastructure should be made with Hadoop framework, data broker service, distributed management system. The Hadoop framework ensures the efficiency and resiliency of computation and optimum scheduling.

The data analytics storage is similar to the Google File System, it is integrated with the data broker which utilizes the pattern to optimize the concepts and storage needs [12, 13].

The analysis may be handled by (i) initial data sets, (ii) analytics workflow and (iii) Predictive Modeling Markup Language (PMML). These feature sum up to better e-learning and scalable distributed data analytics [11].

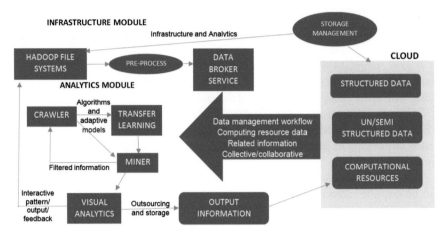

Fig. 4 Data analytics framework

A new tool developed by IBM, Bluemix is very similar to this proposal. Instead of using a cloud platform and interactive interface between company and data scientists, they are just providing online analytical software [14] (Fig. 4).

4   Proposal

The same concept can be applied on a cloud platform. This proposal aims for building a cloud which supports data storage of various types with the platform and the softwares required for data analysis.

It may be so formed that the company can upload their data sets along with the agreement, which can be accessed only by authentic users after signing up. The data analysts need to pay certain amount based on the amount of data they are storing and the space and types of platforms and softwares they are using [15].

The company may also need to pay the cloud providers based on the size of data they are uploading. The company may put up descriptions regarding the data, expected outputs, deadline and stipend for the task (Fig. 5).

The data of these companies can be accessed by the freelancer data analysts after signing up the agreement of the company [12, 13]. They would be permitted to store the data sets on the cloud itself and the processing can also be done using the softwares and platform on the cloud. This would reduce the complex situation of storing such huge data sets on the hard disk (Fig. 6).

It will also increase the remote access of the data from the authentic analysts account from any place. It will reduce the chances of computer crash and data being lost, as the backup will be maintained by the cloud provider, which makes it more reliable.

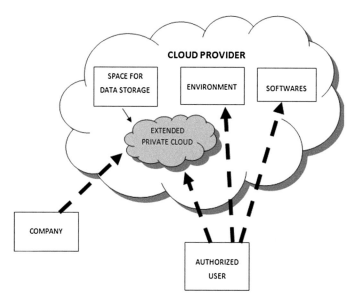

Fig. 5 Cloud architecture

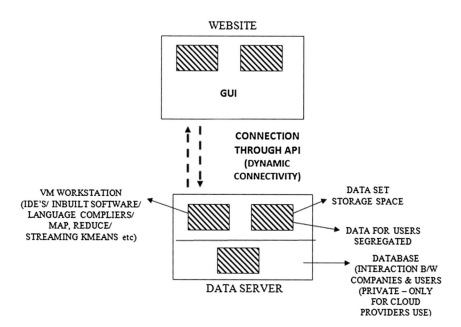

Fig. 6 Use case diagram

5   Conclusion

The motive to be achieved through this proposal is to create a platform on cloud where Data Science, which is one of the most crucial and emergent field, can be provided as a service, similar to any other services on cloud like, Saas, Paas, Iaas etc.

This proposal is suggested keeping in mind the IBM's Bluemix cloud services [14]. The data analytic tools available can be integrated and fed into the Virtual Machine Workstation and dynamically accessed by the users though the website.

6   Discussions and Future Work

The proposal may also be extended in a way that instead of using a single data server, we may use a network of servers, like grid computing or multi core computing, for more efficiency and reliability. With one server holding multiple environments and softwares, there is a possibility of crashing which can be avoided by using cooperative computing technology.

References

1. IBM Point of View: Security and Cloud Computing, Cloud computing White paper, November 2009
2. Ludwig, W., Strunk, O., Westram, R., Richter, L., Meier, H., Buchner, A., Lai, T., et al.: ARB: a software environment for sequence data. Nucleic Acids Res. **32**(4), 1363–1371 (2004)
3. Zissis, D., Lekkas, D.: Addressing cloud computing security issues. Future Gener. Comput. Syst. **28**(3), 583–592 (2012)
4. http://data-informed.com/understanding-data-science-service/
5. Shen, Z., Ma, K.-L., Eliassi-Rad, T.: Visual analysis of large heterogeneous social networks by semantic and structural abstraction. Vis. Comput. Graphics, IEEE Trans. **12**(6), 1427–1439 (2006)
6. http://www.cnet.com/videos/diy-create-your-own-cloud/
7. Zhang, J., Tjhi, W.C., Lee, B.S., Lee, K.K., Vassileva, J., Looi, C.K.: A framework of user-driven data analytics in the cloud for course management. In: Wong, S.L. et al. (eds.) Proceedings of the 18th International Conference on Computers in Education, pp. 698–702. Putrajaya, Malaysia, Asia-Pacific Society for Computers in Education (2010)
8. Brunette, G., Mogull, R.: Security guidance for critical areas of focus in cloud computing v2.1. Cloud Secur. Alliance, 1–76 (2009)
9. Bleiholder, J., Naumann, F.: Data fusion. ACM Comput. Surv. (CSUR) **41**(1), 1 (2008)
10. Guazzelli, A., Stathatos, K., Zeller, M.: Efficient deployment of predictive analytics through open standards and cloud computing. ACM SIGKDD Explor. Newsl. **11**(1), 32–38 (2009)
11. Vinay, K., Girish, M., Siddhi, K., Shreedhar, K.: Study of cloud setup for college campus. Int. J. Adv. Res. Comput. Sci. Softw. Eng. **2**(10), 251–255 (2012)
12. http://www.hongkiat.com/blog/free-tools-to-build-personal-cloud/
13. https://owncloud.org/

14. Smith-Miles, K.A.: Cross-disciplinary perspectives on meta-learning for algorithm selection. ACM Comput. Surv. (CSUR) **41**(1), 6 (2008)
15. https://gigaom.com/2014/03/01/how-to-set-up-your-own-personal-home-cloud-storage-system/

Data Mining Approach for Intelligent Customer Behavior Analysis for a Retail Store

M. Abirami and V. Pattabiraman

Abstract The Occurrence of the recent economic and social changes transformed the retail sector in particular the relationship between the customers and the retail stores changed significantly. In the past the retail industry focused on marketing the products without having detailed knowledge about the customers who availed products. With the proliferation of competitors the retail stores had to target on retaining their customers. To be successful in today's competitive environment retail stores must creatively and innovatively meet their customer needs and expectations. Generic mass marketing messages are irrelevant. This paper put forwards, a new approach of customer classification based on the RFM(Mode) model and also deals with customer data to analyze and predict the customer behavior using clustering and association rule mining techniques.

Keywords Clustering · CRM · CLV · RFM · ARM

1 Introduction

With changing customer preferences and market environment, staying afloat and remaining profitable is the key for success which resulted in the customer relationship management by providing offers. To predict the customer behavior based on the previous transactions and to provide relevant offers to loyal or repetitive customers using data mining techniques. Analysis of the previous transactions of the customers aids in extending the customer lifetime value and provides customer satisfaction business functions such as marketing campaign.

M. Abirami (✉) · V. Pattabiraman
School of Computing Science and Engineering, Vellore Institute of Technology,
Chennai, India
e-mail: abirami.m2014@vit.ac.in

V. Pattabiraman
e-mail: pattabiraman.v@vit.ac.in

© Springer International Publishing Switzerland 2016   283
V. Vijayakumar and V. Neelanarayanan (eds.), *Proceedings of the 3rd International
Symposium on Big Data and Cloud Computing Challenges (ISBCC – 16')*,
Smart Innovation, Systems and Technologies 49, DOI 10.1007/978-3-319-30348-2_23

1.1   Key Success Factors

- Customer Centric—Retailers to change from function based structure to customer centric model. Segment the customers basis on their purchasing behavior and meet the needs of highest valued "Golden" Customers
- Loyalty Programs—Build customer loyalty programs with help of data mining techniques to serve golden customers better and retain them
- Customer Experience—With advent of new fulfillment models, retailers need to improve the shopping experience of the customers
- Data Mining—Few retailers had invested in skills and capabilities to leverage the advantage of smart algorithms and data driven analytic s from big volume of data from multiple sources

This paper aims at classifying the customer based on their visit to the stores (new customer or repeat customer). Purchase behavior of the repetitive customer is identified by application of clustering and associative mining on the customer data which would propose a relevant offer to the right customer at the right time. This would improve the customer satisfaction and builds loyalty thereby increasing the bill value or sales and profits by minimizing the marketing or campaign cost.

(A) *Data Mining*

Data mining is a knowledge discovery process, where we develop insights from huge data which might be incomplete and random. The date is generated from all business functions such as Production, Marketing and Customer Service. Often, we would lost in ocean of data, crunching the date to make meaningful observations calls for data mining techniques. The discovered pattern using Clustering, Classification, Association and Sequential are used in developing marketing campaigns [1].

(B) *Clustering*

Data mining techniques could be broadly classified into two types namely Supervised modeling and unsupervised modeling. In Supervised modeling, the variables could be classified into two types namely, explanatory variables and one dependent variables. The objective of this type of model is to determine the relationship between explanatory and dependent variables. Examples: Regression, Decision Trees. In Unsupervised modeling, there is no distinction between explanatory variables and dependent variables. To perceive information all variables are treated uniformly so that they can be used for grouping and associations. Examples: Clustering, Association, Factor Analysis [2].

Clustering is grouping the objects on the basis of its closeness. Based on the attributes of the input objects grouping is done. The resultant groups are called clusters with high intra cluster similarity and low inter cluster similarity. The clustering methodology dealt with here is the K Means clustering with the RFM model.

1.2   Association Rule Mining

Association rule mining are classified under Unsupervised Models. Association models detect associations between discrete events, products, or attributes. The rules are derived by analyzing historical data to form a rule which comprises the left part of the rule holding a condition: a situation where and when true, the rule applies. The right part of the rule is the conclusion: what tends to be true when the condition (left part) hold true.

Clustering is done to segregate the customers based on their purchase behavior. Association rule mining is done with each cluster separately referring to the full data set to identify the customer specific offers.

2   Literature Survey

A customer segmentation framework based on Customer Lifetime Value (CLV) which provides the companies to estimate the value that the customer generates throughout his life time. An integrated customer segmented framework utilizes this value (CLV) to provide differentiated services/offers to it's customers. This an extension of data mining techniques to deduct a new customer classification model called Customers Segmentation Method Based on CLV.

CLV model is built on three factors namely, Customers value over a time period, customer's length of service, discounting factor. Each factor is separately measured and the model is tested with conventional "Train" and "Test" methodology.

Following assumptions are assumed to identify the repeat customers (Loyal)

- Becomes permanently inactive after initial activity for a short period of time
- Active/Inactive customers classified based on transactions done/not done
- Some customers rapidly moves to inactive

Monetary bill value is independent of the transaction process. This leads to fact that probability of purchase and frequency of future transactions is better than monetary value per transaction. Clusters are formed based on CLV and each segment classifies cluster based on the loyalty [3].

This paper employs clustering techniques in two stages. In stage one, K-Means cluster is applied to form clusters based on RFM's and in Second, demographic data (Age, education, occupation selected basis SOM procedure) is super imposed and new clusters are formed. Customer profile is created with life time value [4].

RFM methodology is used to identify "Golden Customers (High Value Customers)" in the airline industries for providing better flying experiences. Value generated by customers is estimated by the difference between the bill value and absorption cost. The revenue and cost details considered are for a year for an airline company in China [5].

3 Proposed Work

Customer is classified into "New to Store"—first time or repeat customers. The purchase behavior of the first time customer could not be predicted due to lack of historical patterns. For repeat customers, prediction can be done by employing clustering techniques with RFM model and association rule mining.

K-Means—Efficient distance based clustering method to handle large data sets (records/dimensions). This method partitions the data instances into specified number(k) of clusters each of which is represented by the cluster center. The process starts with an initial set of randomly chosen k centers. At the first iteration, each instance is assigned to a clusters based on Euclidean distance between the two—center and the data point. The cluster center gets changed over the iterations. Upon the next iteration the center of each cluster is calculated as the mean of all the instances belonging to that cluster. This procedure is repeated until there is no change in the clusters formed. Clustering is done on the RFM parameters of the transaction.

3.1 RFM Analysis

The Recency, Frequency and Monetary(RFM) parameters are used to analyze customer behavior such as how many days elapsed since his previous purchase (Recency of the customer), the number of purchases a customer does in a particular period of time (Frequency), and how much is the bill value of the customer (Monetary). RFM analysis is widely used in the retail industry to improve customer segmentation and to identify "Golden customers" (profitable) for personalized services and promotions. Retailers could employ clustering models to analyze the RFM components. The clustering model could be supplemented with a classification model such as decision tree as well as Association Rule Mining.

3.2 Algorithm K-Means

Input—Number of clusters to be formed 'K' and the data set D containing 'n' objects. Each object takes recency, frequency and the monetary (taken as mode) parameters.

  Output—A set of K clusters

1. Choose K points as initial centers from the given data set D.
2. Object having Euclidean distance closer to the cluster mean are grouped into the same cluster.

3. Re compute the mean of each newly formed cluster and set the cluster center as the calculated cluster means.
4. The steps 2 and 3 are repeated until there is no change in the objects of the cluster.

This approach used in this paper deals with forming three clusters as

- Customers yielding high profit (provided with huge discounts and offers)
- Customers yielding medium profit (provided with less discounts and offers)
- Customers yielding low profit (for whom, based on their market basket offers are provided to increase their bill value and transactions).

RFM model generally uses absolute monetary value for clustering. This paper aims at proposing the usage of mode of the customer's bill value. As mode indicate the most frequent occurrence on a given series of transactions and indicates the most likely transaction value. Clustering using K Means with RFM (mode), is carried out to segment the customers into three clusters namely high, medium, low.

The input data consists of the transaction ID of the customer and the list of the products purchased for each of the transaction and the bill amount for that transaction from which the recency (number of days elapsed since the last purchase of the customer), Frequency (number of visits), Monetary (bill amount for the transaction) are derived. Based on these RFM numerical inputs the Euclidian distance is calculated to segment the customers as High, Medium and Low.

3.3 Association Rule Mining

Association rule mining is applied to the independently on the transaction data to identify the association between different product purchases. Customers belonging to the low value cluster would be induced to buy more by bundling relevant products, which would lead to increased purchased volume as well as value and thereby higher profitability. Customer's belonging to high value clusters would be offered higher incentives such as cash back or higher discount to build loyalty. The quantum of discount would be lowered for customers in medium clusters. This explains the need of association rule mining integration with the clustering to (Fig. 1).

The figure shows the customers are clustered as high, low, medium valued customers based on their transaction behaviors taken from the customer data. The same data is subjected to the association rule mining to find out patterns of the customer's basket items. After which the results from the clustering and the association are compared to associate the customers with the offers.

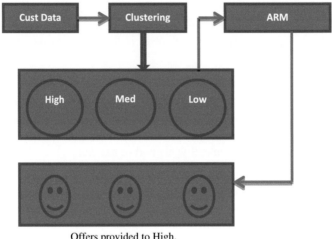

Offers provided to High,
Med,Low clusters based on
their market basket analysis

Fig. 1 Clustering customers and identifying offers

4 Results and Discussion

Extracted around 40,000 unique customers from "Kaggle". This data set consists of almost 350 million rows of transactional data from over 300,000 shoppers.

With help of Rattle (R tool) deployed K means clustering basis RFM methodology. Mode of the transaction value is used in place of the monetary.

K means clustering with RFM (Mean) initially applied to the 40,000 unique customer set and the results are derived. For the same data set K means with RFM (Mode) is applied and the results obtained is compared with the previous result in terms of the following factors:

Table 1 shows that the cluster size varies for RFM (Mean) and RFM (Mode). Cluster one size for RFM (Mode) is 16,097 which is higher than cluster one size of 14,572 whereas cluster two size of RFM (Mode) is lower than that of RFM (Mean).

Cluster one are the golden customers who are provided with the best offers. Cluster 2 and 3 are the medium and low valued customers in which the medium valued customers will get little offers and the low valued customer will get more association based offers based on their basket items.

Time complexity of the K-Means experiment with the mean and mode attributes of the customer purchase remains the same and is equivalent to the existing Kmeans approach. Time complexity of the Kmeans algorithm is calculated based on the total number of operations for an iteration.

Total number of operations for an iteration = distance calculation operations + comparison operations + centroid calculation operations [6].

If the algorithm converges after I iterations,

Table 1 Cluster analysis of RFM (mean) and RFM (mode)

Parameters	RFM (mean)	RFM (mode)
Cluster size	(1) 14,572	(1) 16,097
	(2) 16,101	(2) 14,558
	(3) 9777	(3) 9771
Time taken to cluster (s)	0.23	0.09
Data means	0.20746232428(R)	0.20750022834(R)
	0.00048787448(F)	0.00048767659(F)
	0.00002480102(M)	0.00002472188(M)
Within cluster sum of squares	17.888754	19.58820
	19.590443	16.87176
	9.291215	10.28475
Time taken (s)	0.36	0.16

$$\text{Total number of operations} = [6\text{I}kmn + \text{I}(k-1)mn + \text{I}kmn] \text{ operations} \quad (1)$$

$$\approx O(\text{I} * k * m * n) \quad (2)$$

where n—number of data points, m—number of attributes, k-number of clusters, I—number of iterations.

Number of attributes for the mean and the mode experiment would be same i.e. 3. (R, F, M-mean or mode) and number of clusters $k = 3$. Assuming the minimum number of iterations as 2, the complexity is expressed as $O(18n)$ approximately qual to $O(n)$. Time complexity remains the same while the cluster formation time differs due to the difference in the cluster size which is depicted from the above table.

This methodology enables retailers to mass customize their offerings and optimize their resources to segment, target and build loyalty among their golden customers.

Clustering using RFM (Mode) takes fewer seconds to complete than clustering using RFM (Mean) under similar test conditions.

Figure 2 depicts the visualization of the Bill value of the customer along X axis and the Transaction ID of the customer along Y axis. The difference in shades of colors indicates the clusters to which the customer belongs to.

Figure 3 depicts the spread of the customer's in each cluster based out of the three attributes recency, frequency and monetary.

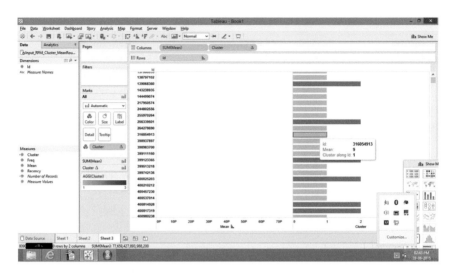

Fig. 2 Visualization of the customer purchase pattern in tableau

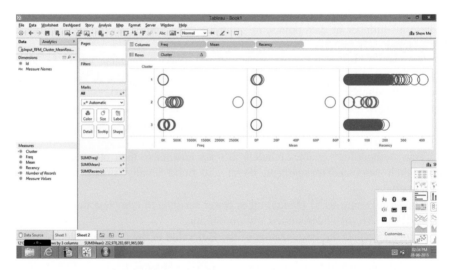

Fig. 3 Visualization of the clusters in tableau

5 Conclusion

Thus from the tool based experimental analysis of K means clustering with RFM (Mode) with the historic data is effective in segmenting the customers basis mode value.

Mean values of transaction though closer to the actual transaction size might not be the right indicator of representing the customer's transaction size. As it would be affected by one off events and customers might be graded at higher levels than actual.

Employing Mode instead of Mean overcomes this drawback and provides retailers accurately the customer's transaction size.

Substituting the Mode values for monetary values in RFM model makes the model effective by removing the effect of outliers and also in identifying the loyalty of the customers through their historic data.

6 Future Enhancements

As the proposed work is on the historic data, the same analysis can be extended in future to the streaming data such as stock markets where timely decisions are crucial.

Products tend to be purchased or browsed together can be analyzed in a distributed system through predictive models. By examining these associations, retailers can determine which joint offers, of pairs or sets of products, are most likely to generate additional sales. This approach can provide certain in-store deployment capabilities, such as posting discounts when products are purchased together.

Similar approach can be extended to the banking domain to segregate the customers to provide offers based on their purchase pattern.

References

1. Wanghualin, Jiangxi: Data mining and its applications in CRM. Second International Conference on Computer Research and Development. IEEE 2010
2. Han, J., Kamber, M., Pei, J.: Data Mining: Concepts and Techniques, 3rd edn. (2011)
3. Chen, Y., Fu, C., Zhu, H.: A data mining approach to customer segment based on customer value. In: Fifth International Conference on Fuzzy Systems and Knowledge Discovery, vol. 4. IEEE (2008)
4. Morteza, N.: A two phase clustering method for intelligent customer segmentation. International Conference on Intelligent Systems, Modelling and Simulation (ISMS). IEEE (2010)
5. Liu, J., Du, H.: Study on airline customer value evaluation based on RFM model. International Conference on Computer Design and Applications (ICCDA), vol. 4. IEEE (2010)
6. http://codeplustech.blogspot.in/2013/05/time-space-complexity-of-basic-k-means.html
7. Liu, F., Wu, K.: Application of data mining in customer relationship management. International Conference on Management and Service Science (MASS). IEEE (2010)
8. Weili, K., Yang, C., Li, H.: The application of improved association rules data mining algorithm apriori in CRM. 2nd International Conference on Pervasive Computing and Applications (ICPCA 2007). IEEE (2007)

Comparison of Search Techniques in Social Graph Neo4j

Anita Brigit Mathew

Abstract Past few decades, relational databases was the dominant technology used in web and business applications where well structured data was widely used. In the era of BigData and social networking, data is in disorganized form. This unstructured data gives importance for relationships between entities and impersonate many-to-many relationships in graph database. This paper brings out the importance of graph NoSQL database Neo4j in social networks and further inquest how multilevel, multikeyword search in social graph Neo4j outperforms the search connected to relational databases. We summarize the current state of technologies existing in multilevel multikeyword search area, explore open issues as well as identify future directions for research in this important field of Big Data and social graphs in Neo4j. On the basis of comparative analysis, we found graph databases that the former retrieve the results at faster pace. Many multilevel or multikeyword search methods on Neo4j was analyzed based on four research questions on five dimensions, but none of them put forward a benchmark model in the integration of multilevel multikeyword search evaluation.

Keywords Relational database · Graph NoSQL database · Neo4j · Multilevel multikeyord search in social graph (MMSSG)

1 Introduction

In social graph entity relationships can be depicted by graphs. Entity are nodes in a social graph where the information of each individual, event, etc. are a stored. In order to retrieve the information at each node the keyword search can be formed. Current search techniques are able to obtain and rank relevant items in social graph provided they are stored in structured form. When the data stored is in structured form relational database responds to a query in a reasonable time. This is because

A.B. Mathew (✉)
NIT Calicut, Kozhikode, India
e-mail: anita_brigit@rediffmail.com

© Springer International Publishing Switzerland 2016
V. Vijayakumar and V. Neelanarayanan (eds.), *Proceedings of the 3rd International Symposium on Big Data and Cloud Computing Challenges (ISBCC – 16')*, Smart Innovation, Systems and Technologies 49, DOI 10.1007/978-3-319-30348-2_24

current search engines takes only the relevancy between the input queries and items into account, where as when the data is in unstructured or semi structured form as in BigData. This resource-centric search engines in relational databases will fail to locate and rank information of real user interests appropriately and users underlying preferences are not considered. This problem can be eliminated to a great extend by the use of graph NoSQL database. Graph NoSQL databases are a powerful tool for storing, representing and querying in social graph compared with relational database. They have an efficient way to search personal information such as documents, biodata, emails, messages etc.

Let us consider a social graph of university as an example. Here the social graph deals with queries based on relationships. The different components of this social graph are student, faculty, laboratory, departments etc. The query model is shown in Fig. 1. Suppose there is a query for finding *"Clive Owen liked subjects"*.

This input query is split into two keywords K1 and K2 (Fig. 1 illustrates this), where K1 holds *Clive Owen* and K2 holds *liked subjects*. First we should find the node information matching K1(*Clive Owen*) from the social graph. Once match obtained, then we go for the relationship of K1 which is K2(*liked subjects*) from the resultant graph obtained from K1. Input Query: K2K1 ← liked subjects Clive Owen.

Relational Database Query to retrieve Input Query:

Step 1: User = getRequestString("K1");
Step 2: Result = "SELECT * FROM Node_data WHERE Users = " + User;
Step 3: SELECT Result FROM Relation_Data WHERE Result LIKE K2;
Step 4: SELECT Result FROM Property_Data WHERE Relation Type LIKE K2;

When the above query is posted in relational databases, a repeated procedure is required to obtain nodes matching K1 which means Step 2 is to be recursively

Fig. 1 Input query

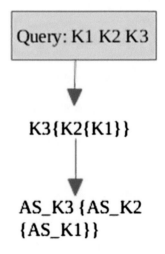

processed till the end of node list. If this list is frequently updated due to the increase in social graph data then the tables also need to be frequently upgraded, also Step 2 query is to be executed at multiple times manually, this is a difficult task. Hence after obtaining the result for K1 in relational database we should compute K2, this again is hard because there could exist many to many relations between nodes and all tables where relationships are stored should be traversed. From above SQL query Step 3 and 4 are repeatedly processed manually until all relation tables accessed.

The above example help us to list the problems faced when we go for relational database search from social graph.

1. The relations in a relational database are handled by joins between tables. To perform many joins between several tables would be a costly work and require a lot of lookahead, but if go for graph NoSQL database it is better due to the linked structure.
2. Relational database performance becomes poor when search is performed and multiple intermediate results are to be stored. Every node in Graph NoSQL database stores relations which can be preprocessed from different distributed environment that does not require to perform join operations manually. It just require traversal through nodes among the distributed environment.
3. Fixed schema is required in relational database but NoSQL graph databases are schema-flexible. Non-restricted NoSQL graph database model of social graph is shown in Fig. 2. When populating through the graph database the edges are made according to the defined relations between nodes as in Fig. 2. An edge is drawn from node Faisal to Anoop as the relation *Work on same topic*.
4. Multiple entities in social graph may read and write at the same time, so there is a need for concurrency of reading and writing with low latency. Relational databases need to perform concurrency control by ensuring ACID properties.

Hence we can state there is a need for graph NoSQL database convention in social graph for multilevel multikeyword search. Next we have to see which graph

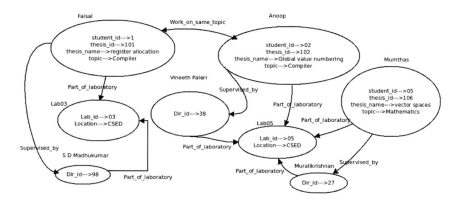

Fig. 2 NoSQL graph database schema of our social graph

NoSQL database for this, Mathew et al. [1] suggests Neo4j. Now we have taken graph NoSQL database Neo4j to store and manage a large social graph with millions of users. This is because sharding is avoided in Neo4j. This database guarantee to break graph so that all the relationships are ensured [2–4]. Relational databases are great when it comes to relatively static and predictable tabular data. But social networks are tend to be less static and predictable, and are ideal candidates for graph NoSQL database Neo4j. Hence there is a need for an interactive and iterative approach to execute Multilevel Multikeyword Search on Social Graph (MMSSG) connected with a graph NoSQL database Neo4j.

Presently there is no in-depth review available in the area of MMSSG with Neo4j. This paper is an effort towards it to provide a systematic in-depth scrutiny. The rest of this paper is organized as follows. Section 2 describes the related terminology and the social graph model. Section 3 introduces the dimensions and its related questions for the review. Section 4 explains about the extracted data related to scrutiny and Sect. 5 explains how search accomplished in Neo4j. Section 6 analyze the existing multilevel or multikeyword search on social graph Neo4j based on the related questions of the review. Section 7 critically analyze all identified survey related to MMSSG Neo4j. Finally, Sect. 8 concludes the paper.

2 Preliminary

This section describes the basic terminology, structure and model of a social graph. The basic terminologies used are:

1. Multikeyword Search—To research for n search terms entered into the search engines when conducting a search.
2. Multilevel Search—Search in nested form where initial search output will be input to next level, in n number of levels.
3. Multilevel Multikeyword Search—To search for multiple terms in a text search engine for multiple levels of a social network connected to a graph database.
4. Social Graph—This is a portrayal of relationships among people, organizations, groups and events in a social network. The idea refers to both the social network and a diagram representing the network called social graph.

The structure of a social graph restricts to an undirected, labeled connected graphs. In this article, we simply call a graph g, defined as a 3-tuple (V, E, L), where V is the set of nodes, E the set of edges, L the set of labels. This is displayed in Fig. 3.

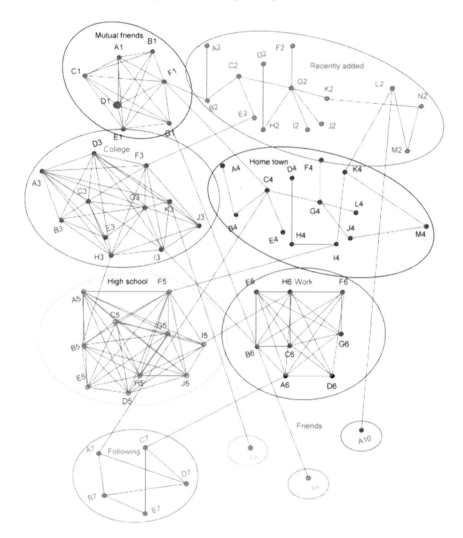

Fig. 3 Social graph model

3 Dimensions of the Survey

The main goal of study is to articulate researchers and practitioners with an organized overview of existing research in the area of search in social graphs. Various dimensions of different parameters chosen varies from Q1 to Q4 and its related questions RQ1 to RQ5 have been identified and are summarized below:

Q1: **What are the multilevel multikeyword search approaches used in social graph Neo4j?**

RQ1: *State the different techniques that are used for multikeyword search in social graph Neo4j?*

RQ2: *State the different techniques that are used for multilevel search in social graph Neo4j?*

RQ3: *Check whether any other search methods are incorporated to execute query in social graph Neo4j?*

Q3: **What could be the input pattern for query in social graph?**

RQ4: *Suggest the important patterns to be considered in a MMSSG Neo4j?*

Q4: **Identify what are the parameters chosen for query search evaluation process?**

RQ5: *What are the hiatus in current strategies used for multilevel multikeyword search?*

Inorder to answer these research questions, we have performed the following process:

Survey: Exhaustive review of the papers published in reputed journals, conferences and workshops to identify the different search techniques used for search in social graph Neo4j.

Categorization: The collected data was categorized based on the following parameters Author, Social Network, Graph Database Support and Search Technique.

Analysis: The categorized data were analyzed based on the techniques used for multilevel multikeyword search.

4 Data Extraction

This section presents the distribution of selected papers over the years and sites of publication. Figure 4 indicates the rise in the study of multilevel multikeyword search approaches over the years in relational databases followed by NoSQL

Fig. 4 Year wise distribution of selected papers that is being reviewed

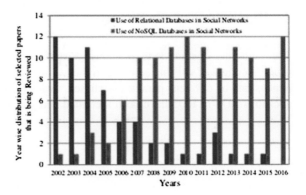

Table 1 Journals/conferences research papers published

Type	Acronym	Description
Journal	TDS	ACM transactions on database systems
	TIS	ACM transactions on information systems
	KDD	ACM transactions on knowledge discovery from data
	ATIS	ACM transactions on information services
	TOW	ACM transactions on web
	DMKD	Data mining and knowledge discovery
	DPD	Distributed parallel databases
	IR	Information retrieval
	ISM	Information systems management
	JDBM	Journal of database management
	JNSM	Journal of network systems management
Conferences	EMCIS	European and mediterranean conference on information systems
	FORTE	IFIP international conference on formal techniques for networked and distributed systems
	ICACCI	International conference on advances in computing, communications and informatics
	ICCIT	International conference on computer and information technology
	ICDCS	International conference on distributed computing systems
	ICDCN	International conference on distributed computing and networking
	ICIS	International conference on information systems
	ICPADS	International conference on parallel and distributed systems
	ICWE	International conference on web engineering
	ICWS	International conference on web services
	IPDPS	IEEE international parallel and distributed processing symposium
	KDD	ACM SIGKDD conference on knowledge discovery and data mining
	VLDB	International conference on very large data bases

databases. This survey incorporates papers published in reputed journals, conferences and workshops. Table 1 presents the lists of popular publications in journals and conferences in MMSSG connected to a graph database. Figure 5 shows the distribution of article publications across the popular publication sites in journals. The blue bars indicate NoSQL databses and red bars indicate relational databases. In Fig. 6 the distribution of article publications across the popular publication sites in conferences are shown. The blue bars indicate NoSQL databses and red bars indicate relational databases.

Fig. 5 Distribution of
selected papers in journals

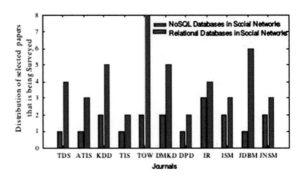

Fig. 6 Distribution of
selected papers in conferences

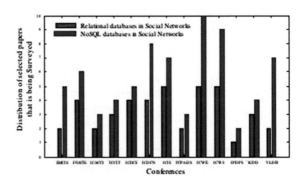

5   How Search Accomplished in Neo4j?

The multilevel multikeyword search is based on traversing the relations between
nodes and provide query information in Neo4j. Cypher query language is used in
Neo4j to traverse the search [5]. In Neo4j there is no schema for any entity, all the
entities are scattered as nodes and the relation among the nodes are the edge-join
relations in the graph. We have taken the graph of NITC students with relations
labeled on the edges. The Cypher queries executed for search are:

(Node A)—[:Relation]-¿ (Node B)
Creation of a node and relationship among the nodes using
Cypher using simple Java code to build.
GraphDatabaseService graphDb = …Get factory
//Create Student1
Node Student1 = graphDb.createNode();
Student1.setProperty = (name, student);
Student1.setProperty = (age, 19);
Student1.setProperty = (year, 1 st);
//Create a relationship representing that they know each other Student1.
createrelationshipTo(Student2, relType.Knows); Here we have nodes say A and B
with properties like A have properties name and age while B has properties like

work_name, rank and occupation, this example show the schema free design of query in graph NoSQL database. This query creates the relationship among the nodes with relation *KNOWS*. This explains that while creating nodes we can define the relationship among the nodes and we can add further more and more relation among the nodes similarly. Now if we want to know all the students1 friends (*KNOWS*) we have following query for this.

```
// Instantiate a query traverser that return Student1s friends
Traverser    friendsTraverser    =    Student1.traverse(
Traverser.order.BREADTH_FIRST,
stopEvaluator.END_OF_GRAPH,
ReturnableEvaluator.ALL_BUT_START_NODE,
RelTypes.KNOWS, Direction.OUTGOING);
// Traverse the node space and print out the result
System.out.prentln(Students    friend:);    for(Node
friend:friendTraverser)    System.out.println(At    depth
friendTraverser.currentPosition(    ).getDepth(    ),
friend.getProperty(name));    Similarly we can find the
relation that returnd the nodes which love Node A
// Create a traverser that returns all friend in love
Traverser    loveTraverser    =    Student1.traverse
(Traverser.order.BREADTH_FIRST,
StopEvaluator.END_OF_GRAPH,
new ReturnableEvaluator( )
public boolean isReturnableNode(TraversalPosition pos)
return    pos.currentNode(    ).hasRelation-
ship(Reltypes.LOVES,Direction.OUTGOING);
RelTypes.KNOWS, Direction.OUTGOING);
```

6   Auditing The Existing Techniques

Here we audit into MMSSG to find some of the related works carried out in the area of search in social graph. Sakr et al. [6] gives a exhaustive study of tremendous approaches and mechanisms of deploying data across distributed environments in NoSQL databases which are gaining a lot of momentum in both research and industrial communities nowadays. They advocate open and future challenges in viewing the need for multikeyword querying, consistency, economical processing of huge scale of data on the NoSQL database. In Yi et al. [7] paper they talk how to search for keywords in a graph that connect to various data sources. They present two case studies of real-world BigData applications of social networks where there is a need for quicker and significant search. Zou et al. [8] talks regarding how graph NoSQL databases come across interesting data management and search problems, such as subgraph search, shortest-geodesic query, attainable verification, and pattern match. They mainly look on pattern match search over huge data graph G. Given a input query how a model graph can be generated from a huge graph. They also suggest methods to reduce the search space significantly, by transforming the vertices into points in a vector space via graph embedding techniques. These techniques helps to convert a pattern match query into a distance-based multi-way join problem over the converted vector space. Several pruning strategies and a join order selection method to process join processing efficiently are also suggested.

Cudr-Mauroux et al. [9] focus on the administration of very huge volume graphs such as social networks. First focus on the main designs used to store the graph (dense/sparse native graphs, triple storage or relational layouts), the access patterns and typical queries monitored (reachability or neighborhood queries, updates various reads, transactional requirements and graph consistency models). Second they talk how to map the data and estimation models to concrete graph management systems, highlighting destination application domains, implementation techniques, scalability and workload requirements. Ugander et al. [10] speaks regarding social graph structure of Facebook users. They focus on features like clustering, pattern matching and finding path distance between friendship relation of users, and come to numerous observations. First they distinguish the global structure of social graph finding whether the social network is completely connected. Second they look how local clustering of graph neighbourhoods because of the sparse nature of Face-book graph. Finally, characterize the patterns in the graph by studying the network user properties. Kumar et al. [11] focus on the structure of large social network online. They present measurements to segment the networks into regions. Each of the region overwhelmingly represent star structure. Traditional RDBMS is used to organize the users of the network in order to perform query process. Khan et al. [12] a neighborhood-based similarity measure to avoid valuable graph isomorphism and edit distance computation is proposed here. Based on this they found an data management propagation model that is able to divert a huge network into modules of multidimensional vectors, where refined indexing and correlation search algorithms are available. The suggested method, called Ness (Neighborhood Based Similarity Search), is suitable for graphs with low automorphism and high noise, which are common in many social and information networks. Martinez-Bazan et al. [13] suggests a highlighting query engine based on an algebra of operations on sets. The new engine combines some regular relational database operations with some extensions oriented to accumulate processing and convoluted graph queries. They study the query plans of graph queries expressed in the new algebra, and find that most graph activities can be effectively expressed as semijoin programs. Morris et al. [14] explore the phenomenon of keyword querying in social graph. They present detailed data on the frequency of different type of keyword querying, and respondents to motivations for asking their social networks rather than using more traditional relational search tools like Web search engines. The above mentioned previous works leads to the knowledge of various techniques used for MMSSG, Table 2 symbolizes this.

7   Critical Analysis

The critical analysis came across during the search based on previous works are:

1. Problem of quadratic growth of BigData. Example—If everybody starts following everybody else on Twitter, the overall load will be too high. We can not

Table 2 Search techniques in social network connected to graph database

Author	Social network	Graph database support	Search technique
Cohen et al. [15]	Twitter	Hypergraph model in HyperGraphDB	B Tree
Zhang et al. [16]	eToro	AllegroGraph, Neo4j	Modified Red Black Tree
Cheng et al. [17, 18]	Facebook	Oracle NoSQL Database, Neo4j	Sequential
Rajbhandari et al. [19]	Facebook, Twitter	Neo4j, Relational	BFS, DFS
Barcel et al. [20]	MyLife, MyHeritage	Neo4j, ArangoDB-Apache	BFS
Nishtala et al. [21, 18]	Facebook	MemCache, Neo4j	Sequential
Fan et al. [22]	Twitter, LinkedIn	Infinite Graph, InfoGrid	Dynamic Hash, Tree Based
Khan et al. [23, 24]	Academia.edu	2-connected graph	Sequential
Mondal et al. [25]	Foursquare, Friendfeed	Neo4j, AllegroGraph	BFS, Heuristics
Gomathi et al. [26]	Small Scale Social Network	Relational Database	Cuckoo
Zhao et al. [27]	Small Scale Social Network	Neo4j, Relational	Tabu

perform analytics on data in Graph database like finding sum, average etc. At least not with a built-in command.

2. If there are millions of nodes in social graph it is difficult to place information of all nodes in a single system.
3. Require efficient methodologies to perform distributed search.
4. Need for query categorization model in order to classify the input query to be executed at each level.
5. Various search techniques available but none supports multilevel multikeyword query search.

8   Conclusion

This paper has reviewed and critically analyzed the different facets and brings on certain interesting insights for Multilevel Multikeyword Search in Social Graph (MMSSG) connected to graph NoSQL database Neo4j compared to relational databases. This review reveals that even after substantial amount of research in the area of MMSSG Neo4j, there is still a deficiency of a benchmark model for evaluation. A defined benchmark model for the integration of multilevel multi-keyword search in social graph will navigate the social graph to understand both users and their relationships through query process. These queries could carry relational databases to their knees. Unlike other NoSQL databases, graph databases are fast at traversing correlations, creating them as a perfect option for our social network.

References

1. Mathew, A.B., Madhu Kumar, S.: Analysis of data management and query handling in social networks using NoSQL databases. In: 2015 International Conference on Advances in Computing, Communications and Informatics (ICACCI), pp. 800–806. IEEE (2015)
2. Holzschuher, F., Peinl, R.: Performance of graph query languages: comparison of cypher, gremlin and native access in Neo4j. In: Proceedings of the Joint EDBT/ICDT 2013 Workshops, pp. 195–204. ACM (2013)
3. Miller, J.J.: Graph database applications and concepts with Neo4j. In: Proceedings of the Southern Association for Information Systems Conference, Atlanta, GA, USA 23–24 Mar 2013
4. Webber, J.: A programmatic introduction to Neo4j. In: Proceedings of the 3rd Annual Conference on Systems, Programming, and Applications: Software for Humanity, pp. 217–218. ACM (2012)
5. Mathew, A.B., Kumar, S.M.: An efficient index based query handling model for Neo4j. IJCST 3(2), 12–18 (2014)
6. Sakr, S., Liu, A., Batista, D.M., Alomari, M.: A survey of large scale data management approaches in cloud environments. Commun. Surv. Tutorials, IEEE 13(3), 311–336 (2011)
7. Yi, X., Liu, F., Liu, J., Jin, H.: Building a network highway for big data: architecture and challenges. Network, IEEE 28(4), 5–13 (2014)
8. Zou, L., Chen, L., Özsu, M.T.: Distance-join: pattern match query in a large graph database. Proc. VLDB Endowment 2(1), 886–897 (2009)
9. Cudré-Mauroux, P., Elnikety, S.: Graph data management systems for new application domains. Proc. VLDB Endowment 4(12) (2011)
10. Ugander, J., Karrer, B., Backstrom, L., Marlow, C.: The anatomy of the facebook social graph. arXiv preprint arXiv:1111.4503 2011
11. Kumar, R., Novak, J., Tomkins, A.: Structure and evolution of online social networks. In: Link Mining: Models, Algorithms, and Applications, pp. 337–357. Springer (2010)
12. Khan, A., Li, N., Yan, X., Guan, Z., Chakraborty, S., Tao, S.: Neigh-borhood based fast graph search in large networks. In: Proceedings of the 2011 ACM SIGMOD International Conference on Management of data, pp. 901–912. ACM (2011)
13. Martinez-Bazan, N., Dominguez-Sal, D.: Using semijoin programs to solve traversal queries in graph databases. In: Proceedings of Workshop on GRAph Data management Experiences and Systems pp. 1–6. ACM (2014)
14. Morris, M.R., Teevan, J., Panovich, K.: What do people ask their social networks, and why?: a survey study of status message q&a behavior. In: Proceedings of the SIGCHI Conference on Human Factors in Computing Systems, pp. 1739–1748. ACM (2010)
15. Cohen, S., Ebel, L., Kimelfeld, B.: A social network database that learns how to answer queries. In: CIDR. Citeseer (2013)
16. Zhang, S., Gao, X., Wu, W., Li, J., Gao, H.: Efficient algorithms for supergraph query processing on graph databases. J. Comb. Optim. 21(2), 159–191 (2011)

17. Cheng, J., Ke, Y., Ng, W.: Efficient query processing on graph databases. ACM Trans. Database Syst. (TODS) **34**(1), 2 (2009)
18. Mathew, A.B., Kumar, S.M.: Novel research framework on sn's NoSQL databases for efficient query processing. Int. J. Reasoning-based Intell. Syst. **7**(3–4), 330–338 (2015)
19. Rajbhandari, P., Shah, R.C., Agarwal, S.: Graph database model for querying, searching and updating. In: International Conference on Software and Computer Applications (ICSCA) (2012)
20. Barceló Baeza P.: Querying graph databases. In: Proceedings of the 32nd Symposium on Principles of Database Systems pp. 175–188. ACM (2013)
21. Nishtala, R., Fugal, H., Grimm, S., Kwiatkowski, M., Lee, H., Li, H.C., McElroy, R., Paleczny, M., Peek, D., Saab, P., et al.: Scaling memcache at facebook. In: nsdi, vol. 13, pp. 385–398 (2013)
22. Fan, W.: Graph pattern matching revised for social network analysis. In: Proceedings of the 15th International Conference on Database Theory, pp. 8–21. ACM (2012)
23. Khan, A., Wu, Y., Aggarwal, C.C., Yan, X.: Nema: fast graph search with label similarity. In: Proceedings of the VLDB Endowment, vol. 6, no. 3, pp. 181–192. VLDB Endowment (2013)
24. Mathew, A.B., Pattnaik, P., Madhu Kumar, S.: Efficient information retrieval using lucene, lindex and hindex in hadoop. In: 2014 IEEE/ACS 11th International Conference on Computer Systems and Applications (AICCSA), pp. 333–340. IEEE (2014)
25. Mondal, J., Deshpande, A.: Stream querying and reasoning on social data. In: Encyclopedia of Social Network Analysis and Mining, pp. 2063–2075. Springer (2014)
26. Gomathi, R., Sharmila, D.: A novel adaptive cuckoo search for optimal query plan generation. Sci. World J. **2014** (2014)
27. Zhao, Y., Levina, E., Zhu, J.: Community extraction for social networks. Proc. Natl. Acad. Sci. **108**(18), 7321–7326 (2011)

A Novel Approach for High Utility Closed Itemset Mining with Transaction Splitting

J. Wisely Joe and S.P. Syed Ibrahim

Abstract Data mining techniques automate the process of finding decisional information in large databases. Recently, there has been a growing interest in designing high utility itemset mining algorithms. Utility pattern growth algorithm is one of the most fundamental utility itemset mining algorithms without candidate set generation. In this paper, a different possibility of designing a lossless high utility itemset mining algorithm is discussed. This algorithm can achieve high utility closed itemsets (HUCI's) even when any number of long transactions present in the database. HUCI's are generated by applying both UP Growth and Apriori concepts with closed itemset mining algorithm on the large database. Too many long transactions in database may affect the efficiency of the algorithm. To resolve this transaction splitting is used. This algorithm adopts transaction weighted downward closure property which guarantees only promising items are high utility items. The proposed algorithm will generate high utility closed itemsets in an efficient way.

Keywords High utility itemset · Closed itemset · Utility mining · Lossless information

1 Introduction

Association analysis [1–3] handles a group of transactions to find association rules that gives the probability of occurrence of an item based on the occurrences of all other items in the transaction. Relationship between the items in datasets are identified and expressed in the form of association rules and frequent itemsets. They provide information in the form of if–then statements. These rules are much useful in business decision making in all the fields like market analysis, healthcare, credit

J. Wisely Joe · S.P. Syed Ibrahim (✉)
SCSE, VIT University, Chennai Campus, Chennai, India
e-mail: syedibrahim.sp@vit.ac.in

J. Wisely Joe
e-mail: wiselyjoe.j2013@vit.ac.in

© Springer International Publishing Switzerland 2016
V. Vijayakumar and V. Neelanarayanan (eds.), *Proceedings of the 3rd International Symposium on Big Data and Cloud Computing Challenges (ISBCC – 16')*,
Smart Innovation, Systems and Technologies 49, DOI 10.1007/978-3-319-30348-2_25

card, census data and fraud detection analysis. The best algorithm for finding such rules and frequent itemsets is Apriori algorithm. It uses two measures called support and confidence to filter interesting rules. Apriori algorithm has some disadvantages too. At times it produces non interesting rules which are not giving fruitful business decisions, very huge number of rules and large number of candidate itemsets.

Frequent pattern growth algorithm [4, 5] plays an important role in frequent itemset mining without candidate itemset generation. This algorithm uses a compact, compressed tree structure for storing itemsets. The database is compressed into a condensed data structure in less number of database scans. This algorithm adopts pattern fragment growth method which avoids candidate set generation and divide and conquer partitioning leads to less usage of search space. This algorithm works well with long and short transactions and frequent patterns. The tree structure used to store compressed frequent pattern information is FP-tree. The same FP-tree is used in many frequent pattern methods like closed and maximal frequent pattern algorithms. Each node in the FP-tree has three fields which includes node name, item count and node link to the next node with same item.

Frequent Itemset Mining algorithms may discard the weight/profit information of each item and number of units of item purchased for every transaction during the process of finding frequent itemsets. Thus they generate many frequent itemsets with low profit and fail to generate less frequent itemsets that have more profit. Utility of item in a transaction is the units of item purchased in that transaction. The main motto of high utility itemset mining is to extract the itemsets with utility value higher than the specified utility threshold, by considering profit, weight, importance, quantity and cost of item or itemsets. High utility itemset mining algorithms proposed in [6–12] resolve these issues in big transactional databases which are the collection of data of day to day operations and process them. They are subjected to an important serious limitation, performance of the algorithm. Extraction of high utility itemsets has a crucial role in many real time applications and is an important research issue in association mining. High Utility itemset mining is much useful in identifying rare but profitable itemsets. In many realtime applications rare itemset mining is mostly used in profitable decision making. If Apriori algorithm is used to compute high utility itemsets, very large number of candidate itemsets are generated in which most of them are irrelevant [1, 2]. Such a large number of non profitable candidate itemsets may degrade the performance of mining process. To avoid this, HUI-Miner(High Utility Itemset Miner) algorithm [13] uses a list called utility list to store the utility information which mines the HUI's in single phase [13]. In this method costly candidate generation is very much reduced whereas the traditional approach ends in irrelevant utility itemsets. In another approach, a kind of FP-tree called UP Tree is used in high utility itemset mining with extra information stored in it. The UP tree construction is same as FP Tree construction except a new measure node utility stored in every node. This method is very much useful when the transaction is too long. This algorithm is called UP-Growth [11], that prunes candidate sets and mines high utility itemsets. It scans the database only twice. This will reduce the memory access in turn reduces the execution time even when database has many number of long transactions.

2 Background

2.1 Problem Definition

The challenges faced by high utility itemset mining algorithms are:

1. High Utility Itemset mining may produce too many HUIs and not all are fruitful.
2. Many profitable HUIs may not be generated by existing algorithms
3. Algorithms used for discovering HUIs may not be efficient in terms of execution time and space.
4. The performance of existing algorithms is not up to the mark when database contains lot of long transactions

In this paper, these challenges are addressed by giving a compact representation of HUIs named high utility closed itemsets (HUCIs), which has both the concepts of closed itemset and high utility itemset mining.

1. A new algorithm called AprioriH-Closed which is an apriori based high utility closed itemset mining algorithm.
2. The new algorithms proposed may produce less number of HUCIs even when the database contains many long transactions.
3. To avoid processing of long transactions present in database, the transactions are splitted into a specified length. The threshold length can be given by the user or it can be automatically calculated by our algorithm based on other transactions present in the database.

2.2 Related Work

In the section, the preliminaries related with high utility itemset mining, transaction truncation and closed itemset mining are discussed in detail and the existing solutions are given.

2.2.1 High Utility Itemset Mining

This section gives an overview of high utility itemset mining.

Definition 1 Let I be a set of items. A database DB is a set of transactions, DB = $\{T_1, T_2, T_3 \ldots T_n\}$ such that for each transaction T_d, $T_d \in I$ and T_d has a unique identifier d called its TransId. Each item $i \in I$ is associated with a positive number EU(i), called its external utility (e.g. unit profit). For each transaction T_d such that $i \in T_d$, a positive number IU(i, T_d) is called the internal utility of i in transaction T_d.

Fig. 1 Database and utility
table

TransId	Transactions	Quantity
T1	(a, c, d, g)	(2, 3, 1, 2)
T2	(b, c, f)	(1, 2,2)
T3	(c, d, e, f)	(2, 1, 2, 3)
T4	(a, b, c, g)	(4, 1, 2, 3)
T5	(d, g)	(2, 3)
T6	(c, e, d)	(3, 4, 1)
T7	(a, b, c, d, e)	(2, 1, 3, 1, 2)

Itemset	a	b	c	d	e	f	g
Utility	2	1	1	4	3	2	5

Every item in database has an utility value known as internal utility which is stored in a table called internal utility table. External utility EU is the purchased quantity of the item which is variable for every transaction.

Definition 2 The utility of item i in transaction T_d, $U(i, T_d)$ is the product of $IU(i, T_d)$ and $EU(i)$, $U(i, T_d) = IU(i, T_d) * EU(i)$.

For example in Fig. 1, $IU(a, T_1) = 2$, $IU(e, T_6) = 4$, $EU(a) = 2$, $EU(g) = 5$, $U(a, T_1) = 2 \times 2 = 4$, $U(d, T_3) = 1 \times 4 = 4$.

Definition 3 The utility of itemset IS in transaction T_d, $U(IS, T_d)$ is the sum of utilities of items present in IS in T_d.

Definition 4 The utility of itemset IS in database DB, $U(IS)$ is the sum of utilities of IS in all transactions which contain IS in the given DB.

For example in Fig. 1, $U(\{a, b\}, T_7) = U(a, T_7) + U(b, T_7) = 2 \times 2 + 1 \times 1 = 5$.
$U(\{a, b\}, T_4) = U(a, T_4) + U(b, T_4) = 4 \times 2 + 1 \times 1 = 9$.
$U(a,b) = U(\{a, b\}, T_4) + U(\{a, b\}, T_7) = 5 + 9 = 14$.

Definition 5 The utility of transaction T_d, $TU(T_d)$ is the sum of utilities of all the items present in T_d.

Definition 6 Total utility of database DB, $TU(DB)$ is the sum of transaction utilities of all the transactions.

For example in Fig. 1, $TU(T_2) = 1 \times 1 + 2 \times 1 + 2 \times 2 = 7$, $TU(T_5) = 2 \times 4 + 3 \times 5 = 23$.

$TU(DB) = 132$

The TU's are listed in Fig. 2.

Definition 7 Transaction weighted Utility of itemset IS in database DB, twu(IS) is the sum of utilities of all transactions contain IS.

Fig. 2 Transaction utility

T1	T2	T3	T4	T5	T6	T7
21	7	18	26	23	19	18

Fig. 3 Transaction weighted
utility

Itemset	a	b	c	d	e	F	G
TWU	65	51	109	99	55	25	70

For example in Fig. 1, TWU (a) = $TU(T_1)$ + $TU(T_4)$ + $TU(T_7)$ = 65.
The TWU's of all 1-itemsets are listed in Fig. 3.

After calculating TWU for all 1-itemsets, all the items in the transactions are reordered from highest TWU to lowest. An user defined minimum utility threshold is set. If TWU(IS) is greater than or equal to that minimum utility threshold, IS is an high transaction weighted utilized itemset. For mining high utility itemsets, UP-Growth algorithm [11] is given by Vincent S. Tseng et al. In this algorithm a new data structure is used to store itemsets in an compact manner. During the construction of initial UP-Tree itself, the unpromising items are removed. Promising item is the one which satisfies minimum utility threshold property. Else the item is an unpromising one. Unpromising items are removed from modified transactions and also removed from TU calculation. Modified TU is called reorganized transaction utility RTU. UP-Growth algorithm has a UP-tree and a header table with all promising items. Every element in table has a pointer to the same item in UP-Tree and the item's TWU.

Global UP-Tree construction with only promising items is same as global FP-Tree construction. Each node in UP-Tree has two values associated with it. The first one is support count of item and the other is the node utility. Follow the same FP-Growth procedure [4] to find local promising and unpromising items in local UP-Trees. The minimum item utilities of unpromising items are discarded from path utilities of the paths during the construction of local UP-Tree. Finally the listed high Utility Itemsets are compared and Potentially high utility itemsets are generated.

2.2.2 Closed Itemset Mining

In this section, definitions and procedures related to closed frequent itemset mining are given. Mining frequent itemsets without data loss is an important problem in many of the real databases that have too many long transactions. Closed Itemset Mining [14–18] is the best way of mining closed frequent itemsets without loss of information. A frequent set is closed if it has no superset with the same frequency. As closed itemset mining algorithm finds the set of all frequent itemsets with exact support, the closed sets are lossless. The formal definition of closed frequent itemset mining algorithm is given below:

Let I = {i_1, i_2,..., i_n} be a set of n distinct items. Let DB denotes a database of transactions. Each transaction in database has a unique identifier TransID and contains set of items. T = {T_1,T_2, …, T_m} is a set of m transactions.

A set of one or more items is called an itemset IS. Support count of an itemset IS, SC(IS) is the number of transactions which has IS as part of it. An itemset is frequent if it's calculated support is more than or equal to minimum support threshold set by the user. The frequent itemset is an maximal set if it is not a subset of any frequent itemset [19]. A set is closed if it has no superset with the same support count of frequency as that of it.

2.2.3   Transaction Truncation

Truncating the transactions has been proposed to address long transaction problem in privacy preserving data mining [20]. If a transaction has items more than certain limit, the extra items are removed from the transaction. If only less number of transactions are long then they have only little impact on the efficiency of the algorithm. In such cases the truncation threshold can be relaxed that the database can have those long transactions for processing without truncation. Frequent items having good support count in 1-itemsets are kept in the transaction itself. Only less frequent items are truncated and only when many transactions have too many items. Different truncation thresholds can be set to different databases during mining and they mostly depends on the number of frequent items. Because only the frequent items are used to generate candidate sets. But the real problem in truncation method is information loss though the truncated are infrequent.

3   Proposed Work

In this section, the proposed method is given that can mine lossless high utility itemsets. This proposed system also uses transaction splitting method to provide the solution for long transactions problem present in high utility itemset mining.

3.1   Lossless High Utility Itemset Mining

Here a novel approach is proposed for lossless high utility itemset mining [11] which joins closed itemset mining [15, 17] with high utility itemset mining. The join order between these two mining constraints does not affect the end results. In either case same list of high utility closed itemsets(HUCI's) are generated. Minimum utility threshold(Min_Util_Thresh) is set by the user and experiments show that the HUCI's are lossless. An Apriori like algorithm is introduced with few UP-Growth approaches [11] to find HUCI's. Like Apriori [1] this proposed algorithm initially scans the database to compute support count of each 1-item, transaction utility (TU) of every transaction, TWU of each 1-item (Fig. 4).

From the calculated TWU, unpromising items are removed from the transactions using Discarding global unpromising items (DGU) [11] strategy and Reduced TWUs (RTWU) are calculated for every transaction only with promising item. That is, the TWU of unpromising items are removed from RTWU calculation also.

Fig. 4 AprioriH-closed
algorithm

```
------------------------------------------------------------------
ALGORITHM: AprioriH-Closed
------------------------------------------------------------------
Input:  DB: The Database; Min_Util_Thresh;
        set of HUCI's
Output: CHUCI:Complete set of HUCI's
1.      CHUCI := Ø
2.      L₁ = 1-HTWUI's in DB
3.      D₁ = DGU(DB,L₁)
4.      L₁ = 1-HTWUI's in D₁
5.      Apriori_Proc(D₁, CHUCI , Min_Util_Thresh , L₁)
------------------------------------------------------------------
```

Fig. 5 Apriori_Proc
procedure

```
------------------------------------------------------------------
PROCEDURE: Apriori_Proc
------------------------------------------------------------------
Input:  Dₖ: The reduced  Database; Min_Util_Thresh;
        set of HUCI's
Output: CHUCI:Complete set of HUCI's
1.      Lₖ₊₁ := Lₖ;
2.      L₁ = 1-HTWUI's in DB;
3.      for(k=1; Lₖ₊₁ ≠ Ø ; k++)
4.      {
5.              Cₖ₊₁ = Apriori_Recur (Lₖ);
6.              Lₖ = FindClosedItemsets(Lₖ, Lₖ₊₁);
7.              CHUCI  = CHUCI ∪ Lₖ₋₁
8.      }
------------------------------------------------------------------
```

By definition, unpromising items and their supersets are not high utility itemsets or closed high utility itemsets. The promising items which has calculated TWU not lesser than Min_Util_Thresh, are included in the set of 1-HTWUI's C_k. In kth iteration, the set of k-HTWUI's L_k is used to produce (k + 1) candidates C_{k+1} using Apriori_Proc().

Now the algorithm removes the lossy items from (k + 1)-HTWUI's by following method. Ariori_Recur() is a method to produce candidate sets C_{k+1} from L_k during kth iteration.

For each candidate X in (k + 1)-HTWUI's, the algorithm checks if there exists any superset of candidate X and its support count is same that of support count of candidate X (Fig. 5). If found, subset is deleted from candidate set as it is not a high utility closed itemset as per our definition given. If such superset is not found, subset is closed and it may be a HUCI.

3.2 Transaction Splitting

To overcome the reduction in efficiency of high utility itemset mining algorithm when it handles too many long transactions, a new strategy is introduced. That splits the long transactions present in database DB instead of truncation. In truncation the items present in transaction after the maximum length threshold (Max_Length) will be eliminated. Thus truncation leads to information loss. In transaction splitting the database is transformed by dividing the long transactions into many sub transactions. Each sub transaction has to meet the Max_Length constraint and database does not loose any items. Thus transaction splitting avoids information loss. For example the database contains a transaction t = {1, 2, 3, 4, 5, 6}, Max_Length = 4 and {1, 2, 3}, {4, 5, 6} are frequent. In truncation the transaction t becomes t1 = (1, 2, 3, 4} and the items 5, 6 are deleted. Also the support of itemset {4, 5, 6} and its subsets will also decrease. Instead of truncation if we divide the transaction into t1 = {1, 2, 3}, t2 = {4, 5, 6}, support of itemsets and its subsets will not get affected much like previous method. If transaction is longer than the Max_Length, it is divided into multiples of Max_Length. Items present at the last sub transaction may have lesser length than Max_Length. They may loose their and their subset's frequency. To overcome this frequency degradation, transaction splitting can be done based on weight or priority. In some cases weight of sub transaction may be increased to compensate the data loss during splitting process. Frequency of items can also be used for splitting transaction. But it is difficult to determine which itemsets are frequent without mining the database as the number of database scans are restricted to two. To solve this, during the first scan itself we can calculate frequency of each item, TWU of each 1-items and order the items in the transactions based on the calculated TWU. Now repeatedly divide the transactions into half until it satisfies Max_Length.

4   Conclusion

This paper proposes a new algorithm AprioriH-Closed to mine high utility closed itemsets (HUCI's) without information loss. Though the algorithm has multiple database scans, this algorithm will take less execution time than the existing algorithms. Because this algorithm will produce less number of candidates through every iteration than others. The second strategy proposed is transaction splitting to solve the problem of 'too many long transactions' exists in existing HUI mining algorithms. This transaction splitting just divides the transaction into sub transactions without loss of items. This strategy can be easily embedded in AprioriH-Closed to produce lossless high utility itemsets even when the database has too many long transactions. AprioriH-Closed with transaction splitting will outperform existing high ultility itemset mining algorithms on real and synthetic datasets.

References

1. Agrawal, R., Imielinski, T., Swami, A.: Mining association rules between sets of items in large databases. In: Buneman, P., Jajodia, S. (eds.) Proceedings of the 1993 ACM SIGMOD International Conference on Management of Data, pp. 207–216, Washington, DC
2. Agrawal R., Srikant R.: Fast algorithms for mining association rules. In: Proceedings of the 20th International Conference on Very Large Data Bases (VLDB 94), pp. 487–499 (1994)
3. Klemettinen, M., Mannila, H., Ronkainen, P., Toivonen, H., Verkamo, A.I.: Finding interesting rules from large sets of discovered association rules. In: Proceedings of the 3rd International Conference on Information and Knowledge Management, pp. 401–408, Nov 1994
4. Han, J., Pei, J., Yin, Y.: Mining frequent patterns without candidate generation. In: Proceedings of SIGMOD'2000, Paper ID: 196
5. Frequent Itemset Mining Dataset Repository [Online]. Available http://fimi.ua.ac.be/data (2004)
6. Erwin, A., Gopalan, R.P., Achuthan, N.R.: Efficient mining of high utility itemsets from large datasets. In: Proceedings of PAKDD, LNAI 5012, pp. 554–561 (2008)
7. Lin, C.-W., Hong, T.-P., Lu, W.-H.: An effective tree structure for mining high utility itemsets. Expert Syst. Appl. 38(6), 7419–7424 (2011)
8. Le, B., Nguyen, H., Cao, T.A., Vo, B.: A novel algorithm for mining high utility itemsets, In: Proceedings of 1st Asian Conference on Intelligent Information Database System, pp. 13–17 (2009)
9. Liu, Y., Liao, W., Choudhary, A.: A fast high utility itemsets mining algorithm. In: Proceedings of Utility-Based Data Mining Workshop, pp. 90–99 (2005)
10. Li, H.-F., Huang, H.-Y., Chen, Y.-C., Liu, Y.-J., Lee, S.-Y.: Fast and memory efficient mining of high utility itemsets in data streams. In: Proceedings of IEEE International Conference on Data Mining, pp. 881–886 (2008)
11. Tseng, V.S., Wu, C.-W., Shie, B.-E., Yu, P.-S.: UP-Growth: an efficient algorithm for high utility itemset mining. In: Proceedings of KDD'10, 25–28 July 2010
12. Chan, R., Yang, Q., Shen, Y.: Mining high utility itemsets. In: Proceedings of IEEE International Conference on Data Mining, pp. 19–26 (2003)
13. Li, M., Qu, J.: Mining high utility itemsets without candidate generation. In: CIKM'12, Maui, HI, USA
14. Hamrouni, T.: Key roles of closed sets and minimal generators in concise representations of frequent patterns. Intell. Data Anal. 16(4), 581–631 (2012)
15. Lucchese, C., Orlando, S., Perego, R.: Fast and memory efficient mining of frequent closed itemsets. IEEE Trans. Knowl. Data Eng. 18(1), 21–36 (2006)
16. Yun, U.: Mining lossless closed frequent patterns with weight constraints. Knowl. Based Syst. 20, 86–97 (2007)
17. Zaki, M.J., Hsiao, C.J.: Efficient algorithms for mining closed itemsets and their lattice structure. IEEE Trans. Knowl. Data Eng. 17(4), 462–478 (2005)
18. Pasquier, N., Bastide, Y., Taouil, R., Lakhal, L.: Efficient mining of association rules using closed itemset lattice. J. Inf. Syst. 24(1), 25–46 (1999)
19. Gouda, K., Zaki, M.J.: Efficiently mining maximal frequent itemsets. In: Proceedings of IEEE International Conference Data Mining, pp. 163–170 (2001)
20. Su, S., Xu, S., Cheng, X., Li, Z., Yang, F.: Differentially private frequent itemset mining via transaction splitting. IEEE Trans. Knowl. Data Eng, 27(7), 1875–1891

Efficient Mining of Top k-Closed Itemset in Real Time

Riddhi Anjankumar Shah, M. Janaki Meena and S.P. Syed Ibrahim

Abstract Analytics of Streaming data has been interesting and one of the profound research areas in the Data Science. Analysis and examination of real time data is one of the major areas of challenge in the BigData Analytics. One of the areas of research being mining of top k-closed frequent Itemsets in real time. Therefore an efficient algorithm MCSET(Mining closed itemsets) is proposed which uses Hash mapping technique to mine efficiently the closed itemsets. Experimental results shows that proposed algorithm has improved the scalability and improved the time efficiency compared to the existing closed association rule mining algorithm of data streams.

Keywords Big data analytic · Streaming data analytics · Association rule mining · Frequent itemsets · Closed itemsets

1 Introduction

Big data "magnitude" has constantly changed from a dozen of terabytes to numerous petabytes of data. Big data is analysis of huge, complex volume of data generated both in structured and unstructured format. Extraction and examining of significant data is one of the challenging tasks today.

One of the approaches of knowledge extraction in big data is data mining which helps to transform the data in understandable format. Association rule mining is one of the efficient and well proposed research technique used in generation of frequent

R.A. Shah (✉) · M.J. Meena · S.P. Syed Ibrahim
School of Computing Sciences and Engineering, VIT University,
Chennai Campus, Chennai, India
e-mail: ridzyshah.29@gmail.com

M.J. Meena
e-mail: Janakimeena.m@vit.ac.in

S.P. Syed Ibrahim
e-mail: Syedibrahim.sp@vit.ac.in

© Springer International Publishing Switzerland 2016
V. Vijayakumar and V. Neelanarayanan (eds.), *Proceedings of the 3rd International Symposium on Big Data and Cloud Computing Challenges (ISBCC – 16'),*
Smart Innovation, Systems and Technologies 49, DOI 10.1007/978-3-319-30348-2_26

itemsets. Frequent itemsets are created based on two criteria (1) Support (2) Confidence. Support indicates how frequently items occur in dataset Support $(A \Rightarrow B) = P (A \cup B)$ whereas confidence indicates number of times statements have been found true. Confidence $(A \Rightarrow B) = P (B/A) = $ Support $(A \cup B)$/Support (A), where A, B are data items.

Association rule Mining of frequent itemsets has mainly two drawbacks (a) if the value of support is large then less or no frequent itemsets are generated in dataset. (b) if the value of the support is less than larger number of frequent itemsets are generated.

To get the more optimized frequent itemsets, association rule can generate frequent closed itemsets and there corresponding rules. Closed frequent itemsets are itemsets in a data set if there exists no superset that has the same support count as its original item set. Efficient Streaming of data is the major part to be researched in bigdata analysis. Real time analysis of data is collected from different applications such as sensor data, online transactions, e-commerce website, traffic transaction data etc. Data Streams are continuous, unbounded large complex data which are unstructured and distributed changing from time to time.

Mining of closed itemsets in data streaming is the new challenge of big data for generation of optimum frequent itemsets. Initially all the data items are examined at once. Secondly, frequent itemsets are classified depending on the support threshold. Lastly closed itemsets are generated of the frequent itemsets.

2   Literature Survey

Significance of Big data is one of the most researched area in statistical analysis in recent years. (Chang and Lee 2004; Chiet 2006; Feu et le 2000; Lee et al. 2006, Jawie Han). Bigdata was proposed mainly to maintain the huge magnitude of dirty and unclean data to extract knowledge from the same. With emerging of large data increasing every second bigdata is going to play a vital role in the industry.

Numerous extraction techniques are there for getting interesting patters in which association rule mining is the most prominent technique for extracting the frequent itemsets from the database. Apriori algorithm is the simplest approach that finds the frequent itemsets by using candidate key step by step. In apriori user defined support and confidence is given and we have to consider that item with the user defined support and confidence. Items are treated as a binary variable who's values are one or zero depends upon whether that item is present in database or not. In real time environment many items may buy simultaneously, and each item has its own percentage, volume and rate. Profit of the product is depend upon user interest. Profitable extraction of items was thus considered for overcoming the drawbacks of persistent items.

FP-Growth algorithm takes different procedure for extracting frequent itemsets without creating candidate itemsets. It builds a FP-tree data structure and mines the profitable itemsets from the FP-tree. It considers the data and finds the support for each item. Removes the item having less support and then sort the frequent itemsets

Fig. 1 Transaction database

TID	Itemset of transaction
T1	{1,2,3,4}
T2	{2,3,4,5}
T3	{1,2,3}
T4	{2,3}
T5	{1,4}

based on their support. It uses the Bottom up algorithm starting from leaves towards the root. It also use divide and conquer approach for finding profitable itemsets. In real time environment the original database is too huge to represent in form of FP-tree which takes lot of memory space to overcome that a new algorithm is being produced using hadoop framework for managing large amount of data (Fig. 1).

Hadoop is a Big Data framework which uses java language allowing distributed processing of huge data sets across different system using an easy language. Growing of e-commerce websites and online transactions of data is increasing in real time. Many research papers have been presented for streaming of data analysis. According to Streaming analysis of various papers the streaming of association rule mining can be divided into three categories First finding of support of candidate itemsets. Secondly, depending on the user specified threshold support the frequent itemsets are classified in every transaction of sliding window. Thirdly, closed itemsets are generated from the frequent itemsets.

Closed itemsets are more customized frequent itemsets which will give more faster, efficient and scalable performance of dataset. Streaming and distributing of data is one of the most efficient practices of websites. Streaming of data can be classified into two categories (1) Static Streaming: Here dataset is stable i.e. in non-changing state and if change the before data becomes stateless i.e. which has no use in the present dataset example: files, photos. (2) Dynamic Streaming: In this streaming data is nit constant it keeps changing time to time for example: video/audio feed. Dynamic data is data which changes with respect to time and here data can be read only once.

3 Existing System

"Interactive Mining of top k closed Itemsets from data Streams" mines the closed frequent itemsets in data streaming using sliding window technique. Window sliding is nothing but the block of limited number of transactions which moves from time to time depending on the time horizon specified by the user. If a transaction data stream is given then, a transaction sliding window sw with w transactions and one positive integer k, the task was to mine the top k closed frequent itemsets. Example 1: Let window size be 3 transactions. Consider a 5 transaction data set where it forms 3 consecutive sliding windows sw1, sw2, sw3.

Initially, the algorithm converts in the bit vector representation which is the most efficient method of converting the data and extracting the information of data streams.

Secondly, they have built a closed itemsets Lattice which is the proposed data structure where every node with the bit vector representation is created who's of each node. For each node the siblings of nx and ny node is generated and bit vector representation is given to the same after generating all he siblings support is computed of each node. If the support of nx! = ny then it is Nx is closed itemset else Ny is the closed Itemset and the closed itemset is stored in the htc(Hash tag closed itemset). After getting the first sliding window top k frequent itemsets the sliding window of 2nd is activated by moving left shift of the bit vector representation of each node and same procedure is followed (Fig. 2).

The insertion and deletion of the data itemsets is also explained for insertion of new item the right most bit(xi) is set to 1 otherwise 0. Insertion uses top down approach to generate candidate closed itemset. If the itemset to be inserted is closed then it is updated in the closed itemset tree and support key is updated in closed itemset lattice otherwise support is subtracted by 1. For deletion of a node or itemset in the transaction the nodes bit vector is left shifted and the closed itemset lattice is reconstructed. After rebuilding the structure the support of the affected closed itemset is subtracted by1. If support after updating is lesser than 1 then the candidate key or node is removed from the tree. Bottom up enumeration is done.

Oldest transaction is deleted from the closed itemset lattice when window slides from sw1 to sw2. The drawbacks of the existing system are (1) Bounded size (size confined to number of transactions) and (2) Less scalable. The efficiency of the existing system decreases with dynamic increase in the size of the dataset. To overcome the following drawbacks new algorithm is proposed using bigdata framework for both static and dynamic streaming which can take up to 1 lac transaction with more scalable results.

Fig. 2 Existing system

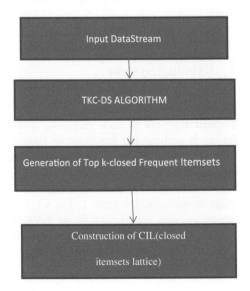

4 Proposed System

The main purpose of introducing the algorithm is to get more scalable and time efficient results for both static and dynamic streaming. Mining of top k closed itemsets is of great interest of researches from past few years to get more interesting pattern combinations of the itemsets. Streaming has emerged since an online era of websites has increased. To mine efficiently large amount of data a simple approach has been proposed using big data framework for static streaming. The algorithm mines the top k-closed frequent itemsets in streaming data.

Problem Definition: A non-empty set of items are taken I = {I1, I2…In} where each row represents different transactions T = {T1,T2…Tn} which are static streaming dataset. User specified support threshold is taken to find the frequent itemsets of the given data set. An itemset Y if it is contained in transaction (z) where the support of each item is denoted by sup(y) specifying number of transactions containing X in dataset. An association rule specifies: Y => Z where Y, Z => I and Y ∧ Z = null set. The support of the rule is denoted by sup(Y => Z) and represented as sup (Y ∪ Z). The result for mining of top k closed frequent itemsets is divided into three parts:

1. Find support for every item and combination of each such item in the database.
2. Depending on the support specified by the user the frequent itemsets are generated.
3. Finally after generation of frequent itemsets closed frequent itemsets are generated by cross checking the support of each level of the dataset support. If the support of the parent i.e. sup (Parent) = Sup (sibling) then sibling is the closed itemsets else parent is the closed itemset.

Example: Consider a Transaction file (Fig. 3). Having 6 transactions initially support count of each item is calculated and the result is stored in key value pair in an output file. Here, for example sup(1)-3, sup(2)-5, sup(3)-1, sup(1,2)-2, sup(1,3)-3, sup(2,3)-3 and sup(1,2,3)-2. After getting the support of each itemset the frequent itemsets are generated depending on the user specified threshold. If the user specified threshold is less than the sup (Ii) then i is the frequent itemset (Fig. 4).

Consider user defined threshold S = 2 then the frequent itemsets are {1, 2,(1, 2), (1, 3),(2, 3),(1, 2, 3)). After getting the frequent itemsets level wise supports are checked to get the closed itemset. The frequent itemsets generated are divided into 3 subsets based on the itemsets (1) containing only 1 element (2) ones containing 1

Fig. 3 Transaction table

Id	Itemset
I1	{1, 2, 3}
I2	{2,3,4,5}
I3	{2,4,5,6}
I4	{1,3,4,5}
I5	{1,2,3,4,5}
I6	{2,4,5,6}

Fig. 4 Proposed system

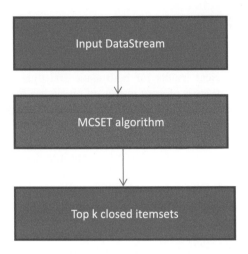

but not 3 (3) ones containing 1, 2 but 3 (4) one containing 1, 3 but not 2 and finally
(5) one containing 1, 2, 3 itemsets. The support of each itemset is calculated
levelwise. Here the closed itemset generated are {(1, 3) and {1, 2, 3}}. The algorithm proposed for getting closed item set is Closet implemented in bigdata
framework. Here the algorithm is experimented for the static streaming data with
transactions up to 50 k. The algorithm is implemented in both hoop single and
multimode framework. The resulting data of top k-closed itemsets increases the
efficiency by increase of scalability, reducing the search space dramatically.

5 Exploratory Results

This segments reports the analysis of the proposed algorithm in contrast to the TKC
(top k-closed itemset) algorithm which was proposed earlier. The algorithm which
is proposed i.e. MCSET (mining closed itemset) which uses hash mapping technique to mine efficiently the closed itemset and has also increased the scalability of
the dataset which accepts data size of more than 50 k and in minimal time. The
algorithm supports high accuracy of output even for vast amount of data.
Experimental results have been performed on various platforms with configurations
like 4 GB RAM, Intel I3, I5 processor, Windows 8 os and Linux(Fedora) environment. The transaction dataset varied from 10,000 to 50,000 itemsets. The
experimental results has shown the time efficiency with the use of BigData
framework with single and Multi node platforms. With increase in the node of
framework the time complexity is reduced. From the experimental result shown in
Fig. 5 Analysis of different hadoop framework nodes is been analysed. Greater the
transaction greater is the time complexity of the dataset in single and multi node.

Fig. 5 Analysis of time with different hadoop nodes

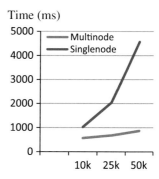

Time (ms)

Whereas the multinode takes lesser time compared to single node since the data here gets distributed in multiple environment the performance time scrutinizes completely.

6 Conclusion and Future Work

This paper proposes the top k closed frequent itemsets in real time streaming of data. An efficient mining algorithm is introduced called MCSET (mining closed itemset) which uses a technique called closet algorithm for generation of closed itemsets. Another major goal of the proposing algorithm is to increase the scalability of data in minimal time. The assessments of the proposed algorithm will prove itself to be one of the efficient mining technique of top k-closed itemset algorithm in streaming analysis of data. The future work will be related to work on dynamic streaming analysis with more concise and generalized algorithm proposal.

References

1. Agarwal, R., Srikant, R.: Quest synthetic DataGenerator. <http://www.Almaden.ibm.com/cs/quest/syndata.html> (1994)
2. Data mining-FP growth algorithm-stack overflow. <stackoverflow.com/questions/5448493/fp-growth-algorithm>
3. Fournier-Viger, P., Wu, C.W., Zida, S., Tseng, V.S.: FHM: Faster high-utility itemset mining using estimated utility co-occurrence pruning. Found. Intell. Syst. **8502**(2014), 83–92
4. Lin, C.-W., Hong, T.-P., Lan, G.-C., Wong, J.-W., Lin, W.-Y.: Efficient updating of discovered high utility item sets for transaction deletion in dynamic databases. Adv. Eng. Inf. **29**(2015), 16–27. <www.elsevier.com/locate/aei>
5. Lin, C.W., Hong, T.P., Lu, W.H.: The pre-FUFP algorithm for incremental mining. Expert Sys. Appl. **34**, 2424–2435 (2008)

6. Lin, J.C.W., Gan, W., Hong, T.-P.: A fast updated algorithim to maintain the discovered high-utility itemsets for transaction modification. doi:10.1016/j.aei.2015.05.003. <http://www.elsevier.com/locate/aei> (2015)
7. Need of temporal database. <stackoverflow.com/questions/.../why-do-we-need-a-temporal-database>
8. Song, W., Liu, Y., Li, J.: Mining high utility itemsets utility by dynamically pruning the tree. Struct. Appl. Intell. 1–15 (2013)

Mining Frequent Itemsets in Real Time

Nilanjana Dev Nath, M. Janaki Meena and S.P. Syed Ibrahim

Abstract Data streams are continuous, unbounded usually come in high speed and have a data distribution which often changes with time. They can be analogous to water stream which keeps on flowing on one direction. Data stream is used in many application domains—one of it being the analysis of data in real time also known as streaming data analytics. They vary to a large extent in case of storing in tradition database which needs the database to be a static one. The store and analyze strategy is not applicable in case of streaming data analytics. In order to analyze, it is required to find a certain amount of associativity among the data stream. Such associativity among the data items leads to the generation of association rules. These association rules are an important class of methods of finding regularities/ patterns in such data. Using association rule mining all the interesting correlation amongst the data can be used to derive. These relationships in the data items, can go on to a large extent in helping larger transactions records in case of making a decision or drawing a conclusion out of it. This paper makes use of Apriori algorithm in data streams which can discard the non-frequent set of data-items and can finally obtain the frequent itemsets from them.

Keywords Big data analytics · Real time analytics · Streaming data · Streaming association rule mining

N.D. Nath (✉) · M.J. Meena · S.P. Syed Ibrahim
School of Computing Sciences and Engineering, VIT University,
Chennai Campus, Chennai, India
e-mail: nilanjanadev.nath2015@vit.ac.in

M.J. Meena
e-mail: Janakimeena.m@vit.ac.in

S.P. Syed Ibrahim
e-mail: Syedibrahim.sp@vit.ac.in

© Springer International Publishing Switzerland 2016   325
V. Vijayakumar and V. Neelanarayanan (eds.), *Proceedings of the 3rd International Symposium on Big Data and Cloud Computing Challenges (ISBCC – 16')*,
Smart Innovation, Systems and Technologies 49, DOI 10.1007/978-3-319-30348-2_27

1   Introduction

In recent years, data size has outgrown to a great extent. This accelerates the need of a sophisticated data mining and data analyzing algorithm that can find out the patterns in data efficiently in lesser time. Here comes the picture of Big Data Analytics that can handle such huge data that is being generated every minute, every second. Big Data Analytics is the current demand for any data-centric companies. It works on both static data or historical data as well as dynamic data or, in other words, data streams. The growth of the huge data is characterised by the four 'V's:

1. **Volume**: The increase in data can be used to describe this term. Huge transaction data items is stored so that they can be analyzed using some heuristics.
2. **Variety**: Data might be structured or unstructured based on their type. These will be ambiguities and inconsistency amongst them. Hence it is required to integrate all these data so they can be used for further processing.
3. **Velocity**: Streaming Data Analytics comes under this. The data which might be unstructured and continuously flowing must be analyzed based on the speed.
4. **Veracity**: The data that is obtained from various sources need to checked for authenticity and validation.

The data streams are mostly generated by mobile sensor networks, social networking sites and various other areas of applications like stock market data, transaction records in banks, all records of sales in a shopping mall and monitoring actions. All such data has to be analysed and information has to be extracted from it in a timely manner. Association rule mining deals with deriving these rules that are necessary to find out the number of occurrences of each items and hence to generate some pattern from them. Streaming Data analytics have various applications most important one being the prediction of sales and revenue for a company, also taking remedies in case of an loss. It can also be used to find the customers trend and hence and be used for drawing the conclusions faster and in an efficient manner. This saves the overall cost for the company and the frauds and other anomalies can be easily detected. All these makes streaming data analytics a very powerful domain for all the enormous data that is being generated in today's world each second.

When the data that is being considered is a real time data, analyzing data becomes a hard nut to crack. Hence in this paper, a time frame has been considered so that each time only certain amount of data has to be analyzed using an efficient algorithm. This time of interest is referred to as a "time window". Time horizon is very much necessary in case of dynamic data streams as each second huge number of data comes into picture. Apache Hadoop is used to handle such huge data stream. It is an open-source software framework written in Java for distributed storage. Hadoop core storage part is known as Hadoop Distributed File System (HDFS). It also has a MapReduce processing part. Using hadoop data across different nodes can be processed, leading to a parallel execution of a large amount of data. "locality

Fig. 1 Real time association
rule mining

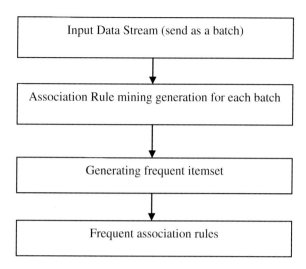

Fig. 1 Real time association
rule mining

of reference" of the data is used which makes the processing faster and efficient. The following items makes up the hadoop framework:

1. HDFS: it is a distributed file system that stores data on systems
2. Hadoop YARN: it is a resource management platform.
3. Hadoop MapReduce: a programming model for large scale processing.

This paper overview can be demonstrated in the given flowchart (Fig. 1).

2 Literature Survey

The mining of data items can actually implies generation of the frequent items and find out the association rules among them. Data mining an important technique for extracting information from databases, has been proposed to solve this problem. There are several functions of data mining, being "classification" is quite important. It has also been used in various areas like finding out relationships among items and doing the companies' risk analysis, predicting all these becomes an important factor.

Association [3, 9] among items or finding the association rules among the data items is first thing that needs to be implemented before any prediction. These can be done by finding out certain amount of *support* and *confidence* [1] between the items. These things are necessary for predicting sales of any supermarket data and also to find out which items should be placed together in a row in accordance with their occurrences with respect to sales. For example an association rule can be like "if 90 % of the customers are buying bread then it is more likely that they'll butter also along with it". Hence it is more likely that bread and butter will be kept in same section or nearby sections in a supermarket.

The problem of mining [11] association rules in large databases can be divided into two subproblems:

(i) finding frequent itemsets;
(ii) generate the association rules among the data items.

The most important and significant algorithm for miningfrequent itemsets is "Apriori", was proposed by Agrawal and Srikant [6]. Parallel to apriori, another algorithm called the FP-Tree [7] can also be used to to find the frequent itemsets. The FP/Growth algorithm follows a merged pattern between the horizontal and vertical model of a database. The transactions that are present in the database are saved in form of a tree data structure and each article contains a pointer pointing towards all transactions inside it. This new pattern of data structure, was named FP-Tree. It had been created by Han et al. [2]. The root of the tree is labelled as "null". The transactions in a FP-Tree are arranged in a reverse order so that the tree size remains small and the frequent items can remain closer to the root. Apriori goes by the notion of generate and test approach. It first generates the itemsets and tests if they are frequent or not with help of a given threshold value.

Let $A = \{i_1, i_2, ..., i_N\}$ be a set of n distinct literals called items. Let D dataset be a collection of transactions over A. Each transaction or record $r \in D$ contains a subset of items, i.e., $r \equiv \{ii, ij, ..., ik\} \subseteq A$. The number of items in r is called its length. The transactions in D can entail different lengths. A subset of items $I \subseteq A$ is called an itemset [8]. The number of times that an itemset I occurs in the transactions is the support count of I, which is denoted as supp(I). Frequent itemsets are defined with respect to the support threshold S; as a result an itemset I is frequent if supp(I) $\geq$ S. The itemset support count is related to the support count of its subsets and supersets. In fact, given two itemsets I and J such that $I \subset J$, the number of times that I occurs is at least the number of occurrences of J because the former is part of the latter.

Therefore, $supp(I) \geq supp(J); \forall I \subset J \subseteq A$.

Also, it is better to classify an itemset A as being maximal frequent if A is frequent, but any $B \supset A$ is not [3]. If |A| is the cardinality of A, then the number of possible distinct itemsets is 2|A|. There are two properties for the search space:

- Downward closure: If there exists subsets of a frequent itemset,then those subsets are also frequent [4, 10]
- Anti-monotonocity: If there exists supersets of an infrequent itemset, then they must also be infrequent.

For our convenience the itemsets are kept sorted in lexicographic order. The number of items in the sets defines its size. So if there are item sets of size k then it is called as k-itemset.

Apriori Algorithm

L_k: be the frequent itemsets
Bk: be the itemsets of k size after generation
C: the highest of all the transaction, i.e., the one which is large enough

Thres: be the minimal threshold of the frequent itemset

1. Find the L_1 first
2. For k = 1 (first pass)
3. While $L_k \neq \emptyset$ and $k < C$
4. Do

   $B_{k+1} = L_k$ join with L_k
   $L_{k+1} = \{I \in B_{k+1} \,|\, \text{supp}(I) \geq \text{Thres}\}$

5. k = k+1
6. end while
7. return U L_K

   In apriori the very first step is to scan the database first. After scanning the database single set items(k = 1), i.e., 1-itemset needs to be found out. All the values in the 1-itemset has to unique hence they are called candidate itemsets. These candidate itemsets are passed through database again till there are no unique 1-itemsets available. These are put in L_k. Every step given in this algorithm comprises of two step. The very first is the generation of (k + 1)-itemsets from the elements present in Lk. And then they are placed in $B_k + 1$. The generation of

Fig. 2 Apriori example

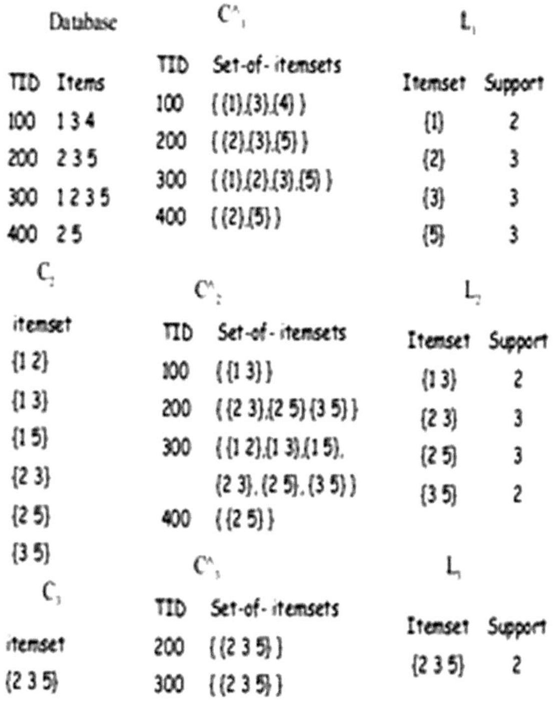

candidate sets is formed by joining L_k with L_k. Here the two itemsets I1, I2 $\in L_k$ are merged and they have (k − 1) items in common to them. The final itemset I = I1 U I2, which has k + 1 items, next is inserted in B_{k+1}.

After this operation again scanning of the database is done and support is calculated. The itemsets whose support is below the threshold Thres is discarded. This process continues till passes covered becomes equal to C. The final ouput comes after frequent itemsets form L_k with k = 1, 2, 3, …. C are merged.

Let us consider a transaction (Fig. 2).

3 Existing System

A window can be regarded as a time frame. The time interval can be regarded as the "frame of interest".. As the window slides new transactions enter the system. We calculate the capacity for that many transactions at a time. In the article "Mining frequent itemsets in data streams within a time horizon" [4] compares various algorithms and its performance within a limited time horizon.

Definition 1: Limited window A window is said to be limited if it has finite lower bound.

As transaction entering in and out of the window will vary and so the itemsets will also vary in accordance with it. Here fixed window is taken into account. The memory capacity is the maximum number of elements the window can hold at that point of time. The flow of the records going in and out of the window is considered same. The capacity of the window changes in accordance with the degree of interest μ(ti) and this can be defined by the mathematical equation

$$C(W) = \sum_{t_i \in W} \mu(t_i)$$

Smooth window is considered here which can be defined as

$$\min(\mu(t_i), \mu(t_j)) \le \mu(t_k) \le \max(\mu(t_i), \mu(t_j)) \forall t_i < t_k < t_j \in W.$$

A smooth window can be made asymptotically limited by

$$C(W) = \lim_{t_i \to \infty} \sum_{t_i \in W} \mu(t_i) = c \in]0, +\infty[$$

Here, the frequent items are generated in way that that it appears as if the frame of interest and data stream are going hand in hand. Transaction- sensitive sliding window are not considered. Real time data flows in and out of the window and the final itemsets are generated after merging with all the smaller subproblems The sliding-window filtering (SWF) algorithm was proposed by Lee et al. [5]. It consists

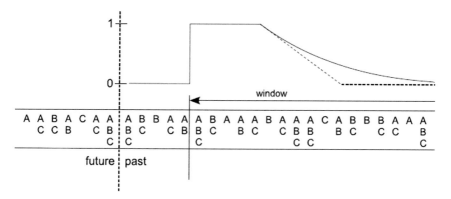

Fig. 3 An asymptotically limited window

Fig. 4 Flowchart of the
existing system

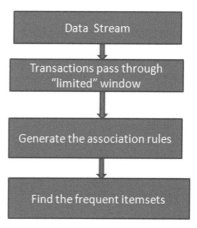

of sequence of partitions. Each of these holds a certain number of transactions. The
transactions that are oldest are discarded so that a new partition makes its way to the
frame of interest. Here as a part of the work, sliding window model has been
considered more appropriate.

The figure depicting asymptotically limited window is shown below (Figs. 3 and 4)

The existing system makes use of only limited windows and hence for large
datasets it might lead to a failure. Hence we go for the proposed system.

4 Proposed System

Finding out the frequent itemsets for smaller data sets can be obtained without
much hassle. But when the data size is huge finding out the frequent itemsets
becomes quite difficult.

Fig. 5 Flowchart of the
proposed system

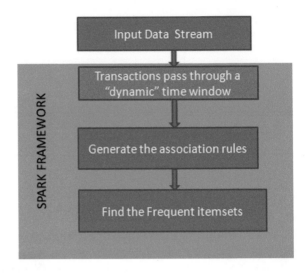

In the proposed system the following things has been done:

1. Input the large dataset
2. From the given data sets the frequency of all the combinations within it has been determined i.e., the support count for each of itemsets is determined individually.
3. A certain threshold has been given
4. From the given threshold and comparing the support for the individual itemsets, the frequent itemsets has been generated (Fig. 5).

"Dynamic window" here corresponds to window size has is not finite. The transactions inside it are distributed across various nodes so that the result is obtained in lesser time and hence it more efficient.

From the frequent itemsets the association rules can be generated. Also the closed itemset can be found but that does is beyond the scope of this work.

5   Experimental Results and Future Work

Basically apriori algorithm is used here but a slight variation. Apriori usually fails to work in case of large datasets since the time complexity becomes almost exponential but here we make use of hadoop's multi node platform and hence results are obtained within a very short period of time. The configurations that had been used here are:

RAM: 32 GB
PROCESSOR: I7
STORAGE: 1 TB

Fig. 6 Time taken by hadoop
single node and multi node

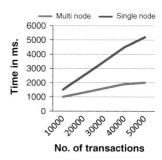

TRANSACTION SET: 50,000 Transactions
FRAMEWORK: HADOOP

Here we used a large dataset of 50,000 transactions with 100 unique items and measure the performance in case of a single node and multi node hadoop platform. The single node consists of only one system i.e., only one computer is needed and the whole transaction record is passed to the system. In multi node, four systems are connected and all the records are distributed across all the four systems. Thus the records gets executed in parallel and results are obtained exceptionally faster. We can depict this using a graph shown in Fig. 6)

Here it is seen that in case of single node when the number of transaction is 10,000 the time taken by multinode is almost same. But as the number of transaction increases by 10,000 each time, there is a huge difference seen between the time taken by single node and multinode analysis. By the time the number of transaction reaches 50,000, the single node configuration will be almost take more than double the time taken by multinode. Hence it is seen that in case of large transaction that is growing each time multinode configuration is more suited. This work can be extended for even larger datasets using spark framework.

6   Conclusion

This work comprises of the generation of frequent itemset from a real time streaming data. Apriori algorithm is used in order to generate the frequent items using which we can analyze the data arriving inside the window in a specified time period and generate the frequent itemsets. The results found with the help of hadoop's multi node system is less time consuming and effective as compared to single node configuration of hadoop framework.

References

1. Agrawal, R., Imielinski, T., Swami, A.N.: Mining association rules between sets of items in large databases. In: Proceedings of the 1993 ACM SIGMOD International Conference on Management of Data, 207–216, Washington, DC, May 1993
2. Han J., Pei J., Yin Y., Mao, R.: Mining frequent patterns without candidate generation: a frequent-pattern tree approach. Data Mining and Knowledge Discovery (2003)
3. Troiano, L., Rodríguez-Muñiz, L.J., Ranilla, J., Díaz, I.: Interpretability of fuzzy association rules as means of discovering threats to privacy. Int. J. Comput. Math. **89**, 325–333 (2012)
4. Troiano, L., Scibelli, G.: Mining frequent itemsets in data streams within a time horizon. Department of Engineering, University of Sannio, Viale Traiano (2013)
5. Lee, H., Lin, C.R., Chen, M.S.: Sliding-window filtering: an efficient algorithm for incremental mining. In: ACM 10th International Conference on Information and Knowledge Management (CIKM 01), 263–270 (2001)
6. Agrawal, R., Srikant, R.: Fast algorithms for mining association rules in large databases. In: Proceedings of the 20th International Conference on Very Large Data Bases (VLDB), Santiago, Chile, Sept 1994
7. Wu, B., Zhang, D., Lan, Q., Heng, J.: An efficient frequent patterns mining algorithm based on Apriori Algorithm and the FP-tree structure. In: Third International Conference on Convergence and Hybrid Information Technology (2008)
8. Szathmary, L., Napoli, A., Valtchev, P.: Towards rare itemset mining. In: Proceedings of the 19th IEEE International Conference on Tools with Artificial Intelligence (ICTAI'07), 305–312. IEEE Computer Society, Washington, DC, USA 305–312 (2007)
9. Mannila, H., Toivonen, H., Verkamo, I.: Efficient algorithms for discovering association rules. In: Proceedings of the AAAI Workshop on Knowledge Discovery in Databases (KDD'94), 181–192. AAAI Press, Seattle, Washington, USA (1994)
10. Li, H.-F., Lee, S.-Y.: Mining frequent itemsets over data streams using efficient window sliding techniques. Expert Syst. Appl. **36**(2), 1466–1477 (2009)
11. Liao, S., Chu, P., Hsiao, P.Y.: Data mining techniques and applications—a decade review from 2000 to 2011. Expert Syst. Appl. **39**, 11303–11311 (2012)

Adaptation of SQL-Isolation Levels to No-SQL Consistency Metrics

Bhagyashri Vyas and Shraddha Phansalkar

Abstract Big-data applications are deployed on cloud data-stores for the augmented performance metrics like availability, scalability and responsiveness. However they assure higher performance at the cost of lower consistency guarantees. The commercial cloud data-stores have unassured lower consistency guarantees which are measured with different metrics. For a traditional application deployed on relational databases with strong and assured consistency guarantees, SQL isolation levels have been used as a measure for the user to specify his/her consistency requirements. Migration of these applications to No-SQL data-stores necessitates a mapping of the changed levels of consistency from SQL isolation levels to No-SQL standard consistency metrics. This work gives insight to user about the adaptation in changed levels and guarantees of consistency from SQL isolation levels to No-SQL consistency metric.

Keywords Migration · Consistency metrics · SQL-isolation level · No-SQL

1 Introduction

In the contemporary cloud applications, data consistency is rationed with No-SQL data stores. The data consistency investigation in the replicated database system implies agreement of values between geographically distributed replicas that is all the replicas should have same status. With the advent of 24×7 availability, however this consensus is very difficult to achieve.

No-SQL data stores have implicit constraints on consistency guarantees to leverage the performance like response time, scalability. Lower levels of consistency are mostly levels of consistency like eventual consistency [1]. Eventual

B. Vyas (✉) · S. Phansalkar
Symbiosis International University, Pune, India
e-mail: Bhagyashri.vyas@sitpune.edu.in

S. Phansalkar
e-mail: shraddhap@sitpune.edu.in

© Springer International Publishing Switzerland 2016   335
V. Vijayakumar and V. Neelanarayanan (eds.), *Proceedings of the 3rd International Symposium on Big Data and Cloud Computing Challenges (ISBCC – 16'),*
Smart Innovation, Systems and Technologies 49, DOI 10.1007/978-3-319-30348-2_28

Consistency and other lower levels of consistency are unassured and they are not directly measurable. Many works are focussed on the measurement of consistency guarantees in No-SQL data-stores [2–4]. Although they define all levels of consistency, NOSQL databases do not provide insights to mapping of these levels to those of ACID transactions in RDBMS.

This is left to the interpretation and understanding of the end-user and requires a manual support. The SQL isolation levels are attributed to SQL transaction and all the data objects in the transaction are observed with the same isolation level. Consistency index (CI) is a data characteristic and is thus co-related to SQL isolation level.

In RDBMS the isolation levels to measure consistency are called as SQL-isolation levels [5]. The isolation levels control the locking and row-versioning behaviour of transaction.

No-SQL data stores do not have these kinds of isolation levels in [5] consistency is measured using a semantic called *Consistency Index* (*CI*) which determines consistency of a data-object with an index value in [0,1]. Consistency index is defined as the ratio of number of correct reads user observes to the total number of reads on a data-object. CI can be considered as a measure of accuracy of read transactions on a data item and can also be used to interpret as an expected value of consistency for a data item.

In this work, we choose Consistency Index (CI) as a No-SQL consistency metric and map it to the SQL-isolation levels to give an insight to the new No-SQL data user about the changed interpretations of the consistency metrics.

In this work we use Amazon SimpleDB as No-SQL data-store in which CI is modelled for every data-object. The different SQL-isolation levels are implemented with efficient use of locks and semaphores. The resultant CI and the performance with respect to response-time is observed for every isolation level. Thus the user is able to gain insight about the choice of the traditional isolation levels, the resultant CI and the consequent performance of the application on the No-SQL data-store.

The rest of the paper is structured in the following sections. In Sect. 2, the related work on adaptation of consistency metrics from SQL isolation levels to NO SQL data store is presented. In Sects. 3 and 4 the standard definition of isolation level and Consistency Index is presented. In Sects. 5 and 6 Implementation set up for measuring the isolation level with CI and the response time of the application is presented. We reproduce definition of CI and redefine the SQL-isolation levels from implementation point of view. In Sect. 7 shows the conclusion.

2   Related Work

Migration from SQL to No-SQL database reveals that No- SQL database does not support ACID transaction. Researchers applied several approaches to improve these consistency anomalies. However all these anomalies are supplemented with consistency measures which are unassured.

In [2] Scalable Transaction Management with Snapshot Isolation based transaction execution, the model is a multi-version based approach utilizing the optimistic concurrency control concepts with SI model helps to provide serializable transaction. The main approach is to decouple the transaction management functions from the storage service and integrate them with the application level processes. Thus it guarantees the scalable consistent transaction on No-SQL data store. The work in the article [3] contributes to how trends have changed in database for optimization problem while focusing on service-oriented architecture. The proposed 28mec architecture inconsistency is handled at application layer because at storage layer there is no access and control entity Authors also talked about minimization of cost designing the system which helps to resolve inconsistencies rather they prevent from inconsistencies. In his work [6], Kraska et al. consider having a middle level layer above the storage, where different consistency models are supported with different performance properties, and then the client can choose dynamically what is appropriate. They build a theoretical model to analyze the impact of the choice of consistency model in terms of performance and cost, propose a framework that allows for specifying different consistency guarantees on data. This paper implements the design of TACT Tunable Availability and Consistency Tradeoffs [7] through the help of three independent metrics a conit-based consistency model was developed to capture the semantics space between strong and optimistic consistency model for replicated of services.

In the work [4], authors put forth a user-perspective of consistency in any replicated database and proposed quantitative measure of data consistency with Consistency Index(CI). CI is user comprehensive consistency metric which is also used to express expected level of consistency.

Although there is a good work on measuring consistency, Objective of work is to establish dependency. This can be further used for finding their level of interactions with consistency guarantees to develop a predictor model for CI.

3   Isolation Levels

In RDBMS, the concept of isolation [5] refers transaction integrity. Database isolation levels have greater impact on ACID properties. Isolation in database systems is all about transaction or queries and the consistency and correctness of data retrieved by queries. There are four levels of isolation according to need of consistency in RDBMS.

(A)   Read uncommitted—This isolation level reads data, that is not consistent with other parts of table and it may or may not be committed. This isolation level ensures the fastest performance but with the dirty reads. No locking mechanism is deployed.

(B)   Read committed—In this isolation level, the rows returned by the query are the rows that were committed when the query was started. And therefore

Table 1 Isolation levels and read phenomena

Read uncommitted	Read committed	Repeatable read	Serializable
No lock	Write lock	Read and write lock	Table lock
Dirty read occur	Non repeatable read and phantom read occur	Phantom read occur	None
Less response time	Average response time	Average response time	High response time

dirty reads do not occur. Lock mechanism is implemented on READ operation.

(C) Repeatable Read—READ and WRITE locks both are implemented. The rows returned by a query are the rows that were committed when the transaction was started. The changes made by other transaction were not present when the transaction was started, and therefore will not be included in the query result.

(D) Serializable—This isolation level specifies that all transactions occur in a completely isolated fashion, meaning as if all transactions in the system were executed serially, one after the other. The DBMS can execute two or more transactions at the same time only if the illusion of serial execution can be maintained. This is highest level of isolation which ensures full consistency in the system. The following table depicts read phenomena (Table 1).

4 Consistency Index

In distributed environment, replication on a data object is carried out for various read and updates operation in the system. Consistency index(CI) [4] of replicated data object can be defined as the proportion of correct reads on the object to the total number of reads on all replicas in the system for an observed period. The value of CI falls in the range of [0,1]. Strong consistency means the CI value is closer to 1. It guarantees serializable level. And 0 shows the weaker level of consistency. So when the value is closer to 0, the CI guarantees are lower. Other isolation levels are categorized between these range. The formula suggests how many correct read operations are performed in the system.

$$CI = \frac{\text{Number of correct reads } (R_C)}{\text{Total number of reads } (R)} \tag{5}$$

Replicas are the major part of the system. Higher the number of replicas, the response time will be increasing. It may also lead to the incorrect read. The frequency of the incorrect read would be depend on the time gap and thread synchronization. The synchronization between all replicas for performing operations on data object should affect the value of CI. Following are the factors for the affecting the CI-

1. Workload
2. Average time gap between an update and read in real time.
3. Number of Replica

5   Implementation

We built experimental set up to execute explicitly all four isolation levels on Amazon SimpleDB [1]. We used TPCC web-application and data-object stock for deploying data. The data object of TPCC schema [8] was implemented on Amazon SimpleDB. TPCC is a popular benchmark for comparing OLTP performance on various software and hardware configurations. It simulates complete computing environment where population of users executes transaction against database. It carries out five transactions as New Order transaction (which creates new order), Payment transaction (execution of payment), Stock level transaction (reads stock level), Delivery transaction (Delivery of items to the customer) and Order Status transaction (Status of Order placed) on entities. We calculate CI on the *stock_qty* attribute of *stock_item* domain in TPCC schema. The *Stock_qty* has semantic transactions like new_order, stock_level transactions which are read and update operations in database.

Amazon SimpleDB is defined database in which tables are called as domains and these domains were replicated at application level. Amazon SimpleDB is a highly available and flexible non-relational data store. Developers simply store and query data items via web services requests and Amazon SimpleDB does the rest. It is a No-SQL document oriented data store with a very simple and flexible schema with weaker consistency guarantees, with little or no administrative burden. SimpleDB creates and manages multiple geographically distributed replicas of your data automatically to enable high availability and data durability. It organizes data in domains within which you can write data, retrieve data, or run queries. Domains consist of items that are described by attribute name-value pairs. Amazon SimpleDB keeps multiple copies of each domain. A successful write guarantees that all copies of the domain will durably persist. Amazon SimpleDB supports two read consistency options i.e. eventually consistent read and consistent read. An eventually consistent read might not reflect the results of a recently completed write. Consistency across all copies of the data is usually reached within a second; repeating a read after a short time should return the updated data. A consistent read returns a result that reflects all writes that received a successful response prior to the read. In our research we have explicitly introduce the mechanism semaphores based locks to achieve the locking as in RDBMS. As the default level of consistency in

SimpleDB is eventual consistency the levels of consistency can be simulated on the top of this level at application layer.

Replication Policy—"Write through policy" is the update policy in which update message is passed to all other replicas. This ensures that updates are safely stored on all the replicas. The write-through policy is good for consistency critical applications.

Time Gap—Arrival of a transaction is random in the system. Also the type of transaction can be a read or update transaction. The relative arrival time of a read transaction from the preceding update is critical determinant for the successful read [4]. Hence we have also varied the time–gap in the different isolation levels and observed its effects on observed value of CI.

At any given time, for different number of replicas, we measured the number of replicas that were updated.

6 Implementation of Isolation Levels

As discussed in Sect. 3 we have 4 isolation levels. The implementation of different isolation levels is done through the simulation of transaction as discussed in Sect. 5. The locking of different object is carried out with semaphore like mutex. The stock level transaction verifies of *stock_qty of item* in TPCC.

1. Read Uncommitted

This is the first isolation level and has very low consistency. It allows all the three concurrency phenomena i.e. dirty read, repeatable read and phantom read. The concurrency is affected by locking implementation in a system. The shared locks mechanism can control concurrency effectively. Read uncommitted level has no shared locks at all, resulting in "dirty reads" the outcome of dirty reads will show the variation in the value of CI. The value is approximately zero implies no or very few correct reads are carried out.

The transactions were simulated with read and write requests continuously without interleaving any delay. The locking mechanism was by passed. Transactions could read from or write to any replicas. There is no acquisition and release of locks. The uncommitted data is read by transactions which would give an arbitrary value of CI on consensus.

Table 2 depicts that time gap between the transactions 0–5000 ms. As discussed in [4] the time gap between an update and read transaction is the critical determinant of the correctness of read operation. This isolation level does not implement any locking and hence the correctness of read is determined by the time gap between read and preceding update.

We varied the Time gap from 0 to 5000 and observed that for a given no of replicas, CI (correctness of read) increased with time gap as shown in Table 2.

Number of replicas: With the increase in the number of replicas, CI decreases. This is because of the increase in availability, correctness is at stake. To control CI

Table 2 CI level of read uncommitted

Number of replicas	Time gap (ms)	Avg response time (ms)	CI
5	3000	11,829	0.1
7	3000	8506	0.14
7	4000	12,290	0.1
7	1000	12,686	0.07
10	5000	14,340	0.1
10	4000	13,098	0.05

for a given number of replicas. We have modified the time gap between a read and an update. CI increases with increases in time gap for a given number of replicas.

Response Time: For a given response time the average response time increases with the delay as discussed in [4]. This is also observed in Table 2. Also the average response time increases with the number of replicas.

2. Read Committed

The SQL standard defines read committed level as second weaker isolation level among four other isolation levels. Many search engines used this as default level. The transaction which is running under read-committed isolations reads only the committed data. Those reads prevents dirty reads. The synchronization of data access would take place through '*acquire*' and '*release*' lock. The behaviour of lock would decide the access of locking acquire and release granularity item of all replicas. According to the definition of read-committed, we implement it as the acquire lock was placed before WRITE operation so that any uncommitted data could not be read by users. We then observe the correct read.

This isolation level has only write lock and hence the correctness of read is determined by time gap between read and preceding update.

We varied the time gap from 0 to 5000 and observed that for a given number of replicas, CI (correctness of read) increased with time gap as shown in Table 3.

For the given number of replicas the average response time increases with the delay. This is also observed in Tables 2 and 3 have same observation and also the average response time increases with the number of replicas. But there is implementation of lock which prevents from dirty read and user will get higher level of consistency. To control CI for a given number of replicas, we have modified the

Table 3 CI level of read committed

Number of replicas	Time gap (ms)	Avg. response time (ms)	CI
5	3000	11,897	0.5
5	4390	12,352	0.9
7	4000	12,382	0.6
7	5000	13,254	0.9
10	5000	15,822	0.9

time gap between between read and update. CI increases with increase in time gap. In this level user is getting high number of correct read as shown in Table 3.

3. **Repeatable Read**

The next strongest level is repeatable read. In this level only single concurrency phenomena known as "phantom read" occur. The repeatable read isolation level provides a guarantee that data will not change for the life of the transaction once it has been read for the first time. Repeatable read can be implemented with one update query on the stock_item and simultaneous insert query on stock_item.

However ecommerce application rarely implies aggregate function. We measure CI for every data object which is an individual stock_item. There is no time gap. The granularity of items of domains and replicas are in resultant of an aggregate query. Hence the results of repeatable read yields a CI of value 1.

4. **Serializable Read**

The last and strongest consistency level is Serializable which prevents from all concurrency phenomena. Serializable read can be implemented with lock on whole domain. It will maintain the illusion of serial execution. So that every domain will work serially. But the response time is higher and we obtain the value of CI as 1

7 Conclusion

The relational databases have already predefined isolation levels although they are clearly defined, they are the database perspective when we adapt the applications to parameters in a given isolation level. Our work demonstrates the adaptation of SQL isolation level to No-SQL CI. The semantics space between strong and weak consistency is gained by explicit implementation of isolation levels on No-SQL data stores. We would further deploy statistical models on isolation levels to predict desired value of CI with parameters in a given isolation level. NO-SQL, user must have a clear insight of adaptation semantics. Hence we chose CI from literature which is user perspective of measuring consistency.

References

1. Amazon SimpleDB Documentation: http://aws.amazon.com/simpledb/. Accessed Oct 2014
2. Vinit, P., Anand, T.: Scalable Transaction Management with Snapshot Isolation for NoSQL Data Storage Systems, IEEE Trans. Serv. Comput. **8**(1), 121–135
3. Daniela, F., Donald, K.: Rethinking cost and performance of database systems. SIGMOD Record **38**(1), 43–48 (2009)
4. Shraddha, P., Dani, A.R.: Predictive models for consistency index of a data object in a replicated distributed database system. WSEAS Trans. Comput. **14**, 391–451 (2015)
5. wikipedia.org/wiki/Isolation_database_systems

6. Kraska, T., Hentschel, M., Alonso, G.: Kossmann consistency rationing in the cloud: Pay only when it matters. **2**, 253−264 (2009)
7. Haifeng, Y., Vahdat, A.: Design and evaluation of a conit-based continuous consistency model for replicated services. ACM Trans. Comput. Syst. **20**(3), 239−252 (2002)
8. Transaction Processing Performance Council: TPC benchmark C standard specification, revision 5.11. http://www.tpc.org/tpcc/. Accessed Oct 2014

Mood Based Recommendation System in Social Media Using PSV Algorithm

R. Anto Arockia Rosaline and R. Parvathi

Abstract Social media plays a vital role in real life events. The overall mood of a tweeter can be assessed through tweets if they are put through sentiment analysis. An analysis is performed on the various tweets and the missing values are found out for the topic opinion. A model for recommendation system based on the positive score associated with each topic has been proposed.

Keywords Mood · Sentiment · Social media · Twitter · Recommendation · Tweets

1 Introduction

Microblogging is so popular because it has a flow of public opinion. The most popular microblogging service is Twitter. From the opinion rich resource Twitter emotional knowledge can be obtained. This emotional tone or the attitude is termed to be Mood. It is a robust way of expressing the sentiment. Using the microblogs as source it is possible to predict the recent styles, can give an alert message, know the public opinion on an event and also can present a recommendation system.

Tweeters may have interests on the same topic but can have different opinions. A recommendation scheme involving both the interests and sentiment need to be presented. Our proposed work gives a recommendation model which involves three techniques. First method is data collection. Here a real time twitter data is collected

R.A.A. Rosaline (✉)
VIT University, Chennai, India
e-mail: antoarockia.rosaline2013@vit.ac.in

R.A.A. Rosaline
Rajalakshmi Engineering College, Chennai, India

R. Parvathi
School of Computing Science and Engineering, VIT University,
Chennai Campus, Chennai, India
e-mail: parvathi.r@vit.ac.in

© Springer International Publishing Switzerland 2016   345
V. Vijayakumar and V. Neelanarayanan (eds.), *Proceedings of the 3rd International Symposium on Big Data and Cloud Computing Challenges (ISBCC – 16')*,
Smart Innovation, Systems and Technologies 49, DOI 10.1007/978-3-319-30348-2_29

from Twitter and the tweets are labeled as positive or negative based on the user topic opinion. Second technique involves predicting the missing values using matrix factorization. The third method involves the generation of the actual recommendation model.

2 Related Work

Nguyen et al. [1] has clustered the communities based on mood. Livejournal nearly offers 132 moods. This way of clustering the people with negative mood needs to be identified. Then those people with mental illness need to provided mental support and they should be closely monitored.

Ren and Wu [2] has presented the low-rank matrix factorization method as the basic model to predict the user-topic opinion prediction. In Twitter, a user's friends and followers are defined as social friends. To model the social context the social friend relationship network is used. Modeling the topical context is based on the hypotheses that if two topics are similar in content the users also will give similar opinion for those two topics.

Dhall et al. [3] has done the work to detect the group's mood based on the structure of the group. The happiness intensities of the individual faces contribute to the perceived group mood.

Tang et al. [4] has proposed a method say MoodCast to predict the emotion in social network. They have proposed an approach to model the emotional states of the user in an efficient manner. Saari and Eerola [5] has predicted mood information in music. The use of dimensional emotion model to represent moods was supported by their analysis of the structure of the tag data.

The problem of social affective text mining has been analyzed by Bao et al. [6]. The connections between online documents and user generated social emotions were found out by them. Turney [7] has presented a simple unsupervised learning algorithm which was used to classify a review as recommended or not recommended. The written review is the input for the algorithm and the output is the classification. Munezero et al. [8] work is to differentiate five terms such as affect, feeling, emotion, sentiment and opinion detection.

3 Techniques

3.1 Data Collection

The process of data collection involves collecting the tweets for U number of tweeters for the T number of topics from Twitter. Table 1 indicates the number of tweets collected for 5 tweeters and 5 topics. Those topics are related to IPL (Indian Premier League) and the tweets are collected from real time twitter data.

Table 1 Total number of tweets (N)

	Topic 1	Topic 2	Topic 3	Topic 4	Topic 5
Tweeter 1	295	409	1	1	8
Tweeter 2	364	464	2	2	6
Tweeter 3	436	567	1	8	22
Tweeter 4	1	2	0	0	0
Tweeter 5	203	343	4	6	5

Topic 1 indicates Chennai super kings, Topic 2 indicates Mumbai Indians, Topic 3 indicates Rajasthan Royals, Topic 4 indicates Royal Challengers Bangalore, Topic 5 indicates Kolkata knight Riders.

The real world dataset which is collected from Twitter can be labeled as positive or negative based on the user topic opinion. Using a credible sentiment analysis program the mood of the user can well be analyzed based on the presence of words and emoticons. The score is on a scale of 1 (neutral) to 5 (very positive) or −1 (neutral) to −5 (very negative). Each entry in Table 2 indicates the positive score obtained from the set of tweets given by a particular tweeter on a particular Topic. Likewise Table 3 indicates the negative score.

3.2 Prediction

As observed in Tables 2 and 3 it is clear that Tweeter 4 has not given any tweets for the Topics 3, 4 and 5. It is necessary to predict the unknown values from the other known values. Matrix factorization (MF) technique has been used to predict the unknown values. MF is used because of having good scalability, accurate prediction and flexibility.

Table 2 Positive score (PS)

	Topic 1	Topic 2	Topic 3	Topic 4	Topic 5
Tweeter 1	1.15	1.14	2	3	1.75
Tweeter 2	1.14	1.33	1.5	2	2
Tweeter 3	2.33	2.34	1	1.63	1.5
Tweeter 4	1	2	0	0	0
Tweeter 5	1.38	1.83	1.25	1	1.2

Table 3 Negative score (NS)

	Topic 1	Topic 2	Topic 3	Topic 4	Topic 5
Tweeter 1	1.17	1.12	1	1	2.25
Tweeter 2	1.15	1.13	1	1	1.67
Tweeter 3	1.74	1.55	1	2.63	1.23
Tweeter 4	1	1	0	0	0
Tweeter 5	1.28	1.16	1	1.33	2.8

Table 4 Predicted missing values using matrix factorization for positive score

	Topic 1	Topic 2	Topic 3	Topic 4	Topic 5
Tweeter 1	1.15	1.14	2	3	1.75
Tweeter 2	1.14	1.33	1.5	2	2
Tweeter 3	2.33	2.34	1	1.63	1.5
Tweeter 4	1	2	1.37	1.86	1.54
Tweeter 5	1.38	1.83	1.25	1	1.2

Consider a set U of tweeters, and a set T of items. Let R of size $|U| * |T|$ be the matrix that contains the positive score the tweeters have assigned to the topics. It is also assumed that there are k latent features. The task is to find two matrices $P(a^{|U|*K}\text{matrix})$ and $Q(a^{|T|*K}\text{matrix})$ such that their product approximates R. Table 4 shows the predicted missing values using MF for positive score.

3.3 Recommendation and Presentation

After the prediction using matrix factorization the volume is computed from the number of tweets given by the tweeter i to topic j and the total number of tweets given by that tweeter. Volume is computed by the equation

$$Volume\ V_{ij} = N_{ij} / \Sigma N_i \tag{1}$$

where i represents the tweeter and j represents the topic.

Then the average volume for each topic is calculated from the volume and the number of tweeters as follows:

$$Average\ Volume\ AV_i = \Sigma V_{ji} / U \tag{2}$$

where i represents the topic and j represents the tweeter.

Then the average positive score for each topic is calculated from the total positive score and the number of tweeters by the equation as follows:

$$Average\ positive\ score\ APS_i = \Sigma PS_{ji} / U \tag{3}$$

where i represents the topic and j represents the tweeter.

The topic that has to be recommended is one that has the maximum value of the product of average positive score and average volume.

$$REC_i = AV_i * APS_i \tag{4}$$

where i represents the topic.

Thus the PSV (Positive Sentiment Volume) algorithm given below gives a suitable recommendation system based on the weighting function sentiment and volume.

PSV algorithm

Require:Let U be the number of tweeters and Let T
be the number of Topics
for i in U do
    for j in T do
    Compute the number of tweets N_{ij}
    end for
end for
for i in U do
    for j in T do
Calculate the positive score PS_{ij} of the tweets
    end for
end for
for i in U do
    for j in T do
    if $PS_{ij}==0$ then
    predict it using matrix factorization
    end for
end for
for i in U do
    for j in T do
        Compute Volume $V_{ij}=N_{ij}/\sum N_i$
    end for
end for
for i in T do
    for j in U do
Compute average volume $AV_i=\sum V_{ji}/U$
Compute average positive score $APS_i=\sum PS_{ji}/U$
Compute $REC_i=AV_i*APS_i$

    end for

end for

max=REC_1

for i in T do

    if $REC_i > max$ then

        max=rec_i

    end if

end for

Recommend max

Table 5 shows the average volume and average positive score obtained for each topic. It is observed that that the topic 2 has the maximum value and so it is recommended.

Table 5 Computing the product of average volume and average positive score

	Topic 1	Topic 2	Topic 3	Topic 4	Topic 5
Average volume (AV)	0.1970	0.2707	0.0012	0.0025	0.0062
Average positive score (APS)	1.4	1.728	1.424	1.898	1.598
Average volume * Average positive score (REC)	0.27584	0.46779	0.00172	0.00492	0.00998

Table 6 Analysis of various algorithms

Algorithms used	Task						Polarity	
	Sentiment analysis	Sentiment classification	Emotion detection	Prediction	Missing data	Unexpected labeling	General analysis	Pos/Neg
Rule-based	✓						✓	
Lexicon-based, semantic			✓				✓	
Graph-based approach		✓						✓
Lexicon-based, semantic		✓						✓
Statistical	✓						✓	
NB, SVM		✓						✓
Lexicon-based, SVM			✓				✓	✓
Semantic	✓							✓
SVM, 1-NN		✓						✓
Lexicon-based			✓				✓	
Unsupervised, LDA		✓					✓	
NB, SVM		✓						✓
SVM		✓					✓	
PSV (Proposed algorithm)	✓	✓	✓	✓	✓	✓	✓	✓

4 Analysis of Other Algorithms

The survey [9] has been carried out on some of the articles on the various categories such as sentiment analysis, sentiment classification and emotion detection. Table 6 shows the survey that has been carried out on various articles which includes the algorithms used, tasks performed and the polarity.

The PSV algorithm performs the tasks such as sentiment analysis, sentiment classification and emotion detection. Unlike the other algorithms PSV is used to handle any missing data from the available data using the matrix factorization technique. PSV also handles any unexpected labeling in an efficient way.

The proposed algorithm is capable of detecting any kind of mood from the tweets and the emoticons. Thus the PSV algorithm is more efficient and provides more accuracy.

5 Results

The real world dataset has been collected from Twitter for the various topics. The tweets collected from the twitter have revealed that the tweeters may have different opinions on the similar topics. Sometimes they may not be interested in tweeting for a particular topic. In that case using matrix factorization their mood can be traced out from the sentiment found in other tweets.

Sentiment found in the tweets is the way by which we can trace their mood. By analyzing what they have tweeted i.e. sentiment and the volume of their tweets it is possible to present a recommendation system. It is also evident that the proposed mood based recommendation system holds good for real time twitter data as the recommendation coincides with the result of the IPL event.

6 Analysis and Discussion

An error analysis has been conducted on the obtained results. The various categories of errors are listed below:

1. Unpredictable mood

From any tweets, it is possible to predict the mood which is a strong form of sentiment expression. It is because of using a credible sentiment analysis program which is used to find out the sentiment from the text and emoticons.

2. Missing data

If any data is missing i.e. the tweeter has not given any tweets on a specific topic, it is possible to predict it from the other data available. So handling of missing data is much simpler.

3. Unexpected labeling

In a tweet if there is a word expressing the happiness say "happy" and also an emoticon to express the sad feeling, the credible sentiment analysis program handles this kind of unexpected labeling in an efficient way. It handles the situation by assigning a positive score to the word happy and a negative score to the emoticon.

7 Conclusion and Future Work

In this paper a recommendation system is presented which is purely based on the user's mood. The basic concept behind this work is that users having similar interests on a topic may have different opinions. The algorithm namely PSV (Positive Sentiment Volume) which is based on the positive sentiment and volume has been used for the recommendation system. Thus this work makes the recommendation more accurate and reliable in Twitter.

As future work there is a plan to undergo a deep sensitivity analysis to incorporate the social context in twitter using undirected weighted graph. Also analysis has to be carried out including the negative score and present a recommendation based on it.

References

1. Nguyen, T., Phung, D., Adams, B., Venkatesh, S.: Mood Sensing from Social Media Texts and its Applications. Springer, London (2013)
2. Ren, F., Wu, Y.: Predicting user-topic opinions in twitter with social and topical context. IEEE Trans. Affect. Comput. 4, 412–424 (2013)
3. Dhall, A., Goecke, R., Gedeon, T.: Automatic group happiness intensity analysis. IEEE Trans. Affect. Comput. 6(1), 13–26 (2015)
4. Tang, J., Zhang, Y., Sun, J., Rao, J., Yu, W., Chen, Y., Fong, A.C.M.: Quantitative study of individual emotional states in social networks. IEEE Trans. Affect. Comput. 3(2), 132–144 (2012)
5. Saari, P., Eerola, T.: Semantic computing of moods based on tags in social media of music. IEEE Trans. Knowl. Data Eng. 26(10), 2548–2560 (2014)
6. Bao, S., Xu, S., Zhang, L., Yan, R., Su, Z., Han, D., Yu, Y.: Mining social emotions from affective text. IEEE Trans. Knowl. Data Eng. 24(9), 1658–1670 (2012)
7. Turney, P.D.: Thumbs up or thumbs down? Semantic orientation applied to unsupervised classification of reviews. In: Proceedings of the 40th Annual Meeting of the Association for Computational Linguistics (ACL), Philadelphia (2002)
8. Munezero, M., Montero, C.S., Sutinen, E., Pajunen, J.: Are they different? Affect, feeling, emotion, sentiment, and opinion detection in text. IEEE Trans. Affect. Comput. 5(2), 101–111 (2014)
9. Medhat, W., Hassan, A., Korashy, H.: Sentiment analysis algorithms and applications: a survey. Ain Shams Eng. J. 5(4), 1093–1113 (2014)

Design of Lagrangian Decomposition Model for Energy Management Using SCADA System

R. Subramani and C. Vijayalakhsmi

Abstract This paper mainly deals with the design and analysis of an Energy Management model using a SCADA (Supervisory Control and Data Acquisition) system. Each power system is restricted by its applicable control authority, forming a decentralized structure by using consistent network. A central optimal power flow problem is decomposed into distributed subproblems to obtain the optimal solution. A new energy management model is designed which enables a flexible and efficient operation of various power plants. Based on the numerical calculations and graphical representations the renewable energy sources in both configurations is independent of the enduring or intermittent main energy resource availability, which leads to effective production.

Keywords Distribution automation · SCADA (Supervisory Control and Data Acquisition) · Transmission · Lagrangian relaxation · Energy management · Optimal power flow

1 Introduction

Electrical energy is fundamental factor of the industrialized and all round progress of any country. Centrally, electricity is generated in huge quantity and transmitted efficiently. Electrical energy is preserved on each step in the procedure of production, transmission, distribution and consumption of electrical energy. It is very difficult to handle the electrical industries which are very complex in the largest industries in the world. Particularly the electrical industries have very challenging

R. Subramani (✉) · C. Vijayalakhsmi
SAS, Mathematics Division, VIT University, Chennai, India
e-mail: subramanivit@gmail.com

C. Vijayalakhsmi
e-mail: vijusesha2002@yahoo.co.in

© Springer International Publishing Switzerland 2016   353
V. Vijayakumar and V. Neelanarayanan (eds.), *Proceedings of the 3rd International Symposium on Big Data and Cloud Computing Challenges (ISBCC – 16')*,
Smart Innovation, Systems and Technologies 49, DOI 10.1007/978-3-319-30348-2_30

problems to be handled through power Engineering, in designing potential power system to distribute large size of electrical energy. These procedures as well as its control play a vital role in the Electrical Power Engineering world. During this automated electricity forward world, it is important that the power utility management has to be taken case for energy efficiency.

SCADA systems are mainly used in huge and superior levels of power systems, to collect the data such as transmission measurements, status information, and control signals to and from Remote Terminal Units (RTUs) must be changed to play their expected functions at the core of the concept of smart grid the grid of the future. A successive SCADA result is more important to effective operation of a utility's most significant and costly distribution, transmission, and generation assets. It is an open loop system. These systems may be collective with a data acquisition system by adding the use of coded signals over communication channels to acquire information about the status of the remote equipment for display or for recording functions. SCADA system is used for real time control and monitoring of electrical network from the remote location. The proposed SCADA is flexible and preserves smart performance, have been dynamically adapted to the context of the power system.

A Distribution Control System (DCS) is a process control system that uses a network to interconnect sensors, controllers, operator terminals and actuators. A DCS typically contains one or more computers for control and mostly use both proprietary interconnections and protocols for communications SCADA may be called Human-Machine Interface (HMI). The basic design of a SCADA system contains multi controller stations and a supervisory control station in limited or in secluded. Throughout the control system consists of Remote Terminal Units (RTU) and Programmable Logic Controllers (PLC) called data collection collective units which utilizes control station, operators can monitor status and provide the corresponding commands to the appropriate devices.

2 Major Concerns of Power System

(i) **Superiority**: Stability of the preferred frequency intensity and voltage;
(ii) **Consistency**: Very low breakdown rate of the system and components;
(iii) **Safety measures**: Robustness—usual situation yet at the back disorder;
(iv) **Permanence**: Preserve synchronism under disorder; (v) **Wealth**: Optimizing the resources, supervision and preservation Costs.

The key features of SCADA systems are given as follows:

(i) New stage in stimulating grid consistency—improved profits; (ii) Practical difficulty identification and decision—high consistency; (iii) Gathering the mandatory electric power authority requirements—improved consumer fulfillment; (iv) Dynamic conscious resolution providing—cost deductions and enlarged proceeds.

3 Applications of a SCADA System in Power Stations

- *Fault location, separation*
- *Load Balancing*
- *Load Control*
- *Remote Metering.*

4 Literature Review

In past research, a complete introduction about Smart SCADA and Automation System in Power Plants was discussed by Desai et al. [1]. Vale [2] proposed a methodology which reveals that Energy contextual energy resource management. Maheswari and Vijayalakshmi [3] implemented a Lagrangian Decomposition Model for Power Management systems by Distribution Automation. Zhong et al. [4] proposed a decomposition method for integrated dispatch of generation and load which reveals that the parallel computing techniques are utilized with IEEE 30-bus and 118-bus to accelerate the computations. Various applications of phasor measurement units (PMUs) in electric power system networks and utilities are discussed by Singh et al. [5]. Logenthiran et al. [6] combines Lagrangian Relaxation (LR) simultaneously Evolutionary Algorithm (EA) to provide a hybrid algorithm for the real time energy environments. Shalini et al. [7] proposed a model which presents the project management phases of SCADA for real time implementation. Leonardi and Mathioudakis [8] focused on the Industrial Control System of the electrical sector, particularly on the Smart Grid which provides that the essential background information on SCADA and utility applications that run typically at the control center of transmission or distribution grid operators.

Felix et al. [9] reviewed the functions and architectures of control centers; their past, present and likely future. Sayeed Salam [10] compressed a unit commitment problem which reveals the generation of thermal and hydro units using Lagrangian Relaxation technique. Vale [11] has proposed a model reveals that the requirements of upcoming power stations operating in the environment of the higher penetration of distributed generation. Dobriceanu and Bitoleanu [12] proposed a model for the water distribution stations using a SCADA system which allows the optimum performance of the pumping system, safety and survival growth in the equipments and installations exploring, and so obtained the efficient energy utility and most favorable supervision of the drinkable water. Roy [13] proposed a complete controller of an electrical power system network using SCADA system. Felix et al. [9] reviewed the functions and architectures of control centers; their past, present and likely future. Ausgeführt [14] combined decomposition algotithm with Meta heuristics for the Knapsack Constrained Maximum Spanning Tree Problems. Nishi [15] anaysed the solution procedure and structure for various Petri Nets using LR technique.

5 Parameters

N—The set of connected nodes for wireless network	Nt—Total study time (hours)
$F_{k,i}$—Input of flow in time slot k	V_i—Voltage magnitude of flow i
F_i^{min}—Minimum input of flow i	V_i^{min}—Lower bounds of voltage magnitude
V_i^{max}—Upper bounds of voltage magnitude	I_{PVi}—Photovoltaic i unit electric power
θ_i—Voltage phase angle of flow i	I_{Vi}—Voltage i unit electric power
θ_i^{min}—Lower bounds of voltage phase angle for each flow i	I_{PVacti}—Photovoltaic i unit rapidly available electric power
θ_i^{max}—Upper bounds of voltage phase angle for each flow i	$I_{WINDacti}$—Wind i unit rapidly available electric power
e_i—per unit transmit power of flow i	$I_{BIOacti}$—Biomass i unit rapidly available electric power
C_i—Cost resource of the system in flow i	I_{Vacti}—Voltage i unit rapidly available electric power
C_{si}—Set up cost of unit i at time slot k	$P_i(k)$—Power output of unit i at time slot k in mega volt (MV)
C_{vi}—Voltage Cost for unit i at time slot k	$R_i(k)$—Allocated reserve power of generation i at time slot k
C_{pvi}—Production cost of photovoltaic unit i at time slot k	P_i^{min}—Minimum power flow transmission line i at time slot k
C_{BIOi}—Production cost of Biomass i unit	P_i^{max}—Maximum power flow transmission line i at time slot k
C_{WINDI}—Production cost of Wind i unit	$P_D(k)$—System load demand at time slot k (MV)
C_{Fi}—Production cost of Fuel thermo-electric i unit	$I_i(k)$—Commitment state (0 or 1) of flow i at time slot k
I_{BIOi}—Biomass i unit electric power	
I_{WINDi}—Wind i unit electric power	
I_{Fi}—Fuel thermo-electric i unit electric power	

6 Optimization Model Formulation

$$\text{Min} \sum_{i=1}^{N} \sum_{k=1}^{Nt} [C_i I_i(k) + C_{vi} I_i(k) + C_{Si} I_i(k) + C_{pvi} I_i(k) + C_{Fi} I_i(k) + C_{WINDi} I_i(k) + C_{BIOi} I_i(k)]$$

(1)

$$\text{Max} \sum_{i=1}^{N} \sum_{k=1}^{Nt} F_{k,i} \tag{2}$$

$$\text{Min} \sum_{i=1}^{N} \sum_{k=1}^{Nt} \left[C_i + C_{vi} + C_{Si} + C_{pvi} + C_{Fi} + C_{WINDi} + C_{BIOi} \right] \tag{3}$$

subject to

$$\sum_{k=1}^{Nt} F_{k,i} + \sum_{i=1}^{N} V_i \geq \sum_{k=1}^{Nt} \sum_{i=1}^{N} P_i^{\min} \tag{4}$$

$$\sum_{i=1}^{N} I_{Fi} + \sum_{i=1}^{N} I_{BIOi} + \sum_{i=1}^{N} I_{WINDi} + \sum_{i=1}^{N} I_{PVi} + \sum_{i=1}^{N} I_{Vi} \geq \sum_{i=1}^{N} I_{DEMAND} \tag{5}$$

$$\sum_{k=1}^{Nt} \sum_{i=1}^{N} F_{k,i} \geq F_i^{\min} \quad \forall i \tag{6}$$

$$R_i(k) + P_i(k) \leq P_i^{\max}(k) \quad i = 1, 2, \ldots, N; \; k = 1, 2, \ldots, N_t \tag{7}$$

$$P_i^{\min}(k) \leq P_i(k) \leq P_i^{\max}(k) \tag{8}$$

$$0 \leq R_i(k) \leq P_i^{\max}(k) - P_i^{\min}(k) \tag{9}$$

$$\sum_{k=1}^{Nt} \sum_{i=1}^{N} P_i I_i(k) \leq P_D(k) \tag{10}$$

$$V_i^{\min} \leq V_i \leq V_i^{\max} \quad \forall i \tag{11}$$

$$\theta_i^{\min} \leq \theta_i \leq \theta_i^{\max} \quad \forall i \tag{12}$$

$$0 \leq I_{Vi} \leq I_{Vacti} \quad \forall i \tag{13}$$

$$0 \leq I_{PVi} \leq I_{PVacti} \quad \forall i \tag{14}$$

$$0 \leq I_{WINDi} \leq I_{WINDacti} \quad \forall i \tag{15}$$

$$0 \leq I_{BIOi} \leq I_{BIOacti} \quad \forall i \tag{16}$$

$$V_i - e_i \theta_i \geq 0$$
$$I_{Fi} \geq 0 \quad \forall i \tag{17}$$

7 Lagrangian Decomposition Model

Lagrangian Relaxation (LR) and Dynamic Programming are extensively used to develop the grade of an industry. While compared to the other mathematical approaches these methods are has a reasonable computation time. Normally the optimal bounds for the optimization problems are obtained by using LR technique. Due to the flexibility in dealing with the different types of constraints the proposed method (LR) is more beneficial. An iterative method used to solve the subproblems; on solving subproblems the Lagrangian multipliers are frequently updated. A heuristic method is applied to adjust the present infeasible solution to the invention of feasible solution. These optimal solutions are evaluated by using duality gap called optimal bounds. Lagrangian relaxation is flexible in incorporating supplementary combination constraints that have not been calculated so far. Each constraint has its Lagrangian multiplier for each period of iteration those are adjoined into the objective of the proposed Lagrangian problem. However, the degree of sub optimality and the number of units are inversely related. These feasible branches are obtained by relaxing one or more constraints with the help of Lagrangian multiplier.

The LR approach replaces the problem of identifying the optimal values of all the decision variables with one of finding optimal values for the Lagrangian multipliers. This has been done in a Lagrangian based heuristic method. The main advantage of Lagrangian-based heuristics is that the production of optimal boundaries of the optimal solution for the main problem. For all standard dataset to the Lagrangian multipliers, the solution lies in the Lagrangian model which is less than or equal to the solution of the original model. Optimization model is designed for the power generation and distribution problem. The central theme of the objective of the proposed model is to minimize the sum of the total operating cost and economic dispatch costs. These factors are subject to demand, reserve, generator capacity and generator schedule constraints. The proposed Lagrangian Decomposition Model (sub-problem) gives the minimum production cost to the Optimization Model.

Generation of power can be distributed through a substation which is a part of an electrical generation, transmission, and distribution system. Substations transform voltage from low to high, or the reverse, or perform any of several other important functions. Between the generating station and consumer, electric power may flow through several substations at different voltage levels. In Power Distribution, Optimal Power Flow (OPF) Problem minimizes the total generation cost with respect to active power and reactive power. Optimizing the reactive power distribution network is a sub problem of OPF which minimizes the active power losses in the transmission network. Transmission network losses can be minimized by installing Distributed Generation capacity (DG) units. These DG units increase the maximum load ability of the system by improving the system voltage profiles on the distribution system. Studies on computational results for the proposed optimization models indicate that the Lagrangian Relaxation technique is capable of solving large problems to the acceptable convergence.

A LR approach is presented for the optimization problems in different areas of power system. Optimization models have been formulated with respect to various parameters and operating conditions of the power system. The solutions of the Optimization models are obtained from Lagrangian Decomposition (LD) models. These LD models are achieved by Lagrangian Relaxation (LR) technique. In this relax-and-cut scheme, the reformulation of the optimization problems which contains addition of the relaxed constraints is associated with the Lagrangian multipliers with respect to the additionally introduced variables which improve the convergence rate dramatically. In order to obtain the optimal bounds of the above objective function LR technique has been implemented. Based on the objective function, the optimal solution is obtained. The difference between the lower and upper bounds is defined as the "gap".

$$L[U, \lambda] = \text{Min} \left[\text{Max} \left[\sum_{i=1}^{N} \sum_{k=1}^{Nt} F_{k,i} + \sum_{i=1}^{N} \sum_{k=1}^{Nt} \lambda_i (F_{k,i} - F_i^{\min}) \right] \right] \tag{18}$$

Subject to

$$P_i^{\min}(k) \leq P_i(k) \leq P_i^{\max}(k) \tag{19}$$

$$0 \leq R_i(k) \leq P_i^{\max}(k) - P_i^{\min}(k) \tag{20}$$

$$V_i^{\min} \leq V_i \leq V_i^{\max} \quad \forall i \tag{21}$$

$$\theta_i^{\min} \leq \theta_i \leq \theta_i^{\max} \quad \forall i \tag{22}$$

$$0 \leq I_{Vi} \leq I_{Vacti} \quad \forall i \tag{23}$$

$$0 \leq I_{PVi} \leq I_{PVacti} \quad \forall i \tag{24}$$

$$0 \leq I_{WINDi} \leq I_{WINDacti} \quad \forall i \tag{25}$$

$$0 \leq I_{BIOi} \leq I_{BIOacti} \quad \forall i \tag{26}$$

$$V_i - e_i \theta_i \geq 0$$
$$\lambda_i \geq 0 \quad \forall i \tag{27}$$

LR technique replaces the main problem with its relevant subproblems called Lagrangian problem. The optimal solution the Lagrangian problem provides the bounds of the objective function of the problem. The optimal solution is achieved by relaxing one or more constraints of the proposed model, and adding these constraints, multiplied by the corresponding Lagrangian multiplier (λ) in the objective function. The objective of the proposed model is to relax (eliminate) the constraints with the purpose of getting relaxed problem. When the Lagrangian Multiplier gets the solution from the relaxed problem, it is very much easier to obtain optimal values of the main problem. The role of these multipliers is to derive the Lagrangian problem towards a solution that satisfies the relaxed constraints.

Based on the solution part of the Lagrangian problem normally the Lagrangian multipliers are attains the solutions from the relaxed constraints. Some of the Lagrangian based algorithms include supplementary heuristics to adapt these infeasible solutions to feasible solution. By this approach, the researchers are able to generate suitable solutions to the original model. The optimal solution is obtained at any point of extremum bound in all the procedure of solving feasible solution. When the optimal gap reaches zero (Based on the model it can take some minimum value) then the optimal solution has been obtained. Or else, when the optimal gap gets suitably small (for example <1 %), the analyst might stop the process as well as be fulfilled that the current feasible solution is in 1 % of optimality. In this paper, Non Linear Integer programming model is designed, the optimal scheduling decision for each power flow is obtained by Lagrangian dual which minimizes the power flow route losses.

8   Conclusion

This paper provides an optimal energy control efficient strategy for the electricity systems by power generation and optimal load management systems. A Lagrangian decomposition algorithm has been formulated as a Non-Linear Integer Programming problem with respect to variety of optimal constraints energy cost, etc. A novel SCADA-based decentralized approach is that proposes for power management in today's electricity forwarded market for efficient energy power management system. The capacity of transmission lines is optimally allocated to individual transactions for maximizing the social welfare. In the efficient power management systems, spot prices and reserved prices in the market that are the significant parameters for the profit based unit commitment problems.

References

1. Desai, P., Mahale, S., Desai, P.: Smart SCADA and automation system in power plants. Int. J. Curr. Eng. Technol. (2014) E-ISSN 2277-4106, P-ISSN 2347-5161
2. Vale, Z.: Distribution system operation supported by contextual energy resource management based on intelligent SCADA. Renew Energy **52**, 143–153 (2013)
3. Maheswari, Vijayalakshmi: Implementation of Lagrangian decomposition model for power management using distribution automation. In: National Conference on Frontiers in Applied Sciences and Computer Technology (FACT'12) 6, 69–76 (2012)
4. Zhong, H., Xia, Q., Kang, C.: An efficient decomposition method for the integrated dispatch of generation and load. IEEE Trans. Power Syst. (2015) (0885-8950 ©)
5. Singh, B., Sharma, N.K., Tiwari, A.N.: Applications of phasor measurement units (PMUs) in electric power system networks. Int. J. Eng. Sci. Technol. **3**(3), 64–82 (2011)

6. Logenthiran, T., Woo, W.L., Phan, V.T.: Lagrangian relaxation hybrid with evolutionary algorithm for short-term generation scheduling. Electr. Power Energy Syst. **64**, 356–364 (2015)
7. Shalini, Sunil Kumar, Birtukan Teshome: Working phases of SCADA system for power distribution networks. Int. J. Adv. Res. Electr. Electron. Instrum. Eng. **2**(5), 2037–2043 (2013)
8. Leonardi, A., Mathioudakis, K.:Towards the smart grid: substation automation architecture and technologies. Adv. Electr. Eng. **2014** (2014) (Article ID 896296)
9. Felix, Moslehi, K., Bose, A.: Power system control centers: Past, present, and future, In: Proceedings of the IEEE. **93**(11), 1890−1908 (Nov. 2005)
10. Sayeed Salam, MD.: Solution to short-temunit commitment problem. In: Energy Systems, pp 255–292, Springer, Berlin (2010)
11. Vale, Z.A., Morais, H., Silva, M., Ramos, C.: Towards a future SCADA. IEEE Xplore, 29 June 2015
12. Dobriceanu, M., Bitoleanu, A.: SCADA system for monitoring water supply networks. WSEAS Trans. Syst. **7**(10) (Oct. 2008)
13. Roy, R.B.: Controlling of electrical power system network by using SCADA. Int. J. Sci. Eng. Res. **3**(10) (Oct. 2012) ISSN 2229-5518
14. Ausgeführt, AM.: A Lagrangian decomposition approach combined with meta heuristics for the knapsack constrained maximum spanning tree problem. A master's thesis, Viena university of technology (2008). ISBN 978-3-540-70807-0
15. Nishi, T.: Lagrangian relaxation approach for solving optimal firing sequence problems by decomposition of timed petri nets. In: SICE Annual Conference 2008. The University Electro-Communications, Japan (2008)
16. Arnold, M., Knopfli, S.: Improvement of optimal power flow (OPF) decomposition methods applied to multi-area power systems. IEEE Trans. (2007). doi: 10.1109/PCT.2007.4538505, pp. 1–6
17. Salam, S.: Unit commitment solution methods. World Acad. Sci. Eng. Technol. **11** (2007)

Twitter Data Classification Using Side Information

Darshan Barapatre, M. Janaki Meena and S.P. Syed Ibrahim

Abstract The important aspect of information gathering is to know what people think about particular topic. With the growth of IT and Internet, resource of opinion-rich text goes on increasing and thus the importance of text classification is increased in order to determine opinion about a subject. The main goal of sentiment analysis is to define sentiment or polarity of a sentence or a piece of text. Polarity means to check whether a particular piece text is positive, negative or neutral. The goal is to extract useful data and present it in a clear way so that one can get a general idea about people's opinions. While doing Text analytics on posts or tweets obtained from social networking site, along with text some useful information is available know as Side Data. Such side data may be found in the form of access behavior of user from web logs, referred links to sites, special symbols, emotions, etc. This side data is helpful in predicting the polarity of the text. Further, the process of classification can be done using SparkR to improve processing speed and to analyze huge amount of data.

Keywords Big data analytics · Text classification · Sentimental analysis · Side data · Distributed processing · Map reduce · SPARK

1 Introduction

Big data is a term used to represent large amount of data in a structured, semi-structured and unstructured form which is difficult to process in traditional database software. Big Data magnitude is constantly changing from dozens of

D. Barapatre (✉) · M. Janaki Meena · S.P. Syed Ibrahim
School of Computing Sciences and Engineering, VIT University,
Chennai Campus, Chennai, India
e-mail: darshan12patre@gmail.com

M. Janaki Meena
e-mail: Janakimeena.m@vit.ac.in

S.P. Syed Ibrahim
e-mail: Syedibrahim.sp@vit.ac.in

© Springer International Publishing Switzerland 2016
V. Vijayakumar and V. Neelanarayanan (eds.), *Proceedings of the 3rd International Symposium on Big Data and Cloud Computing Challenges (ISBCC – 16')*,
Smart Innovation, Systems and Technologies 49, DOI 10.1007/978-3-319-30348-2_31

terabyte to numerous petabytes due to rise of IT and high inclusion of Internet. To gain value from this data, big data analytics is used where it processes huge data and provides value in a timely manner.

Sentiment analysis is the most discussed topic of natural language processing and text mining. Sentiment is a thought, view, attitude, opinion or feeling toward something, such as a person, organization, product or location. It involves extraction of useful information obtained from posts or tweets of social networking sites in order to analyze the polarity of a particular object. It is useful for analyzing trends about a particular topic in the social media, helpful in marketing and research. It has its uses in knowing popularity of new products, customer reviews on political issues, in industrial markets, etc.

Huge amount of side data is present along with the text but extracting meaningful information from it is a real task. Side data is some extra information present with the actual data which can be used to provide the additional information to the classifier for classifying the text in the categories of emotions. Attributes for side data such as user web-logs, links and other non-textual attributes are hidden in the documents. Such attributes provide insight about a particular subject in order to improve opinions and reviews while performing sentiment analysis.

Side data can be found in many forms such as data about the user location or the date and time when the tweets or post has been made. Moreover, if the user log files are considered then this data may contain some information about the links which user are frequently visiting. The analytics may include categorizing the users according to their interests and suggesting websites to them based on their previous interests. In cases, such data can sometimes useful in improving the quality of the classification process, but it can also be a risky when the side-information present is very complex. In such cases, it may also reduce the quality of mining process. The basic approach deals with the process of classification in which attributes and side data gives similar hints about the sentiment of text, and also ignore those aspects which are conflicting.

Main goal of this paper is to study the classification of text data which is obtained using various social networking sites, identifying and categorizing opinions from it to determine whether the user's opinion about a particular subject is positive, negative, or neutral using the side data and to process it using Spark environment.

2 Problem Statement

Accuracy of the twitter data classification may be improved when classifier is constructed along with side data. Moreover, large amount of data can be analyzed using Spark framework.

3   Literature Survey

Big data is a collection of data that is huge and yet growing exponentially with time. Its data is so large and complex that none of data management tools are able to store and process it. It was mainly proposed to maintain huge amount of structured and unstructured data and to extract knowledge from the same.

Subset of this broad field is "Text Mining". It is the process of extracting useful information from large volume of unstructured data usually in the form of text. Text mining uses Data mining techniques for discovering useful patterns for texts. In contrast with data mining, text mining extracts patterns from natural language text rather than databases. It can help organizations in deriving valuable data from text based content such as text documents, E-mails, Social media posts, etc. One such type is "Sentiment Analysis". Sentiment analysis is the task of identifying positive or negative opinions and categorizing these opinions (obtained from piece of text for e.g. blogs, user comments, community web sites, etc.) in order to determine user's opinion about particular topic.

R programming Tool provides the functionality for the data extraction from social networking sites Facebook and Twitter. It provides a package called "Sentiment" which distributes text into six classes namely Joy, Sadness, Fear, Anger, Disgust and Unknown. We can also calculate Polarity (positive, negative or neutral) of the posts. Social media's providing access to its data via Access Token which we have to generate by creating app on developer site [1].

R tool is now integrated with Apache Spark. Spark is an open-source, fast, in-memory data processing engine. SparkR is an R package that allows us to analyze large volume of data and interactively run programs on them from the R tool. It uses Spark's parallel DataFrame abstraction. We can create SparkR DataFrames from "local" R data frames. SparkR supports all the Spark's operations and other analytical functions. They also support converting query results to and from DataFrames. SparkR uses the Spark's parallel engine, which gives us parallel computing environment to scale larger sizes of data. It is a great package which provides flexibility of the R language in order to perform fast computations on very large volume of datasets. Spark computations are distributed over all the nodes available on the Spark cluster, thus it can be used to analyze large volume of data using R [2].

But accuracy of this technique is low because of sarcastic post and difficulty in interpreting the mood of a subject. To improve accuracy, we can use "Side Information" present with the posts or tweets. The side data considered in case of the social networking site is the like count or the comment count or the share count which can give the importance of the comment or Tweet. Given the character limitations on tweets, classifying the sentiment of Twitter messages is most similar to sentence-level sentiment analysis. However, the informal and specialized language used in tweets, as well as the very nature of micro blogging domain make Twitter sentiment analysis a very different task [3].

Fig. 1 Existing system

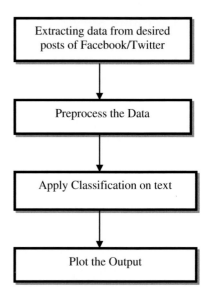

4 Existing System

Figure 1 shows the existing system. In existing system, desired data from social networking sites is extracted. This can be done using R tool by using packages Rfacebook and twitterR. Extracted data is then preprocessed by removing special symbols, numbers, links, spaces, etc. which are not required for analysis. After preprocessing, we apply classification on text. R tool provides a package called "Sentiment" which distributes text into Joy, Anger, Sadness, Fear, Disgust and Unknown. Using Bayes Algorithm, we can define Polarity of text as Positive, Negative or Neutral. Finally, plot the desired output.

5 Proposed System

In proposed system, we have to take care of large data and to access it efficiently. Large amount of data can be extracted through **Flume** and stored on HDFS. To process this large amount of data that we have to use Spark framework. This data is then efficiently processed by Spark on multi-node environment. We try to improve accuracy of classifier using side information. Side information is the additional data present with actual text which can help in improving accuracy. On comparing results of these two, using Side data accuracy of classifier may be improved (Fig. 2).

Fig. 2 Proposed system with
Spark framework

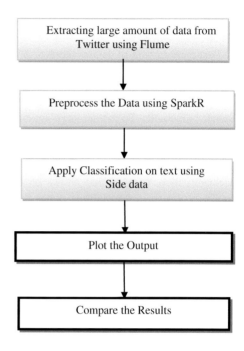

Extracting large amount of data from
Twitter using Flume

Preprocess the Data using SparkR

Apply Classification on text using
Side data

Plot the Output

Compare the Results

6 Dataset

Data can be obtained from social networking sites like Facebook or Twitter using
Flume and **R tool** packages like **Rfacebook** and **twitterR**. The pre-processing of
the data can be done using R packages **tm** and **Snowball**. Classification is then
applied on gathered data.

7 Experimental Analysis

As a starting point for this project, the tweets or posts obtained from the social
networking sites are taken from which we can tell sentiment about a particular
subject is positive, negative or neutral. To do so we can use R tool packages
Rfacebook and **twitterR**. Using these packages, we can get Posts or tweets on
required topic. To access data from Facebook or Twitter, we have to generate app
on respective developer site which provides **Access Token** through which we can
access data. Now we have to preprocess data to make it ready for Sentiment
Analysis by removing blank spaces, symbols, links, numbers, etc. After that we
perform Sentiment Analysis on preprocessed data using package called Sentiment.
It classifies the preprocessed text into six categories Joy, Sadness, Fear, Anger,

Disgust and Unknown. And also polarity as weather posts or tweets is positive, negative or neutral.

For small amount of dataset (approx. 500 kb), it ran in very small time of around 10–20 s. But for the file size of about 100–200 mb it will take very long time. So we have to try the same with Big data platform.

We can use multi-node environment to process large volume of data using **Spark** framework. To extract large amount of data from social networking site we can use **Flume**. Flume lets Hadoop users ingest high-volume data into HDFS for storage. It holds large amount of data and provides easier access. R tool is now integrated with Spark. To provide Spark environment within Rtool R provides a package called **SparkR** that allows us to analyze large volume of datasets and parallelly run them on multiple nodes from Rshell. **SparkR** is based on Spark's parallel Data Frame abstraction. Using this package, we can classification of text can be automatically distributed across all nodes Spark cluster. So this package can be used to analyze terabytes of data using R. Further the work with the side data can be carried out by taking those tweets or posts which are incorrectly classified and using some extra information to classify those tweets and check whether the accuracy can be increased or not.

8 Conclusion

In this project, text classification is being used in order to determine the opinion of a speaker about particular subject. Using SparkR, we improved the processing of large data in multi-node environment. We use side data so that accuracy of the classifier may be improved.

References

1. Access to Facebook API via R. Document of Rfacebook
2. Blog on announcement of SparkR: http://blog.revolutionanalytics.com/2015/06/sparkr-announcement.html
3. Barbosa, L., Feng, J.: Robust sentiment detectionon twitter from biased and noisy data. In: Proceedings of Coling (2010)
4. Sentiment Analysis on Facebook Data: Text analytics Tutorial. A guided example by Manoj Kumar

Part VI
Internet of Things

IoT Based Health Monitoring System for Autistic Patients

K.B. Sundhara Kumar and Krishna Bairavi

Abstract The transition towards the field of Internet of Things is occurring exponentially and with the advent of sensors and automation, a revolution in the healthcare sector is certain. A persistent deficit in social communication and the social interaction among people, especially children is termed as Autistic Spectrum Disorder. Our system provides an automatic monitoring framework, using sensors, for persons affected by autism. The system keeps track of the sensor readings that are obtained from the brain signals of affected persons. The analytics are performed on the collected data and a notification is constantly sent to their care-takers.

Keywords Autism · IoT · Wearable devices

1 Introduction

The ability to continuously obtain values and analyze them has already gathered a lot of attention in the field of health-care. Rapid developments in this field as well as in sensors and automation [1] could prove a boon, especially to the people who cannot be physically present at all times to take care of the ill.

Autistic Spectrum Disorder is a group of complex and heterogeneous developmental disorders involving multiple nervous system dysfunctions, that are generally found in children in early stage of development. There are a lot of studies that have attempted to classify the nervous system abnormalities associated with ASD. This disorder is highly heritable and has high recurrence rates in siblings.

With technology like Internet of Things ruling the roost, wearable devices and wireless body area network sensors are mainly used to curtail mortality from

K.B. Sundhara Kumar (✉) · K. Bairavi
SSN College of Engineering, Chennai, India
e-mail: sundharakumarkb@gmail.com

K. Bairavi
e-mail: krishnabairavi95@gmail.com

© Springer International Publishing Switzerland 2016
V. Vijayakumar and V. Neelanarayanan (eds.), *Proceedings of the 3rd International Symposium on Big Data and Cloud Computing Challenges (ISBCC – 16'),*
Smart Innovation, Systems and Technologies 49, DOI 10.1007/978-3-319-30348-2_32

chronic diseases, by monitoring health parameters. Wearable devices are networked devices that can collect data and track activities. These wearable technologies are among the fast-growing subset of IoT and promise to provide more assistance in the health-care domain.

ASD can be detected by taking Electroencephalograms. EEGs contain very useful information pertaining to different psychological states of brain and it's a very effective tool to understand dynamics related to the brain.

Our proposed system uses a neurological sensor to collect the EEG waveforms of patients and monitors this reading continuously. The system then sends out e-mail alerts to the relevant persons. If the system detects any emergency conditions, it alerts the respective doctor monitoring the patient on a regular basis.

2 Related Work

This section includes the most relevant works on our study. In this literature survey, a comprehensive review on sensors and various techniques of data acquisition from brain waves are discussed. These works are dedicated to autism related studies.

The authors of paper Nabar et al. [2] proposed an optimized framework to achieve an optimal trade-off between the energy consumption and latency in wireless body sensor networks. They proposed a polling based algorithm that resolves the scheduling conflicts among the devices.

Emily Waltz [3] suggests that autism could be diagnosed in early development stages in children using EEG. The study conducted on normal children and the autistic ones prove that there were different coherent factors that distinguished a normal kid from an autistic one. The connectivity pattern of the brain differed in autistic children mostly in the left hemisphere which is responsible for speech, due to which most of the autistic people have difficulties in communicating. This paved way to proper diagnosis in early development stages using EEG, as opposed to diagnosis through abnormal activities of the kid as in the past.

The authors of paper [4] have proposed an automated expert system for seizure and epilepsy detection and classification. Epilepsy is characterized by uncontrolled excessive activity by either a part or all of the central nervous system (CNS) due to the presence of abnormalities. These abnormalities could be detected and classified into different types by using SVM classifier which acts as an automatic classifier for Disease Diagnostic Expert System (DDES) by placing electrodes on the brain using 10–20 electrode system.

The authors of paper [5] give an incisive view that EEG can be used to study brain physiology, as it has high tolerance and temporal ability. It is suitable to study abnormalities present in autism and is a non-invasive technique. Also, spectral power coherence and hemispherical asymmetry were explained along with the EEG resting state abnormalities. These resting state EEG abnormalities help in providing a detailed platform for the study of Autism spectral disorder (ASD).

The authors of paper [6] designed a dry sensor that overcame the drawbacks of the conventional wet sensor that needed pre-requisites like proper abraded skin, and conductive gel that should be applied to patient before starting the recording procedure. The designing of naval dry sensors ameliorated this arduous task of applying gel and abrading the skin which is not possible every time. A 16 channel wearable wireless EEG dry sensor was designed and proposed by Shang-Lin Wu et al., which could trap EEG signals from the brain non-invasively without following any abrading procedure or applying conductive gel. It is cost effective and was made into a wearable cap-like device of varied sizes. Hence it could be worn by all, irrespective of the head size or age. The EEG waveforms are then sent to the computer using brain–computer interface (BCI).

3 Proposed Work

The philosophy of this work is to provide long-term continuous support and instantaneous health information of autism affected children to the medical practitioners and the caretakers. The health information of the autistic patients is collected using head-gear wearable sensor.

The brain readings of the affected persons are then sent to a continuously monitoring server. On occurrence of any anomalies in the readings, alerts are raised and an e-mail is sent to the respective care takers. In case of any emergency, alerts are also sent to the medical practitioner. The individual patient's details can also be stored in a cloud environment [7] for effective scalability and security aspects. Figure 1 shows the proposed system architecture.

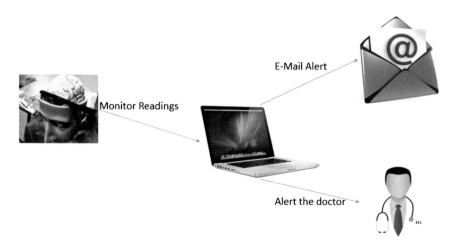

Fig. 1 System architecture

The work flow of this paper is discussed below:

- Collection of brain readings from the sensor
- Process the readings and send alerts to the respective doctor/care taker.

4 Implementation and Discussion

The EEG signals can be sensed from different locations in the brain by using particular electrodes (e.g. 10–20 lead system can be used). FP1, FP2, represent the position of these electrodes that are placed in the head of the subject whose brain readings are to be taken. By selecting the number of channels that should be recorded and the position of the brain from which the signals are supposed to be recorded, we get a graph that displays the amplitude (in micro volts) versus Time (in seconds) relationship of the EEG signal.

The wearable device consists of a sensor along with the lead electrodes that are placed in certain positions, such that once the patient wears the device, brain signals are sensed and can be recorded using the Acquire and Analyze software. The software also has a function by which the position of the brain from which the signals are collected can be changed accordingly.

From the obtained values, these readings can be represented as a graph for better visualization. FP1 and FP2 positions are shown in the Fig. 2. The system then checks whether the obtained readings are normal. Whenever there is any abnormal reading from the patient, an alert is immediately sent to the respective person in-charge. Based on the criticality of the situation, alerts will be automatically sent

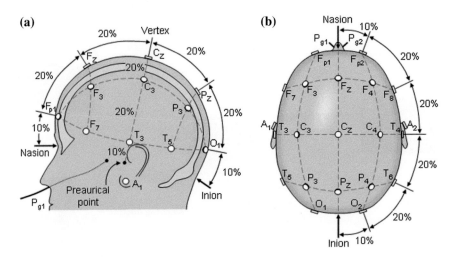

Fig. 2 10–20 lead electrode system

Fig. 3 Signal obtained from electrode FP1 with respect to the reference electrode

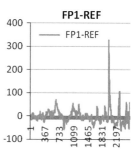

Fig. 4 Signal obtained from electrode FP2 with respect to the reference electrode

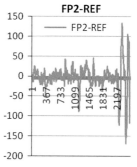

to either the doctor or the caretaker of the patient. By exposing our system to various readings, our system can be trained to predict and give accurate results (Figs. 3 and 4).

Acknowledgements We extend our heartfelt gratitude towards SSN Institutions for providing us with the necessary infrastructure and funding in carrying out this work.

References

1. Benharref, A., Serhani, M.A.: Novel cloud and SOA-based framework for e-health monitoring using wireless biosensors. IEEE J. Biomed. Health Inform. **18**(1):46 (2014)
2. Nabar, S., Walling, J., Poovendran, R.: Minimising Energy consumption in body sensor networks via convex optimization. In: BSN'10: Proceedings of the 2010 International Conference on Body Sensor Networks, pp. 62–67. IEEE Computer Society, Washington, DC, USA (2010)
3. http://spectrum.ieee.org/tech-talk/biomedical/diagnostics/simple-eeg-test-might-help-spot-autism-in-children
4. Pal, P.R., Khobragade, P., Panda, R.: Expert system design for classification of brain waves and epileptic-seizure detection. In: 2011 IEEE Students' Technology Symposium (TechSym), pp. 187–192. IEEE (2011)
5. Wang, J., et al.: Resting state EEG abnormalities in autism spectrum disorders. J. Neurodevelopmental Disord. **5**(1):24 (2013)

6. Wu, S., et al.: Design of the multi-channel electroencephalography-based brain-computer interface with novel dry sensors. In: 2012 Annual International Conference of the IEEE Engineering in Medicine and Biology Society (EMBC). IEEE (2012)
7. Pandeya, P., Voorsluys, W., Niua, S., Khandokerb, A., Buyyaa, R.: An autonomic cloud environment for hosting ECG data analysis services. Future Gener. Comput. Syst. **28**: 147–154 (2012)

Addressing Identity and Location Privacy of Things for Indoor—Case Study on Internet of Everything's (IoE)

Ajay Nadargi and Mythili Thirugnanam

Abstract The "Internet of Everything" (IoE) expresses the delivery together of knowledge, process, people and things to create networked associations additional connected and useful than ever before. IoE encompasses each machine-to-machine (M2M) and web of Things (IOT) technology. The standard security services aren't directly applied on IoT owing to totally different communication stacks and varied standards. Thus versatile security mechanisms square measure required being fictitious, that handle the security threats in such dynamic setting of IoT. Notably, mechanism serves main issues like identity management authentication assessment and optimizing the secure transmission while not revealing their location. This works proposed framework to create user privacy aware by detecting work sensitive content of smart home environment data furthermore as offer privacy preservation for identity and location privacy of IoT devices. Smart Home Environments IoT application is taken into account as a case to look at the privacy and security challenges of smart home environments IoT.

Keywords Internet of things · Internet of everything's · Authentication · Location privacy · Authorization · Network enforced policy · Secure analytics

1 Introduction

Internet of Things (IoT) has turn out as a reliable innovation to enhance the personal satisfaction in smart homes through offering different intuitive, computerized and calm administrations. "Internet of Things (IoT)" may as well be called "Internet of Everything (IoE)". The Internet of Everything (IoE) is changing how people and things connect, how we collect and harness data, and how they all work together to

A. Nadargi (✉) · M. Thirugnanam
School of Computing Science and Engineering, VIT University, Vellore,
Tamilnadu, India

© Springer International Publishing Switzerland 2016
V. Vijayakumar and V. Neelanarayanan (eds.), *Proceedings of the 3rd International Symposium on Big Data and Cloud Computing Challenges (ISBCC – 16')*,
Smart Innovation, Systems and Technologies 49, DOI 10.1007/978-3-319-30348-2_33

377

enable intelligent processes. This is the Internet of Everything (IoE), a paradigm shift that marks a new era of opportunity for everyone, from consumers and businesses to cities and governments.

The Internet of Things (IoT) is a system of all inclusive individual physical articles (or things), their mix with the Internet, and their portrayal in the virtual or advanced world. As appeared in Fig. 1, the IoTs permit individuals and things to be related wherever, whenever, with anybody and anything, in a perfect world utilizing any way/system and any administration.

These IoT applications are categorized in [1]:

- Home and Personal: consist of personal homes [2].
- Enterprise: consist of scales of community [3].
- Utilities: consist of regional and national scale [4].
- Mobile—contains IoT applications stretch out crosswise over multi-space because of dispersed network and scale [5].

In an ordinary smart home environment setting, IoT's delivered by the incorporation of these electronic components are relied upon to sense, activity and convey information gathered from blend of distinctive devices, clients, and PCs associated in the earth with a perspective to reacting with specially designed administrations to clients. Be that as it may, guaranteeing protection and sufficiently giving security in these fundamental administrations introduced through IoT's is a principle issue in smart home situations. A smart home environment is expected to be a modest physical world comprises of various devices on the whole with sensors, shows, computational components, and actuators associating and trading data with clients to present them with agreeable, secured, computerized, and tweaked administrations.

Fig. 1 Schematic view of internet of things

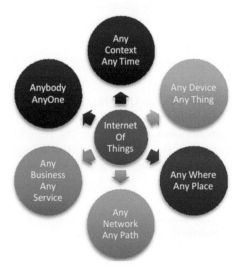

2 Related Works

The essential definitions, patterns and components of IoT are exhibited in paper [6]. It portrays distinctive definitions given by diverse investigates of IoT. The author likewise introduces uses of IoT and the areas which are affected by IoT. Additionally the future examination headings identified with IoT innovation are specified. Then again, paper [7] presents distinctive testing issues while constructing an IoT. It talks about the security issues in IoT. In paper [8] depicts how the setting mindful figuring is valuable for IoT. Setting mindful processing has turned out to be effective in comprehension sensor information. It introduces the fundamental wording in connection mindful processing furthermore demonstrates how this can be material to IoT. The authors introduced review which incorporates strategies, routines, models, applications, frameworks, functionalities and middleware resolutions joined with circumstance mindfulness and IoT. The paper [9] portrays how IoT can be utilized for making Smart City, where the employments of open assets are upgraded. In [10], presented ECC supported mutual authentication protocol for IoT utilizing hash capacities. Mutual authentication is accomplished among terminal hub and stage utilizing mystery key cryptosystem presenting the issue of key stockpiling and administration. The author [1] says in regards to diverse security challenges in Smart City. It depicts the security issues when diverse devices are joined utilizing Internet (IoT). The self guaranteed keys cryptosystem upheld circulated client confirmation plan for WSN is introduced in [11], where just client hubs are verified. However this is not minor answer for IoT. In [12], the author tended to the trouble of authentication and secures correspondence in view of a shared key and is appropriate to restricted area and can't be utilized for expansive district. It manage shared validation however can't be reached out to an asset compelled environment. Effective identification and authentication is exhibited in [13] and depends on the sign properties of the hub yet it is not suitable for portable hubs. The signal's heading is considered as a parameter for node authentication yet it requires more investment to choose the sign bearing with more memory and calculations included. In [14], cluster based authentication is arranged which is chiefly suited for the cutting edge IoT, yet an attacker can get hold of the circulation of cluster key and framework key sets. Era of arbitrary numbers and marks makes significant computational overhead expending memory assets. Portability is imperative part of versatile and remote correspondence and basically out of sight of IoT. With the heterogeneous system topologies like Wi-Fi, LTE and WiMax, authenticated services conveyance with legitimate access control set up on the fly is a major test. Remote Internet Service Provider meandering (WISPr) [15, 16] is a structural engineering, which proposes point by point particulars for permitting between administrators wandering for Wi-Fi customers. WISPr utilizes Remote Authentication Dial as a part of User Service (RADIUS) to give brought together verification and approval. Investigation and security vulnerabilities of RADIUS have been talked about in [17] because of its brought together nature. In [18], the author gave a authentication parameter going amid the handshake. The handshake procedure is tedious and in light of symmetric key cryptography with more memory necessity for vast prime numbers. Security appraisals of EAP have been examined in [19] and investigated numerous shortcoming focuses. Particularly EAP don't address shared confirmation and not impervious to replay attack. Key replication and replay attack on Authenticated Key Agreement (AKA) have been exhibited in [20] which unmistakably demonstrates that there is even a personality is connected with AKA, it is inclined to attack. Relative studies on validation and key assentation techniques for 802.11 remote LANs are displayed in [21]. This best in class in versatile and Wi-Fi environment plainly demonstrates that there is a need of adaptable and secure verification plan. The author [2] expounded outline, execution and assessment of an interruption location framework, called SVELTE, for IoT. Here is intensely related exertion done in the MAGNET venture [22, 23] where insurance affiliations get position with developed correspondence raised and verification stays undetected. Authors offered a dispersed access control key upheld on security layouts aside from attack strife is not uncovered. The authors [24], offered an ECC supported authentication protocol however the primary issue is that it is not Denial of Service (DoS) attack resistant. IoT comprises a number of devices imperviousness to DoS attack is of vital significance.

3 Various Issues Identified in Smart Home Environment

3.1 Security in IoT

One of the central components in securing an IoT foundation is around device identity and mechanisms to authenticate it. Today's authentication and encryption strategies are bolstered on cryptographic suites, for example, Advanced Encryption Suite (AES) for classified Rivest-Shamir-Adleman (RSA) for computerized marks, key transport and information transport and Diffie-Hellman (DH) for key transactions and administration. While the conventions are hearty, they require high register stage an asset that may not exist in all IoT-connected devices. Therefore, authentication and authorization will require proper re-designing to suit our new IoT joined world [25].

3.2 Privacy in IoT

Preservation of privacy has been a concern since the dawn of the Internet. IoT will exacerbate the problem because many applications generate traceable signatures of the location and behavior of the individuals. Privacy issues are particularly relevant in Smart Home, and there are many interesting Smart Home applications that fall within the area of IoT. It can refer to among others the following of smart home equipments, the monitoring of vital insights for types of gear at home or in an assisted living facility. In this environment, it is vital to confirm device ownership and the owner's identity while decoupling the device from the proprietor. Shadowing [25] is a mechanism that has been proposed to achieve this. In essence, digital shadows enable the user's objects to act on her behalf, storing just a virtual identity that contains information about her attributes.

3.3 Security Challenges Within IoT Systems

As plot in the situations and the above attacks in IoT, authentication and privacy are primary security issues which are to be tended to. These components must be considered keeping in mind the end goal to create solid security assurance systems.

- Typically little, modest devices with practically no physical security
- Computing platforms, constrained in memory and compute resources, may not support complex and evolving security algorithms due to the following factors:
  - Restricted security work out capabilities
  - Encryption algorithms need higher processing power
  - Low CPU cycles versus effective encryption
- Designed to operate autonomously in the field with no backup connectivity if primary connection is lost
- Requires secure remote management during and after on boarding
- Identification of endpoints in a scalable manner
  - Individual—e.g., Home Smart Meter
  - Group—e.g., All light bulbs in a room/home
  - Scalability challenges of Individual versus Group
  - Sometimes the location may be more important than the individual identifier (ID)
- Physical Security.

4 Proposed Smart Home Environment Security Framework

To address the highly diverse IoT environment and the related security challenges, a flexible security framework is required. Figure 2 illustrates the security environment from an IoT perspective.

The proposed framework of the secure IoT system is depicted in Fig. 3. IoT secure framework can periodically poll the sensor to retrieve data from it and processing information and actuation. As far as preparing data the framework as of

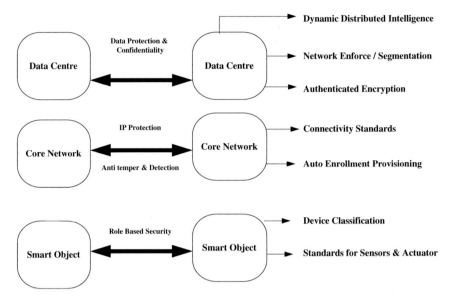

Fig. 2 Proposed IoT security environment

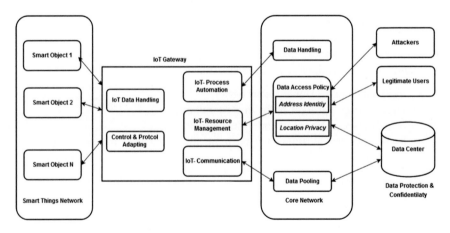

Fig. 3 Secure IoT proposed framework

now takes into access policy approach and for a particular kind of activity. Keeping in mind the end goal to ensure client security, the end-clients must to be in control of access strategies identifying with their own information.

It consist of different parts namely,

1. Smart Things Networks
2. Data Centre
3. Core Networks
4. IoT Gateways.

4.1 Smart Things Networks

Smart things systems comprise of IoT Devices. Equipment segment, element or framework that either measurers properties of a thing/accumulation of things or impacts the properties of a thing/gathering of things or both impacts/measures. Sensors and actuators are devices.

4.2 Data Centre

A Data Centre will act as repositories of data and application. It will be located close to population hubs so they can receive and transmit data to IoT sensors instantly.

4.3 Core Networks

It consists of IoT Data Handling, Data Access Policy and Data Pooling.

4.3.1 Data Access Policy

It is a committed extra part to the Security, Privacy and Trust since it involves usefulness inside the IoT entryway kept up by Trust, Privacy and Security which does not have segments running at this phase of information generation, near the IoT resources or IoT devices. It addresses both authentication and location privacy which is separated in two components:

1. Address Identity
2. Location Privacy.

1. *Addressing Identity* At the heart of this framework is the authentication layer, used to provide and verify the identify information of an IoT entity. When connected IoT devices (e.g., embedded sensors and actuators or endpoints) need access to the IoT infrastructure, the trust relationship is initiated based on the identity of the device.

   How Address Identity works? It includes *Authenticate(UserCredential)* part where *UserCredential* is sort of data utilized by the Authenticate usefulness to check the Users identity to be authenticated (e.g. username–password pair, PIN code, retinal distinguishing proof etc.). At long last, the Authorize segment settles on yielding or denying access to a resource. *Authorise (Authenticate, Resource, ActionType)*. The Authorize function comprises of Authentication,

Resource corresponds to the resources' to be utilized, and *ActionType* assign the activity to be completed upon the resources.

2. *Location Privacy* utilization of soft identities technique, where the authentic identity of the user can be utilized to create different soft identities for particular applications. Each soft identity can be intended for exact application or circumstance without uncovering preventable data, which can manual for security infringement.

The another technique Pseudonymisation alludes to a strategy by which handle that empower identification of a user inside of an information record or Object are substitute by one or more fake identifiers. The reason for existing is to make the Object less confined and these way lower IoT user (e.g. client or patient) restrictions to its utilization.

4.3.2 IoT-Data Handling

It will be vital for Application and Data Service Providers to gather huge total of IoT related information, shaped by a limitless number of IoT assets just about continuously.

4.3.3 Data Pooling

A data pooling is the frequent point in the communication among the end point users, offer synchronization capability of their data. This information is accessible to all end point users in a simple, fast, common and accurate manner.

4.4 IoT-Gateways

It is generally situated at closeness of the devices to be associated. It comprises of IoT-Data Handling, IoT-Resource Management and IoT-communication.

4.4.1 IoT-Communication

IoT-Communication will give normal and bland access to each sort of things regardless of any specialized limitation on interchanges, ordinarily incorporating various conventions and manage intermittence of availability for meandering devices.

4.4.2 IoT-Process Automation

IoT-Process Automation will propose to Services/Application providers general capacities empowering to utilize commitment and arrangement designs that will simplicity programming of programmed procedures including IoT resources.

4.4.3 IoT-Resource Management

Refers to the calculation components (i.e. programming running on devices), make conceivable to assemble data about things and follow up on them.

4.4.4 IoT-Data Handling

The IoT-Data Handling will ready to handle effective information representation positions (together with semantic information) and perform neighboring or close to the IoT assets. They will have the capacity to execute information handling capacities intending to create a littler arrangement of entangled information locally, capacity by a lot of crude information and the keeping the flooding of the backend framework.

5 Conclusions

New inventions that contribute to the Internet of Everything are manifesting themselves in different ways globally, blurring the distinction between the digital and physical worlds. In this paper considered IOT-based 'Smart Home Environment' situation. In this paper, perceived a security's percentage and privacy threats and identified the require for cautious protection and proposed security framework which demonstrates that attack resistant and incorporated methodology for authentication and access control. As particular connections are considered, gaps can likewise be identified and addressed.

References

1. Elmaghraby, A.S., Losavio, M.M.: Cyber security challenges in smart cities: safety, security and privacy. J. Adv. Res. 5(4), 491–497 (2013)
2. Raza, S., Wallgren, L., Voigt, T.: SVELTE: real-time intrusion detection in the internet of things. Ad Hoc Netw, pp 2661–2674. Elsevier (2013)
3. Gubbi, J., Buyya, R., Marusic, S., Palaniswami, M.: Internet of things (IoT): a vision, architectural elements, and future directions. Technical Report CLOUDS-TR-2012-2, Cloud Computing and Distributed Systems Laboratory, The University of Melbourne (2012)

4. Spiess, P., Karnouskos, S., Guinard, D., Savio, D., Baecker, O., Souza, L., Trifa, V.: SOA-based integration of the internet of things in enterprise services. In: Proceedings of IEEE ICWS 2009, Los Angeles, CA, USA (2009)
5. Gluhak, A., Krco, S., Nati, M., Pfisterer, D., Mitton, N., Razafindralambo, T.: A survey on facilities for experimental internet of things research. IEEE Commun. Mag. **49**, 58–67 (2011)
6. Gubbi, J., Buyya, R., Marusic, S., Palaniswami, M.: Internet of things (IoT): a vision, architectural elements, and future directions. Future Gener. Comput. Syst. **29**(7), 1645–1660 (2014)
7. Miorandi, D., Sicari, S., De Pellegrini, F., Chlamtac, I.: Internet of things: vision, applications and research challenges. Ad Hoc Netw. **10**(7), 1497–1516 (2014)
8. Perera, C., Zaslavsky, A., Christen, P., Georgakopoulos, D.: Context aware computing for the internet of things: a survey. IEEE Commun. Surv. Tutorials. **16**(1):414 (2014)
9. Zanella, A., Bui, N., Castellani, A.P., Vangelista, L., Zorzi, M.: Internet of things for smart cities. IEEE Internet Things J. **1**(1):22–32 (2014)
10. Zhao, G., Si, X., Wang, J., Long, X., Hu, T.: A novel mutual authentication scheme for internet of things. In: Proceedings of 2011 IEEE International Conference on Modelling, Identification and Control (ICMIC), pp. 563–566, 26–29 (2011)
11. Jiang, C., Li, B., Xu, H.: An efficient scheme for user authentication in wireless sensor networks. In: 21st International Conference on Advanced Information Networking and Applications Workshops, pp. 438–442 (2007)
12. Balfanz, D., Smetters, D.K., Stewart, P., Wong, H.C.: Talking to strangers: authentication in ad-hoc wireless networks. In: Network and Distributed Systems Security Symposium (NDSS) (2002)
13. Suen, T., Yasinsac, A.: Ad hoc network security: peer identification and authentication using signal properties. In: Proceedings from the Sixth Annual IEEE SMC Information Assurance Workshop (IAW'05), pp. 432–433, 15–17 (2007)
14. Venkatraman, L., Agrawal, D.P.: A novel authentication scheme for ad hoc networks. In: Wireless Communications and Networking Conference (WCNC2000), vol. 3, pp. 1268–1273. IEEE (2000)
15. Best Current Practices for WISP Roaming. WiFi Alliance (2003)
16. RFC 5247: Extensible Authentication Protocol (EAP) Key Management Framework (2008)
17. Feng, J.: Analysis: implementation and extensions of RADIUS protocol. In: International Conference on Networking and Digital Society (ICNDS'09), vol. 1, pp. 154–157, 30–31 (2009)
18. Verma, R.R.S., O'Mahony, D., Tewari, H.: Progressive authentication in ad hoc networks. In: Proceedings of the Fifth European Wireless Conference (2004)
19. El-Nagar, A.M., El-Hafez, A.A., Elhnawy, A.: A novel EAP—moderate weight extensible authentication protocol. In IEEE Seventh International Conference on Computer Engineering (ICENCO2011), pp. 1–6, 27–28 (2012)
20. Yuan, W., Hu, L., Li, H., Zhao, K., Chu, J., Sun, Y.: Key replicating attack on an identity-based three-party authenticated key agreement protocol. In IEEE International Conference on Network Computing and Information Security (NCIS), vol. 2, pp. 249–253, 14–15 (2011)
21. Lei, J., Fu, X., Hogrefe, D., Tan, J.: Comparative studies on authentication and key exchange methods for 802.11 wireless LAN. Comput. Secur. **26**(5):401–409 (2007)
22. Kyriazanos, D.M., Stassinopoulos, G.I., Prasad, N.R.: Ubiquitous access control and policy management in personal networks. In: Third Annual International Conference on Mobile and Ubiquitous Systems: Networking and Services, pp. 1–6 (2006)
23. Prasad, R.: My Personal Adaptive Global NET (MAGNET): Signals and Communication Technology Book. Springer, The Netherlands (2010)
24. Braun, M., Hess, E., Meyer, B.: Using elliptic curves on RFID tags. Int. J. Comput. Sci. Netw. Secur. **8**(2):1 (2008)
25. Secure the Internet of things. http://www.cisco.com/web/about/security/intelligence/iot_framework.html

Intuitive Touch Interaction
to Amputated Fingers

B. Rajesh Kanna, Devarati Kar, Sandra Felice and Nikhil Joseph

Abstract Body parts have an important role in the life of a human being. In today's era, smart phones have also become a part of these body parts virtually. Smart phones and touch devices are of different interaction medium and it has essentially become a part and parcel of our daily life; equipped with touch based screens and thus making the user to interact with the smart phone in a better way. The objective of our project is to use smart phones without using touch feature and use facial features to select the applications, thus making life more comfortable to the end users especially the people with amputated or no fingers. The invention of these smart phones played a vital role in improving the lives of the people with such disabilities. These smart phones will be equipped with both the types of features like touch screen with and without using hands.

Keywords Amputated fingers · Nose tracking · Eye wink · HCI

1 Introduction and Motivation

The mobile phones have come a long way as it was becoming an important means for communication. As the time passed by, these mobiles got updated with different functionalities and this evolution of these newly apt phones were termed as smart phones.

B. Rajesh Kanna (✉) · D. Kar · S. Felice · N. Joseph
School of Computer Science and Engineering, VIT University, Chennai, India
e-mail: rajeshkanna.b@vit.ac.in

D. Kar
e-mail: Devarati.kar2015@vit.ac.in

S. Felice
e-mail: sandra.felice2015@vit.ac.in

N. Joseph
e-mail: nikhil.joseph2015@vit.ac.in

© Springer International Publishing Switzerland 2016   387
V. Vijayakumar and V. Neelanarayanan (eds.), *Proceedings of the 3rd International Symposium on Big Data and Cloud Computing Challenges (ISBCC – 16'),*
Smart Innovation, Systems and Technologies 49, DOI 10.1007/978-3-319-30348-2_34

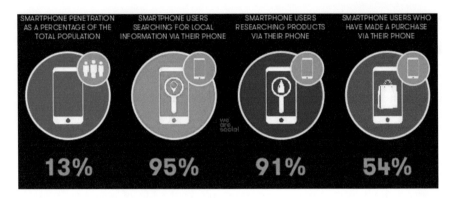

Fig. 1 Mobile phone usage around the world [4]

Figure 1 shows that the smart phones have modified the way we live our lives [4]. About 13 % of the total population use smart phones a lot. 95 % of the smart phone users search for local information via their phone. Besides making phone calls, smart phones provide directions with the help of GPS, pictures with the help of camera, play music and keep track of contacts and other applications of the phone. It has also conjointly advanced in the fields of education and industry providing us the facilities of net banking, e-Commerce, m-Commerce etc. About 54 % have made a purchase via their phone.

However, it is observed from the survey conducted by the USA every year, 50,000 new amputations are recorded each and every year. Ratio of upper limb to lower limb is 1:4 [9]. Figure 2 show the number of smart phones in India grew 54 % throughout in 2014, reaching 140 million in range and reaching 651 million in number [9]. Again, the amount of tablets grew 1.7-fold in 2014, reaching 2 million and reaching 18.7 million. India has become the world's quickest growing marketplace for mobile-based stock commercialism since its launch in 2010. On the National stock exchange of India, the turnover of mobile transactions has inflated by a hundred and thirty percentages over the past year. Thus, sensible phone or

Fig. 2 Amputation risks [7]

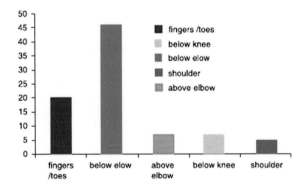

touch-based device has become a crucial part for all types of social elements. But for such screens finger interaction is essential to trigger the user requirements. This modernized technology cannot be directly utilized by people having no fingers or multiple amputated fingers because all interactions on these medium are through fingers. This necessity rules out access for the person with amputation.

Amputation is the removal of a person's limbs or hand through some kind of surgery or by some medical illness. It is used to control pain in the affected area of a person. Causes for amputation may include, Severe injury (e.g.: From a vehicle accident or serious burn), cancerous tumor in the muscle or bone of the limb, Serious infection that does not heal better with antibiotics or other treatment and Thickening of nerve tissue [10].

To explore all the features of touch devices fingers are mandate for operations and there is no alternate methodology emphasized till date for amputated people. Hence, a system should be proposed to bridge this gap. In this proposed work, we investigated how touch-based software applications can be modified to extend the usability for the amputated person. Thus, there is a necessity of a simple interaction with minimum gestures which is able to assist proper functioning operations of touch based phones.

2 Aim and Objective

The aim of our project is to develop a new system to interact with touch devices without fingers designed specifically for people with no fingers or multiple amputated fingers. We will be providing an alternate interaction to make the disabled people INDEPENDENT. Hence, we aim to answer the following research questions:

- Which are the previous methods adopted?
- What are the issues faced by the affected user group?
- Which method or approach should be considered to make more assistance to affected user group?
- Which alternate mechanism works on the existing device making it cost effective?

Further, facial recognition is a very efficient method of interaction, which has become a common feature in smart phones. The human face can be used as a means for security purpose. Moreover, bio-metric face recognition technology has received significant attention in today's world. Face recognition has distinct advantages over fingerprints because of its non-contact process as compared with other biometrics system. Moreover, the facial images can be captured from a distance without touching the person being identified. This does not require interacting with that person. Thus, we are focusing more on looking for an ideal solution that does not involve touch interactions. So, face recognition can be an ideal replacement since it does not require touch as an interactional medium.

3 Literature Survey

There exists a lot of research works on the detection of the facial features which have done by people all over the world. Some papers have been studied by us to know about the techniques and methods used by the researchers to get a clear view of existing methods. Krlak and Strumio [6] describes about a vision-based human computer interface which detects voluntary eye-blinks and interprets them as control commands. Image processing methods like Haar-like features and template matching techniques have been used for the purpose. An eye-blink detection algorithm is applied which consisted of four steps—face detection, eye-region extraction, eye-blink detection and eye-blink classification. Here the voluntary and involuntary eye-blinks were resolved with threshold. If the duration of detected eye-blink is more than 250 ms and less than 2 s, then it is regarded as a voluntary one. The accuracy rate for this came out to be 99 % and the performance tests demonstrate that the user interface was useful for people with disabilities but it required a strenuous training session and time taken is little more.

Kraichan and Pumrin [5] presented a novel user interface with computer. Here also, eye and face detection is done via Haar-like feature and template matching. The center of eye-pair detection is mapped to the mouse coordinates in computer screen where left and right click are controlled by blinking left and right eye respectively. The experimental results suggested that the distance between users and camera is preferable at 60 cm because the eye-center detection achieves 90 % sensitivity. It also contributes 80 % sensitivity or more at the roll and pitch angles of head pose between −15° and +15°.

Gorodnichy and Roy [3] employed nose as facial feature and the face-tracking was carried based on tracking a convex shape nose feature. Recently, it has been proved that the robustness of the feature tracking can be significantly improved if instead of commonly used edge-based features such as corners of mouth, brows, etc., the nose tip is used which laid the basis for this paper. For the tracking of the nose, template matching is used which involves scanning a window of interest with the peephole mask and comparing each thus feature vector with the template vector. The intensity pattern of the tip of the nose is not affected much by the orientation of the nose; which helps in designing the nose feature template. The accuracy was about 90 % on this approach.

Richman [8] described about nose detection in images provided by the Eastman Kodak Company. They proposed a nose detection algorithm which consisted of three components, Preprocessor eliminates regions of the image which do not contain noses through image differencing method. The support vector machine employs a radial basis kernel function and classifies each feature vector generated by the preprocessor. Finally, the post processor is used to convert support vector machine values at each candidate pixel into a description of the number and location of noses in an image. At each pixel, a simple pattern recognition scheme is used to determine which pixels correspond to that of the nose and then a small polygon is drawn around each of the pixels and overlapping polygons to detect the nose. The algorithm is reasonably successful despite costly measures like image rescaling.

Anand et al. [1] proposed to use facial expressions for natural interaction, especially with gadgets like smart phones, tablets, etc. They present a use-case scenario of eBook reader application where the facial expressions are used to control the device while using the application. A facial expression recognition system was introduced which comprised of an action unit (AU) detection module and an AU to emotion mapping module. The eBook reader ran in real-time at 15 frames per second which had a high level of accuracy. Challenges were like in the form of facial occlusions that are static (eyeglasses, facial hair) and dynamic (hand).

Esaki et al. [2] used blinks for quick menu selection on a communicator which was developed for the disabled. The input for the communicator was selected menu was executed using the pupil area which were obtained from the custom made image processor for eye-gaze detection. The voluntary and involuntary blinks were distinguished by the difference of their duration by giving a proper threshold and the results came out to be 100 %.

However all the above methods and techniques were developed and made suitable for PC based systems not to touch based systems.

4   Conceptual Design

The conceptual design of the work is divided into six stages, which are mentioned below, Fig. 3 shows the execution sequences of those stages.

- Design of multiple windows in the phone
- Mapping of front-end camera and menu windows
- Track nose pointer view1

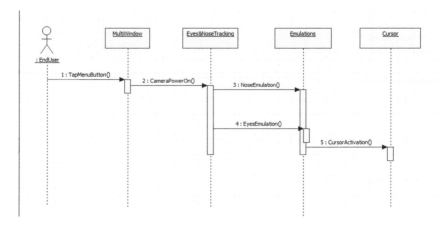

Fig. 3 Proposed system flow diagram

- Emulation of movement direction keys with respect to nose momentum
- Eye blink Tracking
- Emulation touch event with eye blink.

4.1   Design of Multiple Windows and Mapping of Front-End Camera View and Main Menu

First let us know what a multi-window is. Multi-window is a feature in smart phones which allows to run two applications simultaneously on your screen. This feature is limited to supported apps, but does a great job of bringing true multi-tasking to the Smartphone. The multi window feature can be enabled with the help of the following steps:

1. Tap menu button.
2. Tap Settings.
3. Tap Display.
4. Tap the green button to enable multi window.

   Click the home button to go back to the home screen. If you do not see the Multi-Window handle somewhere on the left edge of the screen, you can long-press the back button to open it. It does a great job of bringing true multitasking to the smart phone. The multiple windows decide which windows are visible and also perform transitions and animations of the window when opening or closing an app or rotating the screen. Here, the multi window will be used for keeping one window as the camera for tracking the face of the user and the other will be the main menu for selection. The camera is the first view where the facial expressions are detected and tracked for further navigation. The second view is the actual system where the main menu is there. This is also known as providing simultaneous view on the smart phone and a sample layout is depicted in Fig. 4.

   Mapping means we are coordinating the movement of the facial feature along with pointer on the menu to select an application. Here, the movement of the nose tip viewed from the front end camera view is tracked. Since this movement is mapped with menu pointer in view II, menu pointer moves as per the movement of nosetip.

4.2   Nose Tip Tracking and Emulation of Direction Keys

Touch menu view is presented in the view II, which is navigated with the help of nose by tracking it. Nose movement is coordinated with the movement of four direction keys like up, down, right and left. This direction keys will give the actual position of the application with which we want to work with as shown in Fig. 5.

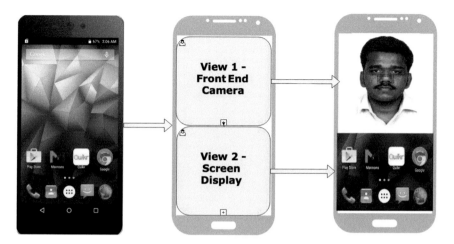

Fig. 4 Provisioning of two views on the touch screen

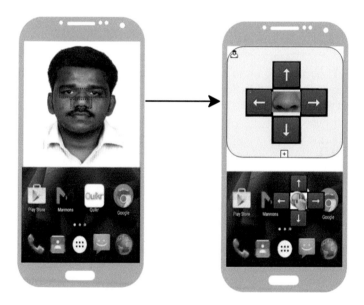

Fig. 5 Emulation of 4-directional key per nose tip motion

4.3 Eye Wink Tracking and Emulation of Tab Event

After the menu is present in front of the user, now it is time for selection of an application and this will be probably done with the eye-wink. In other words, there will be two events event one uses nose as a mouse and the eye-wink as a mode of selection. These events will be passed to another view. Finally, motion feed from

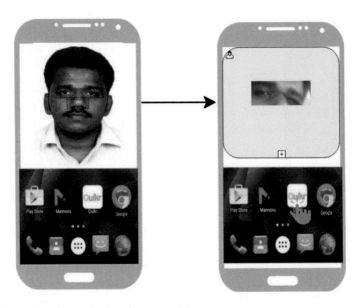

Fig. 6 Emulation of touch tapping from eye wink

View I could be mapped on to View II to interact with menu. Emulation touch event with eye wink: For selection of menu with eye blink, it is actually used as a replacement to touch action. We are using eye wink to select an application and it is illustrated in Fig. 6.

5 Implementation

5.1 Android Studio

The Camera class in Android provides the *startFaceDetection* method to start face detection on an Android phone. Once the method is called the camera will notify Camera. FaceDetectionListener of the detected faces in the required frame. Now, the images are captured by the Android framework through the *android.hardware.camera2 API*. The relevant classes includes: *android.hardware.camera2 and Camera*. The *hardware.camera* package provides the primary API for controlling cameras of devices. It can be used to take pictures when building a camera application. The *Camera* API is for controlling camera on devices. Similarly the *SurfaceView* is a class, used to present a live camera preview to the user.

5.2 OpenCV

OpenCV is a library where we can work with real time images in any platform as in android studio, eclipse, etc. There are certain algorithms in OpenCV with which we can identify faces, videos. The classes of OpenCV used in our paper are for face detection, eye-blink and nose detection which are briefly described below:

ClassMat—It represents a single-channel or multi-channel array which is dense and represented on an n-dimensional space. Moreover, to store matrices and real or complex-valued vectors, vector fields, point clouds, tensors, histograms we use this class of OpenCv.

ClassRect—This OpenCv class can create Template class for 2D rectangles which has the parameters as coordinates.

Class Point—OpenCv class Point can create Template 2D point class which is used for converting point coordinates into the specified type like conversion of floating point coordinates to integer coordinates.

Class Imgproc—This OpenCv class is for image processing.

Class CameraBridgeViewBase—This OpenCv class is a basic class which can combine, bind and implement the camera of the device with the OpenCV library. The important function of this class is to control when can the camera be enabled, processed with the particular function and allow the listener to make any adjustments to the camera frame. The resulting frame can be drawn once the job is done.

Class CascadeClassifier—The OpenCv class Cascade classifier is mainly used for object detection like face detection, textures of a picture, which can be done with the help of Violas Cascade Boosted classifiers using Haar or LBP features.

Class JavaCameraView—This OpenCv class is an implementation of the Bridge View between the OpenCV and Java Camera. The class relays on the functionality available in base class and only implements the required functions: connect Camera—opens Java camera and sets the Preview Callback to be delivered. Disconnect-Camera—closes the camera and stops preview. When frame is delivered via callback from Camera—it processed via OpenCV to be converted to RGB format and then passed to the external callback for modifications if required.

Class VideoCapture—This OpenCv Class is used for capturing videos from cameras and video recordings. C++ API is provided by this class for capturing video from cameras or for reading video files.

Class FeatureDetector—This OpenCv class is used for 2D image feature detection.

6 Conclusion

The authors wish to thank the Ms. Sudha Ramalingam, Managing trustee of ANBAGAM under the Manonmani Trust (NGO), registered office at Husain House, No. 7, Kodi Chetty Street, Chennai for facilitating in indentifying the

Fig. 7 Requirement gathering from amputated person

physically challenged people to gather the exact need of amputated person while smart phone usage, Fig. 7 shows a field discussion and gather feedback from the needy people. As of now, conceptual design of the proposed system was successful done and we are in the process of development and testing the complete system.

References

1. Anand, B., Navathe, B.B., Velusamy, S., Kannan, H., Sharma, A., Gopalakrishnan, V.: Beyond touch: Natural interactions using facial expressions. In: 2012 IEEE Consumer Communications and Networking Conference (CCNC), pp. 255–259, Jan 2012
2. Esaki, S., Ebisawa, Y., Sugioka, A., Konishi, M.: Quick menu selection using eye blink for eye-slaved nonverbal communicator with video-based eye-gaze detection. In: Proceedings of the 19th Annual International Conference of the IEEE on Engineering in Medicine and Biology Society, vol. 5, pp. 2322–2325, 1997
3. Gorodnichy, D.O., Roth, G.: Nouse use your nose as a mouse perceptual vision technology for hands-free games and interfaces. In: Proceedings from the 15th International Conference on Vision Interface. Image Vis. Comput. **22**(12):931–942 (2004)
4. Katz, J.E., Aakhus, M.: Perpetual Contact: Mobile Communication, Private Talk, Public Performance. Cambridge University Press, Cambridge (2002)
5. Kraichan, C., Pumrin, S.: Face and eye tracking for controlling computer functions. In: 2014 11th International Conference on Electrical Engineering/Electronics, Computer, Telecommunications and Information Technology (ECTI-CON), pp. 1–6, May 2014
6. Krlak, A., Strumio, P.: Eye-blink detection system for humancomputer interaction. Univ. Access Inf. Soc. **11**(4), 409–419 (2012)
7. Kulley, M.: Hand prosthetics-statistics. http://biomed.brown.edu/Courses/BI108/BI108_2003_Groups/Hand_Prosthetics/stats.html (2003)

8. Richman, M.S., Parks, T.W.: Detection of noses and faces in consumer images
9. The Economic Times. India to have 651 million smartphones, 18.7 million tablets by 2019. http://articles.economictimes.indiatimes.com/2015-02-03/news/58751662_1_networking-index-mobile-users-population (2015)
10. Wikipedia. Amputation. http://en.wikipedia.org/wiki/Amputation (2011)

A Novel Approach for Web Service Annotation Verification and Service Parameters Validation Using Ontology

S. Shridevi and G. Raju

Abstract Semantic annotations of web services are useful in service discovery only when it reflects accurate service. But the rate of defects in semantic annotations was found to be very high due to the need for human interaction and expertise. The purpose of this work is to verify the semantic annotation of web services. The verification is based on the conventional testing process. In this paper, emphasis is laid on automation of semantic annotations to increase the efficiency and accuracy to provide useful web service through which the required data can be reproduced according to the specification defined by the services. In particular, the study proposed a technique from software testing to address the problem of semantic annotation verification. A software testing has been well proven to be effective for verification of software program behavior according to its specifications. By using a test suite which consists of test cases with specific data values for feeding the execution, therefore the expected outputs are generated as per the specification. As specified by the test cases, the software is executed using the input values. In this way, there is a possibility to identify a defect due to the inconsistency between the output delivered and expected by a test case. The advantages of proposed model are: (1) ease of use when information, (2) automatic compliance to system requirements and specification, (3) improved accuracy, (4) use of ontology based partitioning, (5) efficient defect detection mechanisms.

Keywords Web services · WSDL · Semantic annotation · Ontology · Testing and validation

S. Shridevi (✉)
VIT University, Chennai, India

G. Raju
Kannur University, Kannur, India

© Springer International Publishing Switzerland 2016   399
V. Vijayakumar and V. Neelanarayanan (eds.), *Proceedings of the 3rd International Symposium on Big Data and Cloud Computing Challenges (ISBCC – 16')*,
Smart Innovation, Systems and Technologies 49, DOI 10.1007/978-3-319-30348-2_35

1 Introduction

The rapid development of the Internet in recent years had resulted in the emergence of large number of web services. The Web is now evolving into a distributed device of computation from a collection of information resources such as to find the weather, travel schedules, and the best restaurant while travelling. In addition, Web services also been used as the technology for business process execution and application integration. Although Web services are based on widely accepted standards, the lack of a formal description of the meaning of their functionality and the data exchanged has been a significant roadblock in the realization of integration. As the number of Web services increase, it is important to discover and compose Web services. This task forces the computer to do more of the necessary work where the Semantic web (www.w3.org/2001/sw/) addresses this concern [1] and expected to become a next-generation of the web. Semantic Web, aims at developing a global environment on top of web with interoperable, heterogeneous organizations, humans, data repositories, web services, agents and applications.

Semantic Web Services offer the possibility of highly flexible web service architectures, where new services can be quickly discovered, orchestrated and composed into workflows [2]. The extent of description available in the current WSDL standard leaves room for ambiguous interpretations of the functionality and data of a Web service. To address these issues semantic annotation of Web service elements are proposed, which explicates the exact semantics of the Web service data and functionality elements that are crucial towards the use of the Web service. This is done by annotating the Web service elements with concepts in domain models or ontologies. Since ontologies represent an agreed upon view of the modeled domain, any ambiguity in the interpretation of functionality or data of a Web service is eliminated. The purpose of annotating Web services is to enable unambiguous and automated service discovery and composition.

Yet, semantic annotations of web services are useful in service discovery only when it reflects accurate service [3–5]. But the rate of defects in semantic annotations is found to be very high due to the need for human interaction and expertise. This is because, annotator neither involved in the domain ontology construction and web service development. Presently there are no efficient mechanisms to check the high rate of defects occurring while using semantic annotations. This makes the system prone to disuse and is found to be highly ineffective. Since web service providers supply annotations, therefore one would expect significantly low errors. Yet, the integration of annotation failed to integrate with lifecycle development of web services which resulted in annotation of third parties. In a typical scenario, an annotator is required to analyze the defects and he/she should be involved in the construction of ontology and in the development of web service, therefore there is a possibility for mistaken interpretation of behaviors and concepts while selection of annotation.

The structure of the paper is as follows: Sect. 1 critically reviews the previous studies conducted elsewhere on semantic annotations. This section is followed by brief description of existing system (Sect. 2) and subsequently the proposed system (Sect. 3). Section 4 presents the verification of semantic web service annotation using conventional software testing technique based on the domain ontology. This section also presents the verification process that includes multiple phases. The test cases used in semantic annotation is used to reveal the defects in the web service. The algorithm used in the testing technique reduces the effort of the manual annotator. The execution of algorithm results in normal or abnormal termination, which in turn determines the valid or invalid input. The most outstanding contribution of this work is the foundation of a novel technique, which would cover both the ontology concepts applied for annotation and its effectiveness of the adequacy criteria by applying that in real-world situation (Sect. 5), in the case of online shopping service.

2 Related Work

Recently, Web services related research area are found to be active and the following section reviews the previous major techniques, platforms, standards that have been applied for web services closely related to present research.

2.1 Definitions of Concepts

The definitions of concepts used in this study are described as follows [6]: *Ontology:* This term is understood from a philosophical point of view where it is defined as "the knowledge of what is to be in oneself". In the case, data processing it refers to the structured set of knowledge in a domain. It is an explicit share specification in a specific domain. *Semantic annotation:* A semantic annotation is referent to a relative ontology where it is a link to its semantic description assigned to an entity in the text. *Reasoner:* Based on the specific request it matches service advertisement such as input and outputs. *Matching:* Based on some similarity features, the matching operates two concepts such as inputs, output and precondition and effects. This was done as per the Paolocci et al. [7].

2.2 Semantic Description Languages of Web

Several web service frameworks are promoted for W3C standardization where the most frameworks are evolved from the WSDL-S Specifications. The prominent ones are OWL-S (Ontology Web Language for Services), WSMO (Web service Modeling Language), WSDL-S (Web service Description Language Semantic) and SAWSDL (Semantic Annotation for Web service Description Language). These frameworks are reviewed based on the criteria such as Resources, property, language, annotation, model and matching. Various studies have been conducted previously and the following are the studies that are closely related to our concept. Semantic web access control policies were discussed by Bonatti et al. [8], while Denker et al. [9] developed the security annotations. Ankolekar et al. [10] proposed DAML-S and DAML + OIL ontology to explain the web services capabilities and properties. Further, author also described three aspects of ontology that includes the service grounding, the process model and the service profile, where later provides the information required for an agent to service discovery while service model provides information for an agent to make use of a service. In specifically, author has applied service grounding which connects with low level XML based description of web services. The research although shed light on upper ontology but still it requires human expert for service description, time consuming and does not guarantee for successful deployment. Medjahed et al. [11] proposed an ontology based framework for the automatic composition of web services developed from high-level declarative descriptions and implemented in e-government application. Author in this study defined formal safeguard for meaningful composition through the use of compos ability rules which compares the semantic and syntactic features of web services. Although the service composition is based on the ontology and syntactic and semantic riles, while composite service based on high level declarative descriptions still it failed to address the spatial and temporal availability. Further semantic features did not address before deployment of web service and composite services are not based on composition quality. Lord et al. [2] proposed light-weight semantic discovery architecture and attempted to fits in the larger architecture of the Grid project, specifically for user requirements of bioinformatics. Further this paper provides light weight semantic support and usage of RDF's as backend provides low complexity. Yet, still system required experts in the domain ontology and the proposed failed to provide verification in the annotations therefore, susceptible to be applied to different domains. Patil et al. [12] proposed a Meteor-S web service annotation framework for semi-automatically marking up web service descriptions with ontologies. Author in this study has proposed algorithms to match and annotate WSDL. Although semantic annotations are made with real ontologies with higher quality still the paper failed to provide verifications to check the accuracy of annotation before deployment. Gomadam et al. [13] presented a faceted approach to search and rank web APIs considering attributes or facets of the APIs as found in their HTML descriptions based on their utility and popularity. Although the proposed provides effective search and classification mechanism for Web API's based on the service utilization ranking (serviut), still the proposed system assists human users for annotating web services which demands high costs at the same times there is a possibility of high rate of false positive. Maximilien and Singh [14] addressed the issue related to Quality of Service (QoS) via an agent framework coupled with QoS ontology. The proposed system supported multi-attribute utility and dynamic descriptions of web services and therefore QoS support. However, there is a possibility that proposed system may be vulnerable for attacks such as proofing while agencies data by malfeasant agents. This reflects that there is a lack of guarantee for accurate reflection of web service in semantic annotation. Davidson and Freire [15] applied provenance in scientific workflows to allow for result reproducibility, sharing, and knowledge re-use in the scientific community. Jiang and Luo [16] proposed extended UDDI which takes into account the height factor of ontology tree and local density factor. The findings revealed that the proposed service enhanced the web service discovery efficiency. However, this algorithm ignored quality of services, precondition and effects of web service operations. Boukhadra et al. [17] proposed semantic interoperability in a heterogeneous, distributed architecture based on the automatic dialing services semantic web. Saquicela et al. [18] proposed lightweight semantic annotation of RESTful services using cross-domain ontology and findings showed satisfactory results. Further, several tools like APIHUT, KINO [19], Meteor-S [20], APIHUT [13] are some of the tools that have been proposed to enhance the verification of semantic annotations of web services. The proposed system combines provenance in workflow system with higher quality ontologies. Yet, the proposed model need an efficient way for managing data and failed to combine provenance workflow with database and be visualized.

3 Exisiting System

The existing semantic annotation, described web service parameters partially but do not attempt to describe its transformation, which is critical in web service discovery. In addition, techniques proposed earlier on semantic annotations did showed following drawbacks such as (a) manual annotation needs experts in domain ontology (b) Manual annotation is time consuming task (c) Semantic annotations are created for describing web services under strong assumptions (d) Defect detecting power is very low and (e) Test case generation and execution is expensive [2, 11, 13, 14, 21]. There are dearth of tools and techniques to verify the accuracy of annotation. Further, several studies [22, 23] have been conducted earlier to improve the Quality of Service provided to the end user. The objective is to enhance user experience and enable dynamic detection of requirements and avoid rigidity in creating and reusing information. In almost every system, accuracy seems to have emerged as a very serious threat to its effective implementation. Ontologies enable Web services framework to be machine readable, and this in turn allow software agents to handle more the decision making in completing a task [24], thereby creates effective use to annotations technology [25]. The semantic annotations associate with relevant data structure in the ontology thereby evolving effective defect detection systems [26]. This enables effective use of annotations technology to extract relevant information as per a dynamic specification and service requirements.

4 Proposed System

In this paper, emphasis is laid on automation of semantic annotations to increase the efficiency and accuracy to provide useful web service through which the required data can be reproduced according to the specification defined by the services. In particular, the study proposed a technique from software testing to address the problem of semantic annotation verification. A software testing has been well proven to be effective for verification of software program behavior according to its specifications [27]. By using a test suite which consists of test cases with specific data values for feeding the execution, therefore the expected outputs are generated as per the specification. As specified by the test cases, the software is executed using the input values. In this way, there is a possibility to identify a defect due to the inconsistency between the output delivered and expected by a test case. The advantages of proposed model are: (1) ease of use when information, (2) automatic compliance to system requirements and specification, (3) improved accuracy, (4) use of ontology based partitioning, (5) efficient defect detection mechanisms.

4.1 Semantic Annotation Framework: Architecture

4.1.1 Ontology Based Service Annotator

In this paper, we proposed a semi-automatic semantic annotation framework based on the existing semantic annotation. Various modules enhance the reliability and response time of the web service using the concept of ontology. Thus from the perspective of architecture (Fig. 1) the Ontology based Service Annotator system is divided into main parts domain ontology and creation of web service from the programmer and domain based. The verification of semantic web service module and the end result i.e. the generated output generated is tested using the annotated source code. The process is differentiated in two different sections, which include Domain based and web programmer. In the first part the domain section the focus is given to Ontology Based Service Annotator for the construction and generation of pool instance. In the next phase the input and output is introduced to execute to the test cases. In the second part, the web programmer the WSDL file is used to support the annotation. In the next stage the execution of test is conducted [28].

The main phase of semantic annotation creation includes the Domain ontology construction, test cases, WSDL file, annotation, discovery, test case execution, defection detection, curating and deploying. In the research the main direction of verification of semantic web service module is explored using various semantics, which are presented in different stages.

Stage 1: Creating online shopping and WSDL creation
Stage 2: Ontology construction for online shopping and extending WSDL
Stage 3: Test case generation with legal and illegal values for annotations

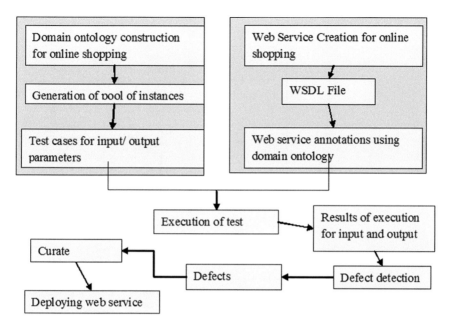

Fig. 1 Architecture of annotation framework

Stages 4: Execution of test cases
Stages 5: Verifying outcome of testing.

   Stage 1: Creating online shopping web service.

   Web service is created for online shopping. The application includes operations such as login, product details, purchase products, product delivery and logout. The operations are described as service descriptions in web service through WSDL file. The WSDL stands for Web Service Description Language, which describes the operations input and output parameters. The operations input and output are based on the methods used in web service. The WSDL files are useful in service discovery and service publishing (Fig. 2).

   Stage 2: Ontology construction for online shopping and extending WSDL.

4.2   Ontology Construction for Online Shopping and Extending WSDL

Ontology is constructed based on the online shopping domain concepts. The ontology consists of classes and properties. The output of WSDL file created from the previous module is given as input for semantic annotation. The WSDL file of the online shopping is semantically annotated using the classes of the ontology.

Fig. 2 Service creation for online shopping

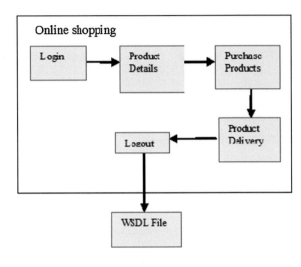

Fig. 3 Ontology annotation

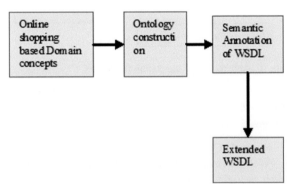

The WSDL file is now semantically annotated with the ontology classes which in turn consist of operations input and output (Fig. 3).

4.3 Annotation Verification Process

In the Annotation Verification Process in the initial phase the i.e. Phase 1 is where the instance of each annotation is prepared which turns into a pool of annotated instances, in Phase 2 the generation of these annotated instances pool the parameters for input test cases are created. Thirdly, these constructed parameters are tested and executed in Phase 3 and in Phase 4 the defects detection in the annotation is carried out from the results parts. Later in Phase 5 the parameters for output test cases are exploited from the executed results and generated output test cases are executed in Phase 6 and in Phase 7 the analysis for identifying the defects is performed. A curator can then correct annotations in the light of the defects discovered (Phase 8). The following section illustrates the same.

Stages 3: Test case generation with legal and illegal values for annotations.

The test cases are created using the automatic process, in which researcher specifies and test the registry of semantic annotation and parameter set. Moreover, the ontology location for semantic annotations is shown below. It is evident that the legal and illegal values are not illustrated in the algorithm for selected parameter (variable pool). Thus the annotated instances should be associated with both legal and illegal values. In addition, during the input parameter testing (op, p) identifies to be associated with a precondition prec and minimize the legal value number during the annotation accuracy verification of input parameter which satisfies the predicate prec. The predicate is utilised to include the constraints to be checked in validating the parameters of the service, thereby reducing the number of test cases for the validation of parameters in the service invoked (Fig. 4).

An operation op has a set of input parameters given by the function inputs(op) and a set of outputs given by outputs(op). A parameter is denoted by a pair op, p where p is the name of the parameter and op is the operation to which the parameter belongs. Especially, while testing the annotation of op, p iff holds(prec, op, p,v), where holds(prec, op, p,v) is true iff prec is satisfied when v is used as, were v stands for legal value. The strong semantics are used to create the test cases. For execution, the annotation should have some pre-condition in strong semantics. Once the pre-condition is satisfied, the annotation is selected as input or output test case. The concept of pre-condition provides high level of accuracy for identifying the test case belongs to legal or illegal.

The algorithm used for generating the test cases is as follows:

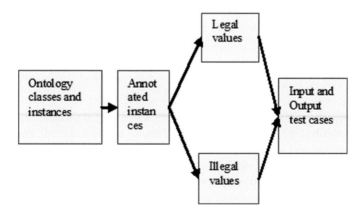

Fig. 4 Test cases generation

```
Input ps_i = input params to be tested
Output ps_i = empty test suite
Begin
For each operation parameter <op, p>∈ ps_i
pool := selected values from pool (should be a
mixture of legal and illegal)
For each v in pool
            If v is legal Select tuple of legal values lv
            for other input params of this operation
            such that the precondition of op holds
            eo := "normal"
If v is illegal
            Select tuple of legal values lv for other
            input params of this operation
            eo := "abnormal"
            tc := <<op, p>,lvU v, eo>
    Add test case tc to ts_i
End
```

Stages 4: Execution of test cases.

The operation is tested with input and output test cases and executed. The execution of test cases terminated with either normal or abnormal termination. In the execution phase the SoapUI [29] is adopted to support the large web service invocations. The algorithm which is used during the executing test Cases for Input Parameters is as follows (Fig. 5).

```
Input: ts_i = test suit
Outputs rs = a null tuple
InputLog = the results of execution of test cases
output Log = the results delivered byoutput
parameter
Begin
For each tc in ts_i
tc = <op, p_i>, vs, eo>
Execute operation op with input params vs
rs := result returned from op
If op terminated normally then
            ao := "normal"
            For each output param op, p of op
                    r := value for <op, p> in rs
                    Add <<op, p>,r> to output Log
Else
            ao := "abnormal"
Add <op, vs, rs, ao> to input Log
End
```

The normal termination describes the values is legal and abnormal termination of test case describes the value is illegal. However, in real time interpreting the web service's normal and abnormal terminations is more complex and in the abnormal terminations case various other factors affect the web service to exit with an exception, other than providing illegal input.

Stages 5: Verifying outcome of testing.

In order to construct the output parameters from the test cases of input parameter execution i.e. the outcome of test case execution are used. Thus to generate the output parameter's Test Cases the following algorithm is used (Fig. 6).

Fig. 5 Test case execution

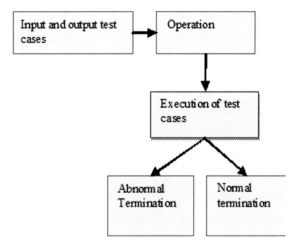

Fig. 6 Outcome verification

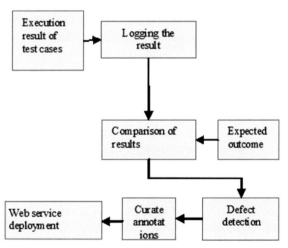

Input ps$_o$ = output params to be tested
Outputs rs = empty test suite
Begin
For each operation parameter <op, p$_i$> $\in$ ps$_o$
os := output values for <op, p$_i$> from output log

  For each ov in os
    Find all services ss with input
    annotated with the same concept
    as <op, p>
    Select tuple of legal values lv for
    other input params of this
    operation
    eo = = expected outcome for lv
$\cup$ ov
    tc = <<op, pi> lv $\cup$ ov, eo>
    Add test cases tc to ts$_o$
End

The generated output parameter's Test Cases is tested and interpreted using the following algorithm.

```
Input ts_o = test suite for outputs
Outputs output Log = the results of execution of
test cases
Begin
For eachtc in ts_o
tc := <<op, p_i>, vs, eo>
            Execute operation op with input params vs
            rs := result returned from op
            If op terminated normally then
                  ao := 'normal'
                  Else
                        ao := 'abnormal'
                  Add <op, vs,rs, ao> to output test
                  Log
      End
```

In addition, the executed results are logged. The logged results are compared with the expected outcome. If the results resemble the expected outcome, then the annotation is verified as valid. If the results contradict the expected outcome, then the annotations are considered as error. The defect is detected by performing above steps. The errors are rectified in semantic annotations.

5 Test Adequacy Criteria

In the semantic annotation process is a explicate Adequacy Criteria or underlying principle is a means which provides form or strategy to compare the different integrating strategies [30]. In the present study the logged results are compared with the expected outcome. If the results resemble the expected outcome, then the annotation is verified as valid. If the results contradict the expected outcome, then the annotations are considered as error. The defect is detected by performing above steps. The errors are rectified in semantic annotations.

6 UML Diagrams

6.1 Sequence Diagram

A sequence diagram is a kind of interaction diagram that shows how processes operate with one another and in what order. It is a construct of a Message Sequence Chart. A sequence diagram shows object interactions arranged in time sequence. It depicts the objects and classes involved in the scenario and the sequence of messages exchanged between the objects needed to carry out the functionality of the scenario. Sequence diagrams typically are associated with use case realizations in the Logical View of the system under development. The Sequence Diagram models

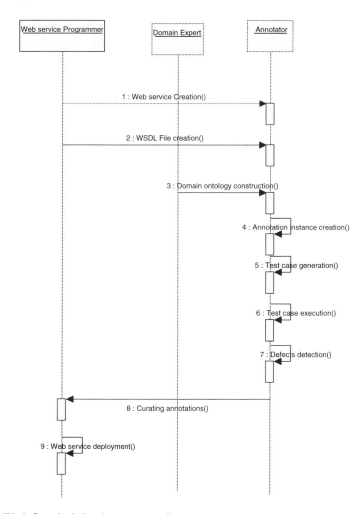

Fig. 7 Work flow depiction thru sequence diagram

the collaboration of objects based on a time sequence. It shows how the objects interact with others in a particular scenario of a use case (Fig. 7).

7 Results

Domain ontology is utilized to generate reduced test cases for the validation of web service parameters. In our work, the basic objective behind adding semantics to WSDL is to enhance the use of metadata that comes in the form of ontology. The ontology adds constraints to all the service parameters thereby during the testing

Table 1 Operation registration

Parameters	Constraints	Input values	WSDL without ontology annotation	WSDL with ontology annotation
Name	Should not be empty, should have only character limited to the size of 20 characters	"Äsdfgh"	125(5^3) exhaustive test runs required for testing all these parameters with 3 constraints each, thereby increasing code development time and testing time	As constraints are in ontology, only 15 test cases as per orthogonal array testing strategy is required, which minimizes the testing time to a very greater extent
Age	Cannot be greater than 100, cannot be negative and only numeric	45		
Address	Alphanumeric	Asd345		
Mobile number	Can be only 9 digits numeric and positive number	123456789		
Email id	Should follow the expression — aaa@aaa.com	aaa@cv.com		

phase the metadata related to the service parameters can be utilized for data validation of the parameters, thereby reducing the code required to be written for data validation and also reducing the number of test cases used in the testing phase of the implementation.

Comparison was made on the present proposed service parameters testing using semantically annotated WSDL and using non- annotated WSDL, i.e., by using the ontology constructed proposed for the domain of services and without using ontology.

In the following Table 1, for the operation registration which has name, age, address, mobile number and email id as the input parameters, nearly 5^3 = 125 test runs are required to test the operation completely with the assumption that each parameter may have 3 constraints. Also the developer has to take care of all these constraints in the data validation code and appropriately take actions for the constraints. This increases, not only the development time of the software and also testing time of the service both from the provider and from the consumer. But if the constraints specified are taken from the ontology for testing, then it can be automatically tested during testing phase and throws the corrective measures to the provider and also to the consumer thereby reducing the test runs to a great optimum level.

7.1 Metrics for Test Case Effectiveness

To measure the effectiveness of a test suite that can detect faults, we have used the harmonic mean (HM) [31], which is independent from the size of a given test suite. The HM is a standard statistical tool to combine a set of numbers, each representing a rate, into an average value [31]. The proposed metric is defined as follows.

Let T be a test suite consisting of n test cases and F be a set of m faults revealed by T. Let TF_i be the first test case in a reordered test suite S of T that reveals the fault i. The harmonic mean of the rate of fault detection (HMFD) of S is defined as follows:

$$HMFD = m/(\frac{1}{TF1} + \frac{1}{TF2} + \frac{1}{TF3} + \cdots + \frac{1}{TFm})$$

Note that a lower HMFD value indicates a more effective result [31]. Let us use an example to clarify the metric further. Suppose that the annotated ontology has three faults and we are given with a test suite with five test cases. Suppose further that the first test cases in a reordered test suite S to detect the fault i are the first, second, and third test case, respectively. The HMFD value for S is, therefore, computed as 3/(1/1 + 1/2 + 1/3), which is 1.64. Similarly the test suite of both input and output parameters of web services were tested and proved to have a lower value which indicates a more effective result.

8 Conclusion

The proposed system provides accurate service results of semantic web service annotation before deploying in public use. Verification process includes conventional testing methodology for testing semantic annotations in validating the service parameters. Annotated instances are taken as test cases and test cases are executed with legal and illegal values. The resulted logged values are matched with expected outcomes. The outcomes may be normal and abnormal termination. If abnormal termination takes place, the annotation is corrected. The system provides high defect detecting power.

References

1. Schrimpsher, D., Etzkorn, L.: A model to predict quality of a reduced ontology for Web service discovery on mobile devices. Knowl. Eng. Rev. **29**(02), 201–216 (2014)
2. Lord, P., Alper, P., Wroe, C., Goble, C.: Feta: a light-weight architecture for user oriented semantic service discovery. Semant. Web Res. Appl. Lect. Notes Comput. Sci. **3532**, 17–31 (2005)

3. Heb, A., Kushmerick, N.: Automatically attaching semantic metadata to Web services. In: Proceedings of IJCAI Workshop Information Integration on the Web (2003)
4. Jannach, D., Shchekotykhin, K., Friedrich, G.: Automated ontology instantiation from tabular web sources-the all right system, web semantics: science. Serv. Agents World Wide Web 7(3), 136–153 (2009)
5. Salih, A.Q.M.: Towards from manual to automatic semantic annotation: based on ontology elements and relationships. Int. J. Web Semant. Technol. 4(3), 21–36 (2013)
6. Talantikite, H.N., Aissani, D., Boudjlida, N.: Semantic annotations for Web services discovery and composition. Comput. Stand. Interfaces 31(6), 1108–1117 (2009)
7. Paolucci, M., Kawamura, T., Payne, T.R., Sycara, K.: Semantic matching of Web service capabilities. In: Proceedings of the First International Semantic Web Conference (2002)
8. Bonatti, P.A., Duma, C., Fuchs, N., Nejdl, W., Olmedilla, D., Peer, J., Shahmehri, N.: Semantic web policies—a discussion of requirements and research issues. In: 3rd European Semantic Web Conference (ESWC). Lecture Notes in Computer Science, vol. 4011 (2006)
9. Denker, G., Kagal, L., Finin, T., Paolucci, M., Sycara, K.: Security for DAML Web services: annotation and matchmaking. In: Proceedings of the 2nd International Semantic Web Conference (2003)
10. Ankolekar, A., Huch, F., Sycara, K.P.: Concurrent execution semantics of DAML-S with subtypes. In: ISWC '02 Proceedings of the First International Semantic Web Conference on the Semantic Web, pp. 318–332 (2002)
11. Medjahed, B., Bouguettaya, A., Elmagarmid, A.K.: Composing Web services on the semantic web. VLDB J. Int. J. Very Large Data Bases 12(4), 333–351 (2003)
12. Patil, A.A., Oundhakar, S.A., Sheth, A.P., Verma, K.: Meteor-s Web service annotation framework. In: Proceedings of the 13th Conference on World Wide Web—WWW '04, p. 553 (2004)
13. Gomadam, K., Ranabahu, A., Nagarajan, M., Sheth, A.P., Verma, K.: A faceted classification based approach to search and rank web APIs. In: 2008 IEEE International Conference on Web Services, pp. 177–184 (2008)
14. Maximilien, E.M., Singh, M.P.: A framework and ontology for dynamic Web services selection. IEEE Internet Comput. 8(5), 84–93 (2004)
15. Davidson, S.B., Freire, J.: Provenance and scientific workflows. In: Proceedings of the 2008 ACM SIGMOD International Conference on Management of Data—SIGMOD '08, p. 1345 (2008)
16. Jiang, B., Luo, Z.: A New algorithm for semantic Web service matching. J. Softw. 8(2), 351–356 (2013)
17. Boukhadra, A., Benatchba, K., Balla, A.: Automatic composition of semantic Web services-based alignment of OWL-S. Available: http://ceur-ws.org/Vol-867/Paper39.pdf. Accessed 09 Apr 2015
18. Saquicela, V., Vilches-Blazquez, L.M., Corcho, O.: Lightweight semantic annotation of geospatial RESTful services. Available: http://oa.upm.es/6968/1/Lightweight_Semantic.pdf. Accessed 09 Apr 2015
19. Ranabahu, A., Parikh, P., Panahiazar, M., Sheth, A., Logan-Klumpler, F.: Kino: a generic document management system for biologists using sa-rest and faceted search. In: 2011 IEEE Fifth International Conference on Semantic Computing, pp. 205–208 (2011)
20. Patil, A.A., Oundhakar, S.A., Sheth, A.P., Verma, K.: Meteors Web service annotation framework. In: WWW, pp. 553–562 (2004)
21. Ankolekar, A., Burstein, M., Hobbs, J.R., Lassila, O., Martin, D.: DAML-S: Web service description for the semantic web. In: Proceedings of the 1st International Semantic Web Conference Sardinia (2002)
22. Li, G., Deng, S., Xia, H., Lin, C.: Automatic service composition based on process ontology. IEEE Xplore (2007)
23. Song, Y., Liu, L., Ren, P.: Web service composition based on the annotated ontology. In: Fifth International Workshop on Education Technology and Computer Science (2009)

24. Maedche, A., Motik, B., Silva,N., Volz, R.: MAFRA—A MApping FRAmework for distributed ontologies. In: EKAW '02 Proceedings of the 13th International Conference on Knowledge Engineering and Knowledge Management. Ontologies and the Semantic Web, pp. 235–250 (2002)
25. Naumenko, A., Nikitin, S., Terziyan, V., Zharko, A.: Strategic industrial alliances in paper industry: XML-vs ontology-based integration platforms. Learn. Organ. **12**(5), 492–514 (2005)
26. Mokarizadeh, S., Küngas, P., Matskin, M.: Ontology learning for cost-effective large-scale semantic annotation of XML schemas and Web service interfaces. EKAW **10**, 401–410 (2010)
27. Meyer, B.: Seven principles of software testing. IEEE Comput. **41**(8), 99–101 (2008)
28. Sabou, M., Pan, J.: Towards semantically enhanced Web service repositories. Web Semant. **5**(2), 142–150 (2007)
29. EVIWARE: SoapUI; the Web Services Testing tool (2015)
30. Eschenbach, C., Gruninger, M.: Formal ontology in information systems. In: Proceedings of the Fifth International Conference (Fois 2008), p. 325. IOS Press, Amsterdam (2008)
31. Ke, Z., Bo, J., Chan, W.K.: IEEE prioritizing test cases for regression, testing of location-based services: metrics, techniques, and case study

Part VII
Artificial Intelligence

Prediction of Protein Folding Kinetic States Using Fuzzy Back Propagation Method

M. Anbarasi and M.A. Saleem Durai

Abstract Protein folding process is extremely vital in major the molecular function. The kinetic order of protein folding chooses whether the molecule reaches its native structure through intermediates or not. They can either fold without stable intermediates (2State/2S) and with stable intermediates (3State/3S). This is generally determined using equilibrium denaturation research and is often time consuming and tedious. Moreover, the unfolding appliance of large number of Proteins available in the PDB are found unknown. Therefore, it created interest and directed us to predict and classify the folding mechanism as two state or three state (Multiple State). We developed the classification models using Fuzzy Back Propagation Network (FBPN) with the known attributes (Protein length (PL), hydrophobicity, hydrophilicity, secondary structural components). The models performed fairly well for predicting two state and three state folding using the well-known variables. The FBPN model produced accuracy of 83 % for cross validation. This method thus can intensely assist as a outline for predicted monomer and dimer structures with unknown folding appliance for further confirmation through investigational studies.

1 Introduction

To comprehend the relationship between auxiliary components of Monomeric and Dimeric proteins and their collapsing sort is observed to be profoundly charming. Proteins can experience collapsing procedure to shape a steady structure without intermediates (2S) or with stable intermediates (MS or 3S) as depicted in Fig. 1. The 2State harmony collapsing system has been accounted for by distinctive gatherings [9, 19–21, 24] as these concentrates plainly speak to the nonappearance of very much characterized middle of the road compliances. Monomeric proteins can overlay with straightforward 2State dynamic or through populated middle of the road state [14].

M. Anbarasi (✉) · M.A. Saleem Durai
School of Computing Science and Engineering, VIT University,
Vellore 632014, India

© Springer International Publishing Switzerland 2016   419
V. Vijayakumar and V. Neelanarayanan (eds.), *Proceedings of the 3rd International Symposium on Big Data and Cloud Computing Challenges (ISBCC – 16')*,
Smart Innovation, Systems and Technologies 49, DOI 10.1007/978-3-319-30348-2_36

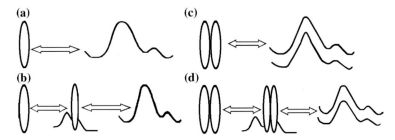

Fig. 1 a Monomer 2S folding. **b** Monomer 3S folding. **c** Dimer 2S folding. **d** Dimer 3S folding

Dimeric proteins additionally fold to their local state with or without intermediates. Henceforth there are very little natural contrasts in the middle of monomers
and dimers in folding [27]. Homodimer protein assumes a noteworthy part in
catalysis and regulation of folding and is very entrancing. Determinations of folding
through existing tests are extremely tedious and dreary. Although, it assumes a
significant part in comprehension basic building design and folding example of
homodimers, likewise some homodimers have been abused as drug targets and are
patented [13, 25]. Heterodimer proteins undertake major part in regulation, catalysis, hindrance, resistance and gathering. Arrangement of two unique proteins
includes a steady interface. Despite the fact that countless studies have been carried
out in understanding and setting up structure folds relationship. The utilization of
information mining strategies and arrangement calculations to develop a choice tree
supporting the characterization of folding energy into 2State and 3State construct up
in light of basic components is accounted for similarly less.

In this paper, we investigate the folding component of monomers and dimers
construct up with respect to their basic information utilizing order calculations. We
portray the improvement and use of diverse characterization calculations like Naive
Bayesian, choice tree, SVM, irregular timberland and neural systems [2, 6, 7, 22, 26]
in view of structure inferred understood variables, for example, Protein Length,
Hydrophilicity, Hydrophobicity and, Secondary auxiliary segments for the order of
folding instruments.

2 Materials and Methods

2.1 Dataset Construction

A dataset of 120 proteins with 20 two State and 20 different State proteins for
monomers; 20 two State and 20 multi state protein for both homodimers and
heterodimers is built (Table 1). The dataset is then arranged with parameters like
PDB ID, Protein length, Secondary basic parts, Hydrophobic and hydrophilic
deposits (Table 2). The procedure of dataset development is delineated in Fig. 2.

Table 1 Proteins for folding state prediction

S. No	Entry name	PDBID	NCBI	Classification	Stoichiometry	References
1	ACLY_HUMAN	3MWD	GI: 298508699	Transferase	Hetero 2-mer-AB	Sun et al. (2010)
2	ACY1_HUMAN	1Q7L	GI: 42543417	Hydrolase	Hetero 2-mer-AB	Lindner et al. (2003)
3	ADX_HUMAN	3N9Y	GI: 335892272	Oxidoreductase, electron transport	Hetero 2-mer-AB	Strushkevich et al. (2011)
4	CP2CI_HUMAN	2CIK	GI: 118137429	Immune system/peptide	Hetero 2-mer-AB	Hourigan et al. (2006)
5	CUBN_HUMAN	3KQ4	GI: 290790132	Transport protein	Hetero 2-mer-AB	Andersen et al. (2010)
6	DPOLZ_HUMAN	3ABD	GI: 288965386	Cell Cycle/replication	Hetero 2-mer-AB	Hara et al. (2010)
7	TIMP3_HUMAN	3CKI	GI: 195927463	Hydrolase, inhibitor	Hetero 2-mer-AB	Wisniewska et al. (2008)
8	FETA_HUMAN	3MRK	GI: 333944222	Immune system	Hetero 2-mer-AB	Gras et al. (to be published)
9	FNTB_HUMAN	1JCQ	GI: 16974883	Transferase	Hetero 2-mer-AB	Long et al. (2001)
10	GELS_HUMAN	1C0F	GI: 7245498	Contractile protein	Hetero 2-mer-AB	Matsuura et al. (2000)
11	GNAI1_HUMAN	1KJY	GI: 21465862	Signaling protein	Hetero 2-mer-AB	Kimple et al. (2002)
12	GNAI3_HUMAN	2IHB	GI: 119390147	Signaling protein	Hetero 2-mer-AB	Soundararajan et al. (2008)

(continued)

Table 1 (continued)

S. No	Entry name	PDBID	NCBI	Classification	Stoichiometry	References
13	ITB1_HUMAN	3G9W	GI: 261824842	Cell adhesion	Hetero 2-mer-AB	Anthis et al. (2009)
14	LMAN1_HUMAN	3A4U	GI: 283135307	Protein transport	Hetero 2-mer-AB	Nishio et al. (2010)
15	PGCA_HUMAN	4MD4	GI: 563320926	Immune system	Hetero 2-mer-AB	Scally et al. (2013)
16	PLCB2_HUMAN	2FJU	GI: 119389489	Signaling protein	Hetero 2-mer-AB	Jezyk et al. (2006)
17	PP2BA_HUMAN	1AUI	GI: 2781188	Hydrolase	Hetero 2-mer-AB	Kissinger et al. (1995)
18	RAF1_HUMAN	1C1Y	GI: 5821936	Signaling protein	Hetero 2-mer-AB	Nassar (1995)
19	RBX1_HUMAN	1LDJ	GI: 21466059	Ligase	Hetero 2-mer-AB	Zheng et al. (2002)
20	SAE2_HUMAN	1Y8Q	GI: 60594165	Ligase	Hetero 2-mer-AB	Lois et al. (2005)
21	PRLR_HUMAN	1BP3	GI: 6729890	Hormone/growth factor	Hetero 2-mer-AB	Somers et al. (1994)
22	SOMA_HUMAN	1A22	GI: 134703	Complex (hormone /receptor)	Hetero 2-mer-AB	Clackson et al. (1998)
23	BDNF_HUMAN	1B8M	GI: 4558267	Growth factor/E43neurotrophin	Hetero 2-mer-AB	Robinson et al. (1999)
24	SMD2_HUMAN	1B34	GI: 6980598	RNA binding protein	Hetero 2-mer-AB	Kambach et al. (1999)

(continued)

Table 1 (continued)

S. No	Entry name	PDBID	NCBI	Classification	Stoichiometry	References
25	POLG_WNV	2GGV	GI: 145579385	Hydrolase	Hetero 2-mer-AB	Aleshin et al. (2007)
26	CLPS_ECOLI	1LZW	GI: 27065491	Chaperone	Hetero 2-mer-AB	Zeth et al. (2002)
27	Q8ZRK4_SALTY	2CO6	GI: 110591065	Fibril Protein	Hetero 2-mer-AB	Remaut et al. (2006)
28	UBA3_HUMAN	3FN1	GI: 225698063	Ligase	Hetero 2-mer-AB	Huang et al. (2009)
29	GNDS_RAT	1LFD	GI: 4930054	Complex (ralgds/ras)	Hetero 2-mer-AB	Huang et al. (1998)
30	CLPS_ECOLI	1MG9	GI: 27065646	Chaperone	Hetero 2-mer-AB	Zeth et al. (2002)
31	LYSC_CHICK	1JTT	GI: 17943062	Immune system, LYSOZYME	Hetero 2-mer-AB	Decanniere et al. (2001)
32	IMM3_ECOLX	1E44"	GI: 14719501	Ribonuclease	Hetero 2-mer-AB	Carr et al. (2000)
33	POC4_YEAST	2Z5B	GI: 166007286	Chaperone	Hetero 2-mer-AB	Yashiroda et al. (2008)
34	IL4_HUMAN	1IAR	GI: 7245490	Cytokine/receptor	Hetero 2-mer-AB	Hage et al. (1999)
35	SPAS_SHIFL	2VT1	GI: 188595908	Membrane protein	Hetero 2-mer-AB	Deane et al. (2008)
36	PRMA_THET8	2NXN	GI: 122921168	Transferase	Hetero 2-mer-AB	Demirci et al. (2007)

(continued)

Table 1 (continued)

S. No	Entry name	PDBID	NCBI	Classification	Stoichiometry	References
37	Q9AJ26_ECOLX	3BZV	GI: 185177894	Membrane protein, protein transport	Hetero 2-mer-AB	Zarivach et al. (2008)
38	O52124_ECOLX	1XOU	GI: 58177543	Structural protein/chaperone	Hetero 2-mer-AB	Yip et al. (2005)
39	3NBB_BOIIR	2H7Z	GI: 114794249	Toxin	Hetero 2-mer-AB	Pawlak et al. (2009)
40	SNTA1_MOUSE	1QAV	GI: 5107634	Membrane protein/E79oxidoreductase	Hetero 2-mer-AB	Hillier et al. (1999)
41	5NT3A_HUMAN	2CN1	GI:110591042	Hydrolase	Homo 2-mer-A2	Wallden et al. (2007)
42	5NTC_HUMAN	2JCM	GI:126031126	Hydrolase	Homo 2-mer-A2	Wallden et al. (2007)
43	5NTD_HUMAN	4H1S	GI: 407944028	Hydrolase	Homo 2-mer-A2	Heuts et al. (2012)
44	ABC3G_HUMAN	3V4K	GI: 374074548	Hydrolase	Homo 2-mer-A2	Li et al. (2012)
45	ACY2_HUMAN	4MRI	GI: 672883368	Hydrolase	Homo 2-mer-A2	Wijayasinghe et al. (2014)
46	ADAT2_HUMAN	3DH1	GI: 198443277	Hydrolase	Homo 2-mer-A2	Welin et al. (to be published)
47	ADH7_HUMAN	1AGN	GI: 1942122	Oxidoreductase	Homo 2-mer-A2	Xie et al. (1997)
48	ADHX_HUMAN	1M6H	GI: 22219399	Oxidoreductase	Homo 2-mer-A2	Sanghani et al. (2002)

(continued)

Table 1 (continued)

S. No	Entry name	PDBID	NCBI	Classification	Stoichiometry	References
49	CP51A_HUMAN	3JUS	GI: 290560261	Oxidoreductase	Homo 2-mer-A2	Strushkevich et al. (to be published)
50	CP7A1_HUMAN	3DAX	GI: 195927532	Oxidoreductase	Homo 2-mer-A2	Strushkevich et al. (to be published)
51	CSK2B_HUMAN	1DS5	GI: 13096757	Transferase	Homo 2-mer-A2	Battistutta et al. (2000)
52	CSN5_HUMAN	4F7O	GI: 443428111	Hydrolase	Homo 2-mer-A2	Echalier et al. (2013)
53	CYGB_HUMAN	1UMO	GI: 55670345	Oxygen storage/transport	Homo 2-mer-A2	De Sanctis et al. (2004)
54	DPEP1_HUMAN	1ITQ	GI: 23200142	Hydrolase	Homo 2-mer-A2	Nitanai et al. (2002)
55	EHMT1_HUMAN	2IGQ	GI: 118138406	Transferase	Homo 2-mer-A2	Wu et al. (2010)
56	ENOA_HUMAN	2PSN	GI: 188595822	Lyase	Homo 2-mer-A2	Hyo et al. (to be published)
57	ENOB_HUMAN	2XSX	GI: 311771970	Lyase	Homo 2-mer-A2	Vollmar et al. (to be published)
58	ENOF1_HUMAN	4A35	GI: 350610930	Isomerase	Homo 2-mer-A2	Muniz et al. (to be published)
59	ENOG_HUMAN	1TE6	GI: 55669906	Lyase	Homo 2-mer-A2	Chai et al. (2004)
60	EXOG_HUMAN	4A1N	GI: 378792600	Hydrolase	Homo 2-mer-A2	Welin et al. (to be published)

(continued)

Table 1 (continued)

S. No	Entry name	PDBID	NCBI	Classification	Stoichiometry	References
61	A4_HUMAN	1AAP	GI: 229659	Proteinase inhibitor (trypsin)	Homo 2-mer-A2	Hynes et al. (1990)
62	CYB5B_HUMAN	3NER	GI: 333944275	Electron transport	Homo 2-mer-A2	Parthasarathy et al. (2011)
63	DICER_HUMAN	2EB1	GI: 160285502	Hydrolase	Homo 2-mer-A2	Takeshita et al. (2007)
64	DSRAD_HUMAN	1QBJ	GI: 5542337	Hydrolase/dna	Homo 2-mer-A2	Schwartz et al. (1999)
65	DTD1_HUMAN	2OKV	GI: 126031644	Hydrolase	Homo 2-mer-A2	Kemp et al. (2007)
66	ITB7_HUMAN	2BRQ	GI: 159162632	Structural Protein	Homo 2-mer-A2	Kiema et al. (2006)
67	NB5R4_HUMAN	3LF5	GI: 300508460	Oxidoreductase	Homo 2-mer-A2	Deng et al. (to be published)
68	PA2GA_HUMAN	1AYP	GI: 1127283	Hydrolase	Homo 2-mer-A2	Oh (1995)
69	PARN_HUMAN	2A1R	GI: 85544056	Hydrolase/rna	Homo 2-mer-A2	Wu et al. (2005)
70	COPG_STRAG	2CPG	GI: 6573448	Gene regulation	Homo 2-mer-A2	Gomis-Ruth et al. (1998)
71	RARC_BPP22	1ARR	GI: 493829	Gene regulating protein	Homo 2-mer-A2	Bonvin et al. (1994)
72	ROP_ECOLX	1ROP	GI: 230307	Transcription regulation	Homo 2-mer-A2	Banner et al. (1987)

(continued)

Table 1 (continued)

S. No	Entry name	PDBID	NCBI	Classification	Stoichiometry	References
73	RCRO_LAMBD	5CRO	GI: 3318783	Gene regulating protein	Homo 2-mer-A2	Ohlendorf et al. (1998)
74	HMFB_METFE	1BFM	GI: 1310900	Histone protein	Homo 2-mer-A2	Starich et al. (1996)
75	VE2_HPV31	1A7G	GI: 4929856	Transcription regulation	Homo 2-mer-A2	Bussiere et al. (1998)
76	G5P_BPF1	1VQB	GI: 1942350	DNA binding protein	Homo 2-mer-A2	Skinner et al. (1994)
77	DBH_THEMA	1B8Z	GI: 7245986	DNA binding protein	Homo 2-mer-A2	Christodoulou et al. (1998)
78	FIS_ECOLI	1ETY	GI: 11514671	Transcription activator	Homo 2-mer-A2	Cheng et al. (2000)
79	ZN174_HUMAN	1Y7Q	GI: 159163639	Transcription	Homo 2-mer-A2	Ivanov et al. (2005)
80	POL_HV1BR	1A8G	GI: 3402073	Hydrolase/hydrolase inhibitor	Homo 2-mer-A2	Ringhofer et al. (1999)
81	3HAO_HUMAN	2QNK	GI: 157836012	Oxidoreductase	Monomer	Bitto et al. (to be published)
82	8ODP_HUMAN	4N1T	GI:618855031	Hydrolase/hydrolase inhibitor	Monomer	Gad et al. (2014)
83	ABC3C_HUMAN	3VOW	GI: 407943679	Hydrolase	Monomer	Kitamura et al. (2012)
84	ABLM2_HUMAN	1V6G	GI:159163205	Metal binding protein	Monomer	Miyamoto et al. (to be published)
85	ACACA_HUMAN	2YL2	GI: 340707530	Ligase	Monomer	Muniz et al. (to be published)

(continued)

Table 1 (continued)

S. No	Entry name	PDBID	NCBI	Classification	Stoichiometry	References
86	ACACB_HUMAN	2DN8	GI:225734325	Ligase	Monomer	Ruhul Momen et al. (to be published)
87	ACE2_HUMAN	1R42	GI: 42543475	Hydrolase	Monomer	Towler et al. (2004)
88	ACE_HUMAN	1O86	GI:409187881	Hydrolase/hydrolase inhibitor	Monomer	Natesh et al. (2003)
89	ACOC_HUMAN	2B3X	GI:88192218	Lyase	Monomer	Dupuy et al. (2006)
90	ADA10_HUMAN	3V35	GI: 402550140	Oxidoreductase	Monomer	Zheng et al. (2012)
91	ADAM8_HUMAN	4DD8	GI: 390136390	Hydrolase /hydrolase Inhibitor	Monomer	Hall et al. (2012)
92	ADA_HUMAN	3IAR	GI:255918005	Hydrolase	Monomer	Ugochukwu et al. (to be published)
93	ADK_HUMAN	1BX4	GI:6435729	Transferase	Monomer	Mathews et al. (1998)
94	ADPPT_HUMAN	2BYD	GI:75766415	Transferase	Monomer	Bunkoczi et al. (2007)
95	CP2A6_HUMAN	1Z10	GI: 75765660	Oxidoreductase	Monomer	Yano et al. (2005)
96	CP2AD_HUMAN	2P85	GI: 146387690	Oxidoreductase	Monomer	Smith et al. (2007)
97	CP2B6_HUMAN	3IBD	GI: 284055614	Oxidoreductase/oxidoreductase Inhibitor	Monomer	Gay et al. (2010)
98	CP2C8_HUMAN	1PQ2	GI: 42543301	Oxidoreductase	Monomer	Schoch et al. (2004)
99	CP2C9_HUMAN	1OG2	GI: 34811351	Electron transport	Monomer	Williams et al. (2003)
100	CP2CJ_HUMAN	4GQS	GI: 414145769	Oxidoreductase	Monomer	Reynald et al. (2012)
101	ADXL_HUMAN	2Y5C	GI: 374414449	Electron transport	Monomer	Webert et al. (to be Published)

(continued)

Table 1 (continued)

S. No	Entry name	PDBID	NCBI	Classification	Stoichiometry	References
102	CSH1_HUMAN	1Z7C	GI: 93278530	Hormone/growth factor	Monomer	Walsh et al. (2006)
103	DNLI3_HUMAN	1IMO	GI: 159162489	Ligase	Monomer	Krishman et al. (2001)
104	DPOLA_HUMAN	1K0P	GI: 33357211	Transferase	Monomer	Evanics et al. (2003)
105	FBXL5_HUMAN	3U9J	GI: 377656349	Protein binding	Monomer	Shu, C.et al. (2012)
106	FKBP8_HUMAN	2AWG	GI: 78101474	Signaling protein inhibitor	Monomer	Walker et al. (to be published)
107	FOXK2_HUMAN	1JXS	GI: 253722939	DNA binding protein	Monomer	Liu et al. (2002)
108	GNAT1_HUMAN	3RBQ	GI: 335892462	Protein transport/lipid binding protein	Monomer	Zhang et al. (2011)
109	ITB2_HUMAN	1L3Y	GI: 159162632	Cell adhesion	Monomer	Beglova et al. (2002)
110	KCMA1_HUMAN	2K44	GI: 238828058	Membrane protein	Monomer	Unnerstale et al. (2009)
111	KDM4A_HUMAN	2GF7	GI: 99032485	Metal binding protein	Monomer	Huang et al. (2006)
112	LN28B_HUMAN	4A4I	GI: 400260697	RNA binding protein	Monomer	Mayr et al. (2012)
113	LRP1_HUMAN	1CR8	GI:4139440	Lipid binding protein	Monomer	Huang et al. (1999)
114	MMP9_HUMAN	1GKC	GI: 21465442	Hydrolase/hydrolase inhibitor	Monomer	Rowsell et al. (2002)
115	MOCS3_HUMAN	3I2V	GI: 254575045	Transferase	Monomer	Bacik et al. (to be published)
116	MSS4_HUMAN	1FWQ	GI: 10835450	Metal binding protein	Monomer	Yu et al. (1995)
117	MT2_HUMAN	1MHU	GI:1431705	Metallothionein	Monomer	Messerle et al. (1990)

(continued)

Table 1 (continued)

S. No	Entry name	PDBID	NCBI	Classification	Stoichiometry	References
118	MT3_HUMAN	2F5H	GI: 109157553	Metal binding protein	Monomer	Wang et al. (2006)
119	MYPC3_HUMAN	1GXE	GI: 33356956	Cytoskeleton	Monomer	Idowu et al. (2003)
120	NGLY1_HUMAN	2CCQ	GI: 110590932	Glycoprotein	Monomer	Allen et al. (2006)

Table 2 Structural features of proteins to classify the folding rates

S. No	PDBID	Length	H	S	C	Hydrophobic	Hydrophilic	States
1	3MWD	0.46	0.44	0.46	0.33	0.46	0.44	MS
2	1Q7L	0.2	0.2	0.15	0.25	0.19	0.2	MS
3	3N9Y	0.53	0.61	0.26	0.25	0.49	0.55	MS
4	2CIK	0.29	0	0	0	0.21	0.36	MS
5	3KQ4	0.42	0.44	0.17	0.25	0.4	0.43	MS
6	3ABD	0.23	0.17	0.2	0	0.21	0.25	MS
7	3CKI	0.27	0.22	0.33	0.33	0.21	0.32	MS
8	3MRK	0.31	0	0	0	0.23	0.38	MS
9	1JCQ	0.48	0.51	0.17	0.33	0.45	0.48	MS
10	1C0F	0.4	1	0.63	0.58	0.37	0.4	MS
11	1KJY	0.35	0.22	0.48	0.17	0.29	0.39	MS
12	2IHB	0.34	0.17	0.39	0.17	0.27	0.4	MS
13	3G9W	0.23	0.44	0.35	0.75	0.19	0.26	MS
14	3A4U	0.26	0.51	0.07	0	0.25	0.27	MS
15	4MD4	0.19	0	0	0	0.17	0.2	MS
16	2FJU	0.17	0.66	0.7	0.58	0.17	0.18	MS
17	1AUI	0.57	0.41	0.39	0.5	0.51	0.61	MS
18	1C1Y	0.16	0.17	0.13	0.17	0.14	0.19	MS
19	1LDJ	0.85	0.22	0.04	0.17	0.66	1	MS
20	1Y8Q	0.37	0.39	0.28	0.17	0.34	0.38	MS
21	2CN1	0.31	0.24	0.3	0.25	0.27	0.33	MS
22	2JCM	0.61	0.49	0.57	0.33	0.51	0.68	MS
23	4H1S	0.58	0.51	0.63	0.33	0.54	0.59	MS
24	3V4K	0.2	0.27	0.17	0	0.16	0.24	MS
25	4MRI	0.33	0.44	0.22	0.17	0.3	0.35	MS
26	3DH1	0.19	0.15	0.11	0.08	0.16	0.21	MS
27	1AGN	0.4	0.49	0.39	0.25	0.4	0.38	MS
28	1M6H	0.4	0.49	0.41	0.25	0.42	0.37	MS
29	3JUS	0.5	0.34	0.5	0.33	0.45	0.53	MS
30	3DAX	0.54	0.41	0.52	0.33	0.47	0.58	MS
31	1DS5	0.35	0.2	0.3	0.42	0.3	0.4	MS
32	4F7O	0.27	0.22	0.15	0.25	0.24	0.29	MS
33	1UMO	0.19	0	0.2	0.17	0.19	0.19	MS
34	1ITQ	0.4	0.39	0.43	0.58	0.36	0.42	MS
35	2IGQ	0.3	0.44	0.43	0.5	0.24	0.35	MS
36	2PSN	0.47	0.29	0.46	0.33	0.46	0.46	MS
37	2XSX	0.47	0.32	0.43	0.33	0.47	0.45	MS
38	4A35	0.48	0.34	0.5	0.5	0.45	0.49	MS
39	1TE6	0.48	0.34	0.48	0.33	0.46	0.47	MS
40	4A1N	0.36	0.27	0.3	0.58	0.3	0.4	MS

(continued)

Table 2 (continued)

S. No	PDBID	Length	H	S	C	Hydrophobic	Hydrophilic	States
41	2QNK	0.3	0.2	0.43	0.08	0.28	0.31	MS
42	4N1T	0.15	0.1	0.24	0.17	0.15	0.16	MS
43	3VOW	0.19	0.17	0.11	0.17	0.14	0.24	MS
44	1V6G	0.06	0.05	0.13	0.17	0.06	0.07	MS
45	2YL2	0.6	0.51	0.39	0.5	0.58	0.57	MS
46	2DN8	0.08	0	0.15	0	0.08	0.09	MS
47	1R42	0.68	0.78	0.26	1	0.57	0.76	MS
48	1O86	0.65	0.76	0.22	0.42	0.56	0.71	MS
49	2B3X	1	0.98	1	0.67	1	0.93	MS
50	3V35	0.36	0.34	0.33	0.17	0.32	0.38	MS
51	4DD8	0.93	0.2	0.09	0	0.89	0.91	MS
52	3IAR	0.39	0.59	0.24	0.17	0.36	0.41	MS
53	1BX4	0.37	0.41	0.35	0.17	0.33	0.39	MS
54	2BYD	0.34	0.34	0.3	0.17	0.3	0.38	MS
55	1Z10	0.52	0.49	0.26	0.5	0.49	0.52	MS
56	2P85	0.52	0.49	0.24	0.58	0.5	0.52	MS
57	3IBD	0.52	0.54	0.22	0.67	0.51	0.5	MS
58	1PQ2	0.52	0.54	0.28	0.5	0.47	0.55	MS
59	1OG2	0.52	0.51	0.33	0.42	0.49	0.52	MS
60	4GQS	0.52	0.51	0.28	0.67	0.49	0.53	MS
61	1BP3	0.19	0.24	0.02	0.33	0.15	0.23	TS
62	1A22	0.24	0.05	0.28	0.25	0.18	0.31	TS
63	1B8M	0.11	0.05	0.2	0.08	0.07	0.15	TS
64	1B34	0.11	0.07	0.11	0.08	0.11	0.11	TS
65	2GGV	0.03	0.02	0.11	0	0.03	0.04	TS
66	1LZW	0.09	0.1	0.09	0	0.09	0.1	TS
67	2CO6	0.11	0.02	0.26	0	0.11	0.12	TS
68	3FN1	0.08	0.1	0.17	0.17	0.07	0.1	TS
69	1LFD	0.16	0.12	0.26	0.17	0.13	0.19	TS
70	1MG9	0.09	0.1	0.09	0	0.09	0.1	TS
71	1JTT	0.12	0.2	0.15	0.42	0.09	0.15	TS
72	1qe44	0.07	0.12	0.09	0.25	0.06	0.08	TS
73	2Z5B	0.14	0.07	0.15	0	0.13	0.16	TS
74	1IAR	0.12	0.15	0.04	0.25	0.07	0.17	TS
75	2VT1	0.08	0.07	0.09	0.17	0.07	0.09	TS
76	2NXN	0.26	0.29	0.33	0.08	0.33	0.19	TS
77	3BZV	0.03	0.12	0.09	0	0.02	0.05	TS
78	1XOU	0.08	0.05	0	0	0.06	0.11	TS
79	2H7Z	0.05	0	0.13	0.08	0.03	0.09	TS
80	1QAV	0.07	0.1	0.15	0.17	0.08	0.07	TS

(continued)

Table 2 (continued)

S. No	PDBID	Length	H	S	C	Hydrophobic	Hydrophilic	States
81	1AAP	0.03	0.05	0.11	0.08	0.03	0.05	TS
82	3NER	0.07	0.17	0.07	0.17	0.06	0.1	TS
83	2EB1	0.2	0.22	0	0.08	0.16	0.23	TS
84	1QBJ	0.06	0.1	0.07	0	0.05	0.08	TS
85	2OKV	0.21	0.15	0.11	0	0.17	0.25	TS
86	2BRQ	0.08	0.54	0.26	0.33	0.1	0.06	TS
87	3LF5	0.07	0.17	0.09	0.17	0.06	0.08	TS
88	1AYP	0.11	0.17	0.09	0.08	0.06	0.17	TS
89	2A1R	0.47	0.41	0.28	0.58	0.38	0.53	TS
90	2CPG	0.02	0.05	0.02	0	0.02	0.03	TS
91	1ARR	0.03	0.07	0.02	0.08	0.03	0.04	TS
92	1ROP	0.04	0.07	0	0	0.03	0.06	TS
93	5CRO	0.04	0.1	0.07	0.08	0.04	0.05	TS
94	1BFM	0.05	0.07	0	0	0.06	0.05	TS
95	1A7G	0.06	0.1	0.09	0.08	0.04	0.09	TS
96	1VQB	0.07	0.05	0.22	0	0.07	0.07	TS
97	1B8Z	0.07	0.07	0.07	0.08	0.08	0.08	TS
98	1ETY	0.08	0.1	0.04	0.17	0.07	0.1	TS
99	1Y7Q	0.08	0.12	0.02	0	0.06	0.11	TS
100	1A8G	0.08	0.02	0.22	0.08	0.1	0.07	TS
101	2Y5C	0.09	0.12	0.13	0	0.1	0.1	TS
102	1Z7C	0.19	0.2	0.02	0.25	0.14	0.24	TS
103	1IMO	0.07	0.1	0.11	0.08	0.07	0.08	TS
104	1K0P	0	0.07	0.04	0.08	0.01	0.01	TS
105	3U9J	0.15	0.2	0	0	0.12	0.19	TS
106	2AWG	0.1	0.1	0.15	0.17	0.11	0.1	TS
107	1JXS	0.08	0.1	0.11	0	0.06	0.11	TS
108	3RBQ	0.19	0.1	0.2	0.17	0.16	0.22	TS
109	1L3Y	0.02	0	0.13	0.17	0.02	0.03	TS
110	2K44	0	0.05	0	0	0.01	0	TS
111	2GF7	0.11	0.07	0.15	0	0.09	0.12	TS
112	4A4I	0.07	0.05	0.13	0.08	0.08	0.07	TS
113	1CR8	0.02	0	0	0	0.01	0.04	TS
114	1GKC	0.16	0.17	0.67	0.25	0.17	0.15	TS
115	3I2V	0.12	0.15	0.13	0	0.12	0.12	TS
116	1FWQ	0.11	0.05	0.24	0.17	0.11	0.12	TS
117	1MHU	0	0.02	0.02	0	0	0.02	TS
118	2F5H	0.01	0.05	0.04	0.17	0	0.03	TS
119	1GXE	0.13	0	0.22	0.08	0.11	0.15	TS
120	2CCQ	0.08	0.07	0.04	0.08	0.08	0.09	TS

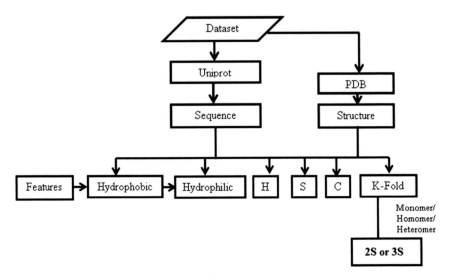

Fig. 2 Flow diagram depicting the process of dataset construction

The PDBID, length and auxiliary basic data of proteins were extricated from uni-prot (http://www.uniprot.org). Protein arrangement length alludes to the aggregate number of buildups in every protein in the dataset and its pertinence has been refered to in writing as a noteworthy component for anticipating protein folding [4, 28]. Auxiliary structure is the local spatial plan of polypeptide's spine, which incorporates α-helices (H), β-sheets (B) and c-curls (C) likewise contribute basically in protein folding. Utilizing protparam apparatus of EXPASY, (http://www.expasy. ch/cgi-receptacle/protparam) the hydrophilic (D, E, R, K, H, S, T, N, Q, W, C, Y) and hydrophobic (V, F, M, L, A, I, P, G) deposits in monomers and dimers have been distinguished. K-fold (http://folding.path.uab.edu/k-fold/K-Fold.html) server was utilized to foresee the folding motor states (two state or different state) of the proteins. The server is based on bolster vector machines (SVM) utilizing direct bit capacities, and it predicts the states, logarithm of folding rates, and diagram demonstrating contacts recurrence and contact request per buildup [1, 10, 23, 29]. After dataset construction then we needs to perform following steps which is given in the Sections "Step I–Step IV".

2.2 Fuzzy Back Propagation

Back propagation learning algorithm is one of the most significant improvements in neural networks. This Back propagation network (BPN) algorithm is applied to multilayer feed-forward networks involving of handling elements with continuous differentiable activation functions the architecture is given in Fig. 3. For the given

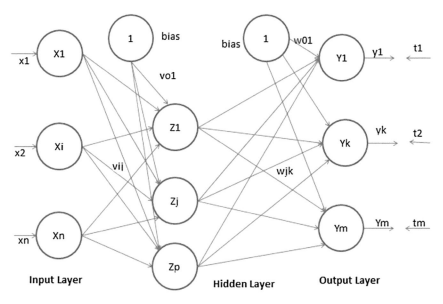

Fig. 3 Architecture of BPN

set of input, output pairs this algorithm delivers a procedure for changing the weights in a BPN. The basic idea used for weight updation is the gradient-descent method as used in simple perceptron network. This method is used to find a local minimum of functions. It starts with an initial guess of the result and takes the gradient of the function at that point. The result is treaded in the negative direction of the gradient repeat the process. Suppose x_k and x_{k+1} are two successive values in the sequel. Then we have $x_{k+1} = x_k - \theta. f^1(x_k), \theta > 0$ is a small number that forces the algorithm to make small jumps. For suitable choices of θ it is guaranteed that. In this method the error is propagated back to the hidden unit. It is different from other networks when it is calculating the weights during learning period. When the number of hidden layers is bigger the complication increases. The error is usually measured in the output layer and there is no information about the error in the hidden layers.

The fuzzy logic method we choose in a fuzzy neural network was used with an artificial neural network in this study. Fuzzy logic can tell the logical meaning used by people in a more straight forward way. Logical decision-making was performed according to the language rules proposed by experts, which can solve non-linear problems that cannot be addressed using different classification methods. The combination of fuzzy logic with a neural network can facilitate self-adaptation through a training function and acquire an algorithm for the information expressed as fuzzy or precise data. The fuzzy BP method uses weights which are trained automatically. This can avoid the difficulties in telling discrete knowledge and processes. The attachment of these two methods can compensate for the incompleteness of a neural network in fuzzy data processing and the insufficiency of pure fuzzy logic in learning functions. To construct a fuzzy neural network structure, this algorithm adds a fuzzification layer above the input layer of the neural network and a Defuzzification layer after the output layer, which results in the fuzzification of the input information and Defuzzification of the output information, respectively. The input and the completeness level of the final output in fuzzy logic, provided by higher minds in this field, were used as the values taken as input and required outputs of the learning samples for the neural network. With the aid of the higher learning and attached memory capacity of the artificial neural network, the neural network is trained for its learning function to generate acquired fuzzy rules that are stored in the network in the forms of input weights and thresholds. Thus, our fuzzy neural network obtains the capabilities of analysing fuzzy questions and making diagnoses and achieves an effective combination of fuzzy logic with a neural network.

2.2.1 Steps for Fuzzy BPN

Step I: Given input we need to convert it into fuzzification
Step II: Apply Fuzzy Back propagation algorithm for the given training data set
    i. Initialize the weights
    ii. Repeat
        1. For each training pattern
           a. Train on that pattern
        End
      Until the error is acceptably low.
Step III: Testing algorithm for BPN
Step IV: Calculate the classifier accuracy
Step by Step procedure

Step I

There are 120 date set given of which 90 are taken for training and 30 are taken for testing. For the given dataset we have to convert it into fuzzification that is based on intuition method using triangular membership function based on Eq. 1 the following membership functions is generated which is given in Fig. 4.

$$f(x, a, b, c) \begin{cases} 0, & x \leq a \\ \frac{x-a}{b-a}, & a \leq x \leq b \\ \frac{c-x}{c-b}, & b < x \leq c \end{cases} \qquad (1)$$

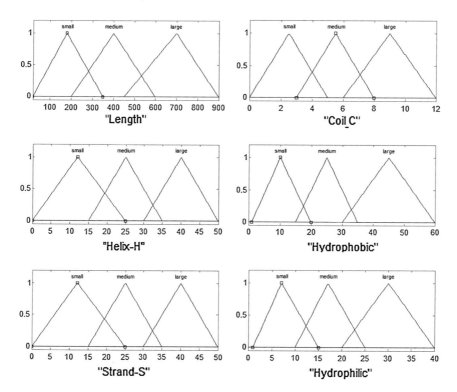

Fig. 4 Membership functions

Step II

Algorithm for Fuzzy Backpropagation

1. Randomly generate the initial weight sets for the input hidden layer and the output hidden layer, where the weight set of the bias is always zero.
2. Perform step 2–15 when stopping condition is false.
3. Let (J_i, K_i), $i = 1, 2, 3, 4...N$ be the N input-output set that our algorithm will train.
4. Let iter denote the no. of iterations (mostly in thousands). Set the counter for the number of iterations and number of pattern to be trained to zero.
   i.e. count_iter = 0, i = 0.
5. Assign values for momentum factor $\acute{\eta}$ and learning rate α.
   Initialize $\Delta W\,(1-1) = 0$, $\Delta W'(1-1) = 0$.
6. Get the another pattern that is subsequent. Assign O = J, p = 1, 2, 3,...,n; O for bias = 1;
7. Calculate $O' = f(NET)$, $j = 1, 2, 3$, where
   $NET_{kp} = CT\left(\sum_{p=1}^{n} Wkp.Okp\right)$ for input hidden layer.
8. Now compute for output hidden layer.

$O'' = f(\text{NET}), j = 1, 2, 3$, where $\text{NET}_{kp} = CT(\sum_{p=1}^{n} W_{kp}.O'_{kp})$

9. Compute change of weights for hidden output layer as follow:

$$\Delta R_{rp} = \left(\frac{\partial Er}{\partial Wpri}, \frac{\partial Er}{\partial Wprk}, \frac{\partial Er}{\partial Wprj} \right)$$

where

$$\frac{\partial Ep}{\partial Wrpx} = -\left(F_{rp} - O''_{rp} \right) O''_{rp}.\left(1 - O''_{rp}\right).(-1\backslash 3).O''_{rp};$$

$$\frac{\partial Ep}{\partial Wrpy} = -\left(F_{rp} - O''_{rp} \right) O''_{rp}.\left(1 - O''_{rp}\right).(-1\backslash 3).O''_{rp};$$

$$\frac{\partial Ep}{\partial Wrpz} = -\left(F_{rp} - O''_{rp} \right) O''_{rp}.\left(1 - O''_{rp}\right).(-1\backslash 3).O''_{rp};$$

10. $\Delta W'(1) = -\dot{\eta}.\Delta E + \alpha.\Delta \Delta'(1-1)$.
11. Now calculating for input hidden layer.

$$B_{pkx} = -\left(U_{kx} - O''_{kx}\right).O''_{kx}.(1 - O_{kx}).1$$
$$B_{pky} = -\left(U_{kx} - O''_{kx}\right).O''_{kx}.(1 - O_{kx}).(-1/3)$$
$$B_{pkz} = -\left(U_{kx} - O''_{kx}\right).O''_{kx}.(1 - O_{kx}).(1/3)$$

Calculate

$$\Delta R_{rp}' = \left(\frac{\partial E'_r}{\partial W'_{pri}}, \frac{\partial E'_r}{\partial W'_{prk}}, \frac{\partial E'_r}{\partial W'_{prj}} \right)$$

where
$$\frac{\partial Er}{\partial Wrpx} = \sum_{G} \partial W''_{rp}.O''_{rp}.(1 - O''_{rp}).1.O'_{rp};$$

$$\frac{\partial Er}{\partial Wrpy} = \sum_{G} \partial W''_{rp}.O''_{rp}.\left(1 - O''_{rp}\right).(-1/3).O'_{rp};$$

$$\frac{\partial Er}{\partial Wrpz} = \sum_{G} \partial W''_{rp}.O''_{rp}.\left(1 - O''_{rp}\right).(1\backslash 3).O'_{rp};$$

12. Compute $\Delta W(1) = -\dot{\eta}.\Delta E + \alpha.\Delta W(1-1)$.
13. Now update the weights for the input hidden layer and the output hidden layer by adding the change in weight to their correspondence layer.
14. Now repeat the algorithm for the next parameter and increment the counter.
15. Check for the stopping condition may be certain number of cycles reached or when actual output is equal to the target output or Repeat till you get a good accuracy.

STEP III (Testing Algorithm for BPN)

1. Initialize the weights. (The weights are taken from the training phase i.e. step II)
2. Perform steps from 2 to 4 for each input vector
3. Set the activation of input for x_i, i = 1,2,...n
4. Calculate the net input to hidden unit and its output $NET_{kp} = CT(\sum_{p=1}^{n} Wkp.Okp)$ and $O'' = f$ (NET)
5. Compute the net input and the output of the output layer (USE SIGMOIDAL FUNCTIONS AS ACTIVATION FUNCTIONS)
   Note: Weight updating not to be applied during testing
6. Now apply the algorithm till step 3 to 5 for testing of data.

Step IV

1. Compare the labels for each test case and note the observation in the confusion matrix.
2. Calculate the accuracy.

   The weights obtained are as follow:

PATERN WEIGHT:-
2.078286183224225 0.7511085752689755 2.0881811383261715 -1.9618997965468586 1.9440714650006676 0.2780081602807892 24.557631540110393 -14.314374743015883
1.154556323277402 2.0834299102048064 1.7419405531774859 -1.6168937529149702 2.0828707161729407 0.7170146241377209 18.938080253063795 -12.10467696180085
1.03020408589074644 -2.0610864072474545 2.433348762472708 -1.556731391215433 1.0898404735102492 0.49461445426753264 30.62517097520484 -15.790261580418248

hidden WEIGHT1:-

3.422711048634155 0.289054779395270 3.4846387279044853
8.53233933465267   0.785412693643350 5.764523035857522
1.140771264634562 0.479425357088316 -2.694848850637537
1.2204598923015932 0.73620937751006536 1.328274394953029

3   Results and Discussion

We predicted the performance of fuzzy back propagation network algorithms by statistical measures like sensitivity, specificity and accuracy. These metrics count how the test was good and consistent. Data are divided into two blocks one is for training and another is for testing phase. Table 4 show that the classifier accuracy for folding rate prediction with accuracy 83 %. Each column of the matrix represents to the examples in an expected class, while every column speaks to the cases in an actual class. Here rows and columns mean the quantity of false positives, false negatives, true positives, and true negatives. The confusion matrix for the fuzzy BPN is predicted with 13 true positive values and 2 for false positive values which are given in Table 3. That is out of 15 TS, 13 were predicted correctly. Similarly the matrix is predicted with false negative values of 12 and false positive values of 3 for

Table 3 Confusion matrix

	TS	MS
TS	13	2
MS	3	12

Table 4 Performance evaluation

Algorithm	Sensitivity	Specificity	Accuracy
FBPN	0.867	0.8	0.83

MS, which is out of 15, 12 were predicted correctly. Sensitivity denoted as the true positive rate that the percentage of positive tuples that are correctly classified. Specificity is the true negative rate that the percentage of negative tuples that are correctly classified.

$$\text{Sensitivity} = \frac{TP}{P} \quad \text{Specificity} = \frac{TN}{N}$$

where TP is true positive, P is positive, TN is true negative and N is negative.

Accuracy measure correctly classified the test tuple in the given condition as it is defined as

$$\text{Accuracy} = \textit{Sensitivity}\frac{P}{P+N} + \textit{Specificity}\frac{N}{P+N}$$

Different attributes were collected to create a final dataset like Protein length, Hydrophobicity, Hydrophilicity, Secondary structural elements to understand their role in deciding the folding state. In Fig. 5 we have two phases training and testing

Fig. 5 Flowchart for FBPN

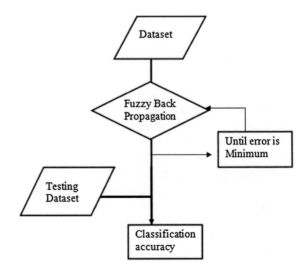

the data set of 120 was spitted into 90 and 30 where 90 was taken as training data and 30 was taken as testing data, the following results were followed and are plotted on the confusion matrix. The features chosen were the most important attributes to distinguish between two state and three state folding of monomers, homodimers and heterodimers. The importance of hydrophobicity in folding kinetics has been documented in different. It is a significant fact that hydrophobicity plays critical role in determining the properties of amino acids, peptides and proteins [3, 11, 15]. As it can be seen 12 true positive values matched, 3 true negative value matched rest 4 false positive and 11false negative value matched.

4   Conclusion

In this paper we predicted protein folding states using Fuzzy BPN with the variables like Protein length, Strand, Helix, Coil, Hydrophobicity, Hydrophilicity, In this study, we have constructed a dataset of 120 proteins with two state and multiple state folding kinetics and investigated the relationship of folding component and structural data using Fuzzy BPN algorithms. Our outcomes therefore affirms the significance of hydrophobicity in folding as all the three classes of proteins like Monomers, Homodimers and Heterodimers demonstrates extensive appropriation of hydrophobic buildups concerning various state folding than in two state folding. The precision for predicting protein folding states when apply the Fuzzy BPN kept running with 83 % which makes it the best model to apply in such conditions and is novel to anticipate lung disease tumour sorts in light of protein traits folding conditions of monomers and dimers. This fuzzy BPN has been implemented in java. The curiosity of this study is the way that, it is interestingly it has been researched the part of protein components to focus folding states in all the three classes of monomers, homodimers and heterodimers.

References

1. Capriotti, E., Casadio, R.: K-Fold: a tool for the prediction of the protein folding kinetic order and rate. Bioinformatics **23**(3), 385–386 (2007)
2. Cheng, J., Randall, A.Z., Sweredoski, M.J., Baldi, P.: SCRATCH: a protein structure and structural feature prediction server. Nucleic Acids Res. **33**, W72–W76 (2005)
3. Cranz-Mileva, S., Friel, C.T., Radford, S.E.: Helix stability and hydrophobicity in the folding mechanism of the bacterial immunity protein Im9. Protein Eng. Des. Sel. **18**(1), 41–50 (2005)
4. Fasman, G.D.: Prediction of Protein Structure and the Principles of Protein Conformation, pp. 1–95. Plenum Press, Berlin (1989)
5. Fawzi, N.L., Chubukov, V., Clark, L.A., Brown, S., Head-Gordon, T.: Influence of denatured and intermediate states of folding on protein aggregation. Protein Sci. **14**, 993–1003 (2005). doi:10.1110/ps.041177505. (A Publication of the Protein Society)

6. Galzitskaya, O.V., Ivankov, D.N., Finkelstein, A.V.: Folding nuclei in proteins. FEBS Lett. **489**(113), 118 (2001)
7. Galzitskaya, O.V., Garbuzynskiy, S.O., Ivankov, D.N., Finkelstein, A.V.: Chain length is the main determinant of the folding rate for proteins with three-state folding kinetics. Proteins **51** (2), 162–166 (2003)
8. Gong, H., Isom, D.G., Srinivasan, R., Rose, G.D.: Local secondary structure content predicts folding rates for simple, two-state proteins. J. Mol. Biol. **327**(5), 1149–1154 (2003)
9. Greene Jr, R.F., Pace, C.N.: Urea and guanidine hydrochloride denaturation of ribonuclease, lysozyme, alpha- chymotrypsin, and beta-lacto globulin. J. Biol. Chem. **249**(17), 5388–5393 (1974)
10. Gromiha, M.M., Selvaraj, S.: Comparison between long-range interactions and contact order in determining the folding rate of two-state proteins: application of long-range order to folding rate prediction. J. Mol. Biol. **310**(1), 27–32 (2001)
11. Hart, W.E., Istrail, S.C.: Fast protein folding in the hydrophobic-hydrophilic model within three-eights of optimal. J. Comput. Biol. **3**(1), 53–96 (1996)
12. Huang, J.T., Cheng, J.P., Chen, H.: Secondary structure length as a determinant of folding rate of proteins with two- and three-state kinetics. Proteins **67**, 12–17 (2007)
13. Ivankov, D.N., Garbuzynskiy, S.O., Alm, E., Plaxco, K.W., Baker, D., Finkelstein, A.V.: Contact order revisited: influence of protein size on the folding rate. Protein Sci. **12**(9), 2057–2062 (2003)
14. Jackson, S.E.: How do small single-domain proteins fold? Fold Des. **3**, 81–91 (1998)
15. Jane Dyson, H., Wright, Peter E., Scheraga, Harold A.: The role of hydrophobic interactions in initiation and propagation of protein folding. PNAS **103**, 13057–13061 (2006)
16. Li, L., Gunasekaran, K., Gan, J.G., Zhanhua, C., Shapshak, P., Sakharkar, M.K., Kangueane, P.: Structural features differentiate the mechanisms between 2S (2 state) and 3S (3 state) folding homodimers. Bioinformation **1**(2), 42–49 (2005)
17. Lin, G.N., Wang, Z., Dong, X., Cheng, J.: SeqRate: sequence-based protein folding type classification and rates prediction. BMC Bioinf. **11**(S1), 3–8 (2010)
18. Lulu, S., Suresh, A., Karthikraj, V., Arumugam, M., Kayathri, R., Kangueane, P.: Structural features for homodimer folding mechanism. J. Mol. Graph. Model. **28**, 88–94 (2009)
19. Pace, C.N.: Determination and analysis of urea and guanidine hydrochloride denaturation curves. Methods Enzymol. **13**, 266–280 (1986)
20. Pace, C.N., Scholtz, J.M.: Measuring the Conformational Stability of a Protein. Protein Structure: A Practical Approach, pp. 299–321. Oxford University Press, New York (1997)
21. Pace, C.N., Grimsley, G.R., Scholtz, J.M., Shaw, K.L.: Protein Stability. In: eLS. Wiley, Chichester (Feb 2014). doi:10.1002/9780470015902.a0003002.pub3
22. Pal, N.R., Chakraborty, D.: Some new features for protein fold prediction. In: Proceedings of ICANN/ICONIP 03. Istanbul, Turkey, pp. 1176–1183 (2003)
23. Plaxco, K.W., Simons, K.T., Baker, D.: Contact order, transition state placement and the refolding rates of single domain proteins. J. Mol. Biol. **227**(4), 985–994 (1998)
24. Saito, Y., Wada, A.: Comparative study of GuHCl denaturation of globular proteins. I. Spectroscopic and chromatographic analysis of the denaturation curves of ribonuclease A, cytochrome c, and pepsinogen. Biopolymers **22**, 2105–2122 (1983)
25. Schulke, N., Varlamova, O.A., Donovan, G.P., Ma, D., Gardner, J.P., Morrissey, D.M., Arrigale, R.R., Zhan, C., Chodera, A.J., Surowitz, K.G., Maddon, P.J., Heston, W.D., Olson, W.C.: The homodimer of prostate-specific membrane antigen is a functional target for cancer therapy. Natl. Acad. Sci. **100**(2), 12590–12595 (2003)
26. Shamim, M.T., Anwaruddin, M., Nagarajaram, H.A.: Support vector machine-based classification of protein folds using the structural properties of amino acid residues and amino acid residue pairs. Bioinformatics **23**(24), 3320–3327 (2007)

27. Tanaka, T., Suh, K.S., Lo, A.M., De Luca, L.M.: p21WAF1/CIP1 is a common transcriptional target of retinoid receptors: pleiotropic regulatory mechanism through retinoic acid receptor (Rar)/retinoid X receptor (Rxr) hetrodimer and Rxr/Rxr homodimerS. J. Biol. Chem. **282**, 29987–29997 (2007)
28. Wishart, D.S.: Tools for protein technologies. Genomics Bioinform. Biotechnol. **2**, 326–342 (2001)
29. Zhou, H., Zhou, Y.: Folding rate prediction using total contact distance. Biophys. J. **82**(1Pt 1), 458–463 (2002)

Decision Making Using Fuzzy Soft Set Inference System

U. Chandrasekhar and Saurabh Mathur

Abstract Soft set theory has brought with it various modified versions of machine learning techniques. One such is fuzzy soft set theory. The paper discusses certain problem solving and decision making approaches available in this concept. We also propose fuzzy soft inference system in line with fuzzy inference system. Defuzzification of parameterized fuzzy soft sets is also discussed.

1 Introduction

Lot of work has happened in soft set theory since it's introduction by Molodtsov [1]. Researchers have chosen fuzzy soft set as one of the betterment for dealing with uncertainties. The following definitions help us to better understand the subsequent discussions.

2 Preliminaries

2.1 Soft Set

Definition According to Cagman et al. [2], a soft set F_A over U is a set defined by a function f_A representing a mapping

$$f_A : E \to P(U) \quad such\ that\ f_A(x) = \emptyset \quad if \ x \in A$$

U. Chandrasekhar (✉) · S. Mathur
School of Information Technology and Engineering (SITE), VIT University,
Vellore, TN, India
e-mail: u.chandrasekhar@vit.ac.in

S. Mathur
e-mail: saurabhmathur96@gmail.com

© Springer International Publishing Switzerland 2016
V. Vijayakumar and V. Neelanarayanan (eds.), *Proceedings of the 3rd International Symposium on Big Data and Cloud Computing Challenges (ISBCC – 16')*,
Smart Innovation, Systems and Technologies 49, DOI 10.1007/978-3-319-30348-2_37

Here, f_A is called approximate function of the soft set F_A, and the value $f_A(x)$ is a set called *x—element* of the soft set for all $x \in E$. It is worth noting that the sets $f_A(x)$ may be arbitrary, empty, or have nonempty intersection. Thus a soft set over U can be represented by the set of ordered pairs

$$F_A = \{(x, f_A(x)) : x \in E, f_A(x) \in P(U)\}$$

2.1.1 Decision Making with Soft Sets

In *Soft matrix theory and its decision making*, Cagman et al. [3], describe a decision making algorithm as follows

- Input the soft set (F, E).
- Input the set P of choice Parameters.
- Compute the corresponding resultant soft set (S, P) obtained from (F, E) with respect to P.
- Compute the score of o_i, $\forall i$ by adding all the values for each parameter.
- The decision is taken is favour of the object with maximum score.

2.2 Fuzzy Set

Definition Let U be a universe. A fuzzy set X over U, according to Zadeh [4], is a set defined by a function μ_X that represents a mapping

$$\mu_X : U \in [0, 1]$$

Here, μ_X called membership function of X, and the value μ_X is called the grade of membership of u. The value represents the degree of u belonging to the fuzzy set X. Thus, a fuzzy set X over U can be represented as follows,

$$X = \{(\mu_X(u) \,|\, / \, u) : u \in U, \mu_X(u) \in [0, 1]\}$$

2.3 Fuzzy Soft Set

In Sect. 2.1, the approximate function of a soft set is defined from a crisp parameters set to a crisp subsets of universal set. But the approximate functions of fuzzy soft sets are defined from crisp parameters set to the fuzzy subsets of universal set.

Definition Let U be an initial universe, E be the set of all parameters, $A \subseteq E$ and $\gamma_A(x)$ be a fuzzy set over U for all $x \in E$. Then, a fuzzy soft set Γ_A according to Maji et al. [5] and Cagman et al. [6] over U is a set defined by a function $\gamma_A(x)$ that represents a mapping

$$\gamma_A : E \to F(U) \quad \text{such that } \gamma_A(x) = \emptyset \quad \text{if } x \notin A$$

Here, $\gamma_A(x)$ is called fuzzy approximate function of the fuzzy soft set Γ_A, and the value $\gamma_A(x)$ is a fuzzy set called x-element of the fuzzy soft set for all $x \in E$. Thus, a fuzzy soft set Γ_A over U can be represented by the set of ordered pairs

$$\Gamma_A = \{(x, \gamma_A(x)) : x \in E, \ \gamma_A(x) \in F(U)\}$$

2.3.1 fs Aggregation

The fs aggregation score, according to Cagman et al. [6] is defined by

$$\mu^*_{\Gamma_A}(u) = (1\backslash|E|) \sum_{x \in E} \mu_{c\Gamma A}(x)\mu_{\gamma A(x)}(u)$$

2.3.2 Decision Making with Fuzzy Soft Sets [6]

- Input the fuzzy soft set Γ_A over U.
- Find the cardinal set $c\Gamma_A$ of Γ_A.
- Find aggregation score for each alternative in Γ_A.
- Find the best alternative having maximum membership grade.

2.4 Fuzzy Parametrized Soft Set

Definition Let U be an initial universe, $P(U)$ be the power set of U, E be the set of all parameters and X be a fuzzy set over E with the membership function μ_X: $E \to [0, 1]$. Then, a fuzzy parameterized soft set according to Cagman et al. [2] F_X over U is a set defined by a function f_X representing a mapping

$$f_X : E \to P(U) \quad \text{such that } f_X(x) = \emptyset \quad \text{if } \mu_X(x) = 0$$

Here, f_X is called approximate function of the fuzzy parameterized soft set F_X, and the value $f_X(x)$ is a set known as x-element of the fuzzy parameterized soft set for all $x \in E$. Thus, an fuzzy parameterized soft set F_X over U can be represented by the set of ordered pairs

$$F_X = \{(\mu_X(x)/x, f_X(x)) : x \in E, \ f_X(x) \in P(U), \ \mu_X(x) \in [0, 1]\}$$

2.4.1 The fps Aggregation Operator

Cagman et al. [2] define the fpfs aggregation score computed by applying the fpfs aggregation operator [2] as

$$(1\backslash|E|) \sum_{x \in E} \mu_x(x) f_X(x)$$

2.4.2 Decision Making with Fuzzy Parameterized Soft Sets

- Using fpfs aggregation operator, the fpfs aggregation score is computed for each object.
- Decision is in favor of object with maximum score.

2.5 Fuzzy Parametrized Fuzzy Soft Set

Definition Let U be an initial universe, E be the set of all parameters and X be a fuzzy set over E with the membership function $\mu_X: E \rightarrow [0, 1]$ and $\gamma_X(x)$ be a fuzzy set over U for all $x \in E$. Then, a fuzzy parametrized fuzzy soft set Γ_X according to Cagman et al. [2] over U is a set defined by a function $\gamma_X(x)$ that represents a mapping

$$\gamma_X : E \rightarrow F(U) \quad \text{such that} \quad \gamma_X(x) = \emptyset \quad \text{if} \quad \mu_X(x) = 0.$$

Here, γ_X is called fuzzy approximate function of the fpfs-set Γ_X, and the value $\gamma_X(x)$ is a fuzzy set known as x-element of the fpfs-set for all $x \in E$. Thus, an fpfs-set Γ_X over U can be represented by the set of ordered pairs

$$\Gamma_X = \{(\mu_X(x)/x, \gamma_X(x)) : x \in E, \ \gamma_X(x) \in F(U), \mu_X(x) \in [0, 1]\}$$

2.5.1 The fpfs Aggregation Operator

The fpfs aggregation score as defined by Cagman et al. [2] is computed by applying the fpfs aggregation operator and is calculated as

$$(1\backslash|E|) \sum_{x \in E} \mu_x(x) \mu_{\gamma X(x)}(u)$$

2.5.2 Decision Making with Fuzzy Parameterized Fuzzy Soft Sets

Algorithm is similar to Fuzzy Parameterized Sets case, the only difference being that the values will be floating point numbers in [0, 1] instead of binary numbers. The algorithm is given by Cagman et al. [2] as

- Using fpfs aggregation operator, the fpfs aggregation score is computed for each object.
- Decision is in favor of object with maximum score.

3 Previous Work

3.1 α Cut on a Fuzzy Soft Set

Let X is a non-empty set, $F(X)$ denotes the set of all fuzzy sets of X. Let A be a fuzzy set in X, that is, Then according to Neog et al. [7] and Bora et al. [8], the non-fuzzy set or crisp set

$$A_\alpha = \{x \in X | \mu_A(x) \geq \alpha\}$$

is called the α cut of the fuzzy soft A.
    If the equation is replaced with

$$A_\alpha = \{x \in X | \mu_A(x) > \alpha\}$$

then A_α is called a strong α cut.

3.2 Fuzzy Soft Set Approach to Decision Making

Roy et al. [9], proposed the following algorithm for decision making using fuzzy soft sets

Algorithm

- Input the fuzzy-soft-sets (F, A), (G, B) and (H, C).
- Input the parameter set P as observed by the observer.
- Compute the corresponding resultant-fuzzy-soft-set (S, P) from the fuzzy soft sets (F, A), (G, B), (H, C) and place it in tabular form.
- Construct the Comparison-table of the fuzzy-soft-set (S, P) and compute r_i and t_i for o_i, $\forall i$.
- Compute the score of o_i, $\forall i$.
- The decision is S_k if, $S_k = max_i S_i$.
- If k has more than one value then any one of o_k may be chosen.

Example Consider the problem of recruitment. A firm wants to recruit graduates in three departments: Technical, Administration and Human Resources, on the basis of their technical, leadership and, communication skills.

Let $U = \{o_1, o_2, o_3, o_4\}$ be the set of candidates.
Let E be the fuzzy set of all possible skills that may be possessed by a candidate.
Let A, B, C be three subsets of E such that

- A denotes level of technical skill and consists of linguistic variables {high, medium, low}
- B denotes leadership skills and consists of linguistic variables {normal, extraordinary}
- C denotes Communication Skills and consists of linguistic variables {good, normal, bad}

Assuming the fuzzy soft set (F, A) denotes 'Candidates with Technical skills',

U	$low(a_1)$	$medium(a_2)$	$high(a_3)$
o_1	0.5	0.2	0.1
o_2	0.1	0.8	0.1
o_3	0.1	0.2	0.6
o_4	0.2	0.25	0.3

(G, B) denotes 'Candidates with Leadership skills' and,

U	$normal(b_1)$	$exta\text{-}ordinary(b_2)$
o_1	0.2	0.4
o_2	0.3	0.4
o_3	0.9	0.1
o_4	0.2	0.6

(H, C) denotes 'Candidates with Communication skills'.

U	$low(c_1)$	$medium(c_2)$	$high(c_3)$
o_1	0.6	0.1	0
o_2	0.2	0.6	0.1
o_3	0	0.1	0.5
o_4	0.3	0.4	0.3

The problem is to select the most suitable candidate to be recruited in the technical, administration and human resources departments.

Let R be the set of parameters required from A and B such that
$R = \{e_{21}, e_{22}, e_{31}, e_{32}\}$ and $e_{ij} = a_i$
and $b_j = \max(a_i, b_i)$. R is given as

$$
\begin{array}{c|cccc}
U & e_{21} & e_{22} & e_{31} & e_{32} \\
\hline
o_1 & 0.2 & 0.2 & 0.1 & 0.1 \\
o_2 & 0.3 & 0.3 & 0.1 & 0.1 \\
o_3 & 0.2 & 0.2 & 0.1 & 0.1 \\
o_4 & 0.2 & 0.2 & 0.3 & 0.3
\end{array}
$$

Let P be the set of parameters required from E = (A and B), and C such that

$$P = \{e_{21} \wedge c_2, e_{22} \wedge c_3, e_{32} \wedge c_3\}$$

P is given as

$$
\begin{array}{c|ccc}
U & e_{21} \wedge c_2 & e_{22} \wedge c_3 & e_{32} \wedge c_3 \\
\hline
o_1 & 0.1 & 0. & 0. \\
o_2 & 0.3 & 0.1 & 0.1 \\
o_3 & 0.1 & 0.2 & 0.1 \\
o_4 & 0.2 & 0.2 & 0.3
\end{array}
$$

The Comparison table is constructed as

$$
\begin{array}{c|cccc}
U & o_1 & o_2 & o_3 & o_4 \\
\hline
o_1 & 3 & 0 & 1 & 0 \\
o_2 & 3 & 3 & 2 & 1 \\
o_3 & 3 & 1 & 3 & 1 \\
o_4 & 3 & 0 & 1 & 3
\end{array}
$$

Row Sum

$$
\begin{array}{c|cccc}
U & o_1 & o_2 & o_3 & o_4 \\
 & 4 & 9 & 8 & 7
\end{array}
$$

Column Sum

$$
\begin{array}{c|cccc}
U & o_1 & o_2 & o_3 & o_4 \\
 & 12 & 4 & 7 & 5
\end{array}
$$

Score = Row Sum − Column Sum

$$
\begin{array}{c|cccc}
U & o_1 & o_2 & o_3 & o_4 \\
score & -8 & 5 & 1 & 2
\end{array}
$$

Since, o_2 has highest score, decision is made in the favor of candidate 2. Chaudhuri et al. [10] proposed an alternative to this algorithm in Solution of the Decision Making Problems using Fuzzy Soft Relations in which by use of suitable fuzzy soft relation [10], the computation of comparison table, row and column sums can be avoided.

4 Proposed Work

4.1 Defuzzyfication of fpfs-Set

- The existing algorithms allow an object to be selected if it compensates its score with respect to one parameter with a larger score with respect to other parameters.
- There are times when making large computations is either not feasible or extremely costly. In such situations, the problem needs to be simplified.

α cut can be applied on fpfs-sets to convert it to an fps-set. Let U be an initial universe, E be the set of all parameters and X be a fuzzy set over E with the membership function

$$\mu_X : E \to [0, 1]$$

and $\gamma_X(x)$ be a fuzzy set over $U \forall\ x \in E$. Let Γ_X be an fpfs-set over U

$$\Gamma_X = \{(\mu_X(x)/x, \gamma_X(x)) : x \in E,\ \gamma_X(x) \in F(U),\ \mu_X(x) \in [0, 1]\}$$

The an α cut on Γ_X is defined as the fuzzy parameterized soft set

$$\Gamma_{X\alpha} = \{(\mu_X(x)/x, f_X(x)) : x \in \Gamma_X,\ \mu_X(x) \in [0, 1]\}$$

where, f_X called approximate function of the fuzzy parameterized soft set $\Gamma_{X\alpha}$, is given by the mapping

$$f_X : E \to P(U) \quad \text{such that } f_X(x) = \emptyset \text{ if } \mu_X(x) > \alpha$$

Example Consider the following variation of Molodtsov's house example. Consider the following fuzzy parametrized fuzzy soft set A describing a set of houses and their properties.

A	expensive	wooden	beautiful	good-surroundings
o_1	0.3	0.4	0.6	0.9
o_2	0.3	0.9	0.3	0.5
o_3	0.4	0.5	0.8	0.7
o_4	0.8	0.2	0.4	0.8
o_5	0.7	0.3	0.6	0.5
o_6	0.9	0.2	0.4	0.3

with weights

A	expensive	wooden	beautiful	good-surroundings
weights	0.5	0.3	0.4	0.6

Say the buyer has the following requirements

- The house should be beautiful at least to a certain extent.
- The house should not be in bad surroundings.
- There are no budget constraints.
- There is no limit to which the house may be wooden.

When we apply a α cut using α values,

A	expensive	wooden	beautiful	good-surroundings
α	0	0	0.6	0.5

We get the reduced the fuzzy parametrized soft set

A	expensive	wooden	beautiful	good-surroundings
o_1	1	1	1	1
o_2	1	1	0	0
o_3	1	1	1	1
o_4	1	1	0	1
o_5	1	1	1	0
o_6	1	1	0	0

with weights

A	expensive	wooden	beautiful	good-surroundings
weights	0.15	0.3	0.4	0.6

5 Solving Decision Making Problems with Fuzzy Soft Set Inference Systems

5.1 Algorithm

- Input parmeter fuzzy soft sets (F_1, A_1), (F_2, A_2), ..., (F_n, A_n)
- Input fuzzy rules of form

  If A is A_0 and B is B_0 then C is C_0

- Compute the output fuzzy sets corresponding to the rules and input.
- Defuzzify the output fuzzy sets using a suitable method.
- Make the decision on the basis of the defuzzified.

5.2 Example

Continuing the recruitment example from Sect. 3.2.2, this time the problem being computing suitability of all candidates, for all the three departments. The output fuzzy sets will be

- P denoting suitability for Technical department.
- Q denoting suitability for Administrative department.
- R denoting suitability for Human Resources department.

each having linguistic variables {high, medium, low}.
   The rule set can be

- If level of technical skill is *medium*, leadership skills are *extraordinary* and, communication skills are *good* then, suitability for Administrative department is *high*, suitability for Technical department is *medium*, suitability for Human Resources is *Medium*
- If level of technical skill is *high*, leadership skills are *normal* and, communication skills are *bad* then, suitability for Administrative department is *low*, suitability for Technical department is *high*, suitability for Human Resources is *Low*

  Similarly $3 * 2 * 3 = 18$ rules can be framed, as defined by functional expert.
   On solving the above rules with given inputs, the output fuzzy soft sets will be obtained similar to

Suitability for Technical Department

$$
\begin{array}{c|ccc}
U & low & medium & high \\
o_1 & 0.7 & 0.1 & 0 \\
o_2 & 0 & 0.9 & 0.1 \\
o_3 & 0.1 & 0.25 & 0.6 \\
o_4 & 0.2 & 0.4 & 0.35
\end{array}
$$

Suitability for Administrative Department

$$
\begin{array}{c|ccc}
U & low & medium & high \\
o_1 & 0.1 & 0.4 & 0.3 \\
o_2 & 0.5 & 0.2 & 0.1 \\
o_3 & 0.6 & 0.25 & 0.1 \\
o_4 & 0.1 & 0.15 & 0.3
\end{array}
$$

Suitability for Human Resources Department

$$
\begin{array}{c|ccc}
U & low & medium & high \\
o_1 & 0.2 & 0.2 & 0.6 \\
o_2 & 0 & 0.9 & 0.1 \\
o_3 & 0.3 & 0.2 & 0.1 \\
o_4 & 0.2 & 0.3 & 0.5
\end{array}
$$

Depending on the number of candidates required for each department and priority of the department all the candidates are selected into respective departments. We could also use comparison tables to calculate the fitness level of each candidate into respective department.

Comparison tables for each of the outputs are constructed as
Suitability for Technical Department

$$
\begin{array}{c|cccc}
U & o_1 & o_2 & o_3 & o_4 \\
o_1 & 3 & 1 & 1 & 1 \\
o_2 & 2 & 3 & 1 & 1 \\
o_3 & 2 & 2 & 3 & 1 \\
o_4 & 2 & 1 & 2 & 3
\end{array}
$$

$$
\begin{array}{c|ccc}
U & row\text{-}sum & column\text{-}sum & score \\
o_1 & 6 & 9 & -3 \\
o_2 & 7 & 7 & 0 \\
o_3 & 8 & 7 & 1 \\
o_4 & 8 & 6 & -2
\end{array}
$$

Suitability for Administrative Department

$$
\begin{array}{c|cccc}
U & o_1 & o_2 & o_3 & o_4 \\
\hline
o_1 & 3 & 2 & 1 & 2 \\
o_2 & 1 & 3 & 1 & 2 \\
o_3 & 1 & 2 & 3 & 2 \\
o_4 & 2 & 1 & 1 & 3
\end{array}
$$

$$
\begin{array}{c|ccc}
U & row\text{-}sum & column\text{-}sum & score \\
o_1 & 7 & 7 & 0 \\
o_2 & 7 & 8 & -1 \\
o_3 & 8 & 6 & 2 \\
o_4 & 7 & 9 & -2
\end{array}
$$

Suitability for Human Resources Department

$$
\begin{array}{c|cccc}
U & o_1 & o_2 & o_3 & o_4 \\
\hline
o_1 & 3 & 1 & 2 & 1 \\
o_2 & 1 & 3 & 2 & 1 \\
o_3 & 2 & 2 & 3 & 1 \\
o_4 & 2 & 2 & 2 & 3
\end{array}
$$

$$
\begin{array}{c|ccc}
U & rowtext\text{-}sum & columntext\text{-}sum & score \\
o_1 & 7 & 8 & -1 \\
o_2 & 7 & 8 & -1 \\
o_3 & 8 & 9 & -1 \\
o_4 & 9 & 6 & 3
\end{array}
$$

Assuming the requirements to be 2 people for Technical, and, 1 each for management and Human Resources, the following selection can be made

- Even though o_3 has the highest score in both technical and administrative departments, he will be offered a position for administrative department since his score is higher there.
- o_1 and o_2 will be offered a position for technical department since they are the next best after o_3.
- o_4 will be offered a position at administrative department *Human resources department*.

Thus, the proposed system is more generic and allows for greater flexibility than the existing fuzzy soft systems.

6 Conclusion

In this paper we defined the concept of α cut on fuzzy soft sets as well as fuzzy parameterized fuzzy soft sets. We also proposed a fuzzy inference system that can be used for fuzzy soft sets as well. The advantage of this mechanism is that multiple choices can be made simultaneously by comparative analysis. The future work can be extended for intuitionistic fuzzy soft sets.

References

1. Molodtsov, D.: Soft set theory first results. Comput. Math. Appl. **37**:19 (1999)
2. Cagman, N., et al.: Fuzzy parameterized fuzzy soft set theory and its applications. Turk. J. Fuzzy Syst. **1**(1):21 (2010)
3. Cagman, N., et al.: Soft matrix theory and its decision making. Comput. Math. Appl. **59**:3308 (2010)
4. Zadeh, L.A.: Fuzzy sets. Inf. Control **8**(3):338
5. Maji, P.K., et al.: Soft set theory. Comput. Math. Appl. **45**:555 (2003)
6. Cagman, N., et al.: Fuzzy soft et theory and its applications. Iran. J. Fuzzy Syst. **8**(3):137 (2011)
7. Neog, T.J., et al.: Some new operations of fuzzy soft sets. J. Math. Comput. Sci. **2**(5):1186 (2012)
8. Bora, M., et al.: Some new operations of intuitionistic fuzzy soft sets. Int. J. Soft Comput. Eng. (IJSCE). **2**(4):2231 (2012)
9. Roy, A.R., et al.: A fuzzy soft set theoretic approach to decision making problems. J. Comput. Appl. Math. **203**:540 (2007)
10. Chaudhuri, A., et al.: Solution of the decision making problems using fuzzy soft relations, CoRR (2013). http://arxiv.org/abs/1304.7238

Part VIII
Software Engineering

Reverse Logistic: A Tool for a Successful Business

B. Neeraja, Arti Chandani, Mita Mehta and M. Radhikaashree

Abstract The present century customers are Hi-Tech. They want to get products and services with latest inputs at their door step. They are not too brand loyal. With the changing demand for updated models related products they feel to be at par with their neighbourhood. It is not that always Customers are brand loyal consumer. Gone are the days when a family would be brand loyal to one company product for generations together. The 21st Century customers are hi-tech. They expect the best and more than they expect from the product and manufacturer. Therefore it is the duty of the manufacturer that he should always be updated with the demands of their customers. Before their competitors pitch into their market zone they are to serve their customers with better products and services. Research has proved that due to marketing and delayed supply of goods and services many Companies have lost a set of customers. It should be noted that one time failure by the producer in delivering the product/service would make him lose 10 times the value of sales from his total sales. It is therefore very important that should always have a track of the orders placed and the delivery of the product. Producer should always have a track of the orders placed and the delivery of the product. The present day customer is too keen on the after sales services too. Earlier the customer had to wait for days or at times even weeks for buying a automobile or luxury goods but now the time for placing an order and delivering the product has come down to hours or at the maximum days. This is because of the development in technology. If the companies try to work out on reverse logistics concepts they can do wonders and reduce more

B. Neeraja (✉) · M. Radhikaashree
Dr. MGR University, Maduravoyal, Chennai
e-mail: neerajavijay93@gmail.com

M. Radhikaashree
e-mail: rathihagayathri@yahoo.com

A. Chandani · M. Mehta
Symbiosis Institute of Management Studies, Symbiosis International University,
Range Hills Road, Khadki, Pune
e-mail: arti.chandani@sims.edu

M. Mehta
e-mail: director@sims.edu

© Springer International Publishing Switzerland 2016
V. Vijayakumar and V. Neelanarayanan (eds.), *Proceedings of the 3rd International Symposium on Big Data and Cloud Computing Challenges (ISBCC – 16'),*
Smart Innovation, Systems and Technologies 49, DOI 10.1007/978-3-319-30348-2_38

cost and time in serving the customer. In this paper the author's try to insist on the concept of cloud computing for reverse logistics. By using Keiretsu concept f or saving time and fast delivery of products. This paper try to concentrated on tie up of various transport companies who could utilize their resources in cost cutting, time saving and effective utilization of resources for serving their customer in a better way by implementing keiretsu concept.

Keywords Supply chain management · Reverse logistics · Customer satisfaction · Keiretsu

1 Introduction

Supply Chain Management is the planning and control of the entire supply chain, from production to transportation, to storage and distribution, through to sales, and back again to production (Bonacich and Wilson 2005, p. 68). Supply Chain management has evolved from a push to a pull system. Supply Chain Management is the reorganization of a business' supply-side activities to maximize customer value and to gain a competitive advantage in the market. SCM represents the intention of the suppliers to develop and implement supply chains methods that are more feasible and affordable to the Indian scenario. Supply chains cover everything from the point of receiving the raw material from suppliers to the stage of the product reaching the end user. Entry of new competitors are shortening product and service lifecycles they are driving companies with traditional, infrastructure-intensive supply chains to adopt innovative strategies and solutions to enhance their competitiveness. As a result of information technology development manufacturers had to see alternatives for their logistic issues to fulfil and substantiate their supply chains options which would become more dynamic, more affordable, and more capable of supporting the capacity, vision and mission and financial objectives of the company. Currently companies are moving relatively cautiously towards leveraging cloud technologies in their supply chains, studies show that interest in the potential of cloud computing is already strong. Historically, supply chain operations have proven to be adept at adopting and capitalizing on innovative technology solutions, and cloud computing will be no exception (Fig. 1).

Defining Logistics: The word logistics is derived from the French word loger that means art of transport, supply, and quartering of troops. Logistics can be defined as a concept designed for military troops so as to ensure careful planning and implementation of supplying weapons, food, medicines, and other necessary items in the battle field. Initially the concept of logistics included planning and implementing the strategy of procurement of inputs for the production process to make goods and services available to the consumers. But the present scenario is demanding for the concept of reverse logistics which is a challenging issue for business houses. Especially with the products where the disposal is to be planned and done in a systematic way, reverse logistics has been gaining increasing

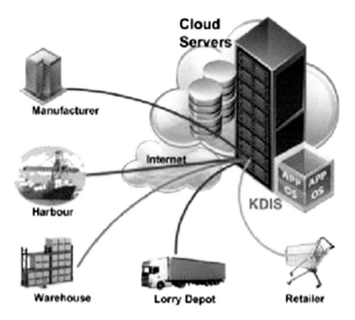

Fig. 1 Supply chain management structure

attention and awareness in the supply chain community, both from business house, manufacturers, customers and researchers. This can be attributed to increased regulatory pressure such as extended-producer responsibility, consumer expectations and societal sustainability demands, as well as to the intrinsic value that can be regained from Collected products (De Brito 2003) (Fig. 2).

Reverse logistics is the method of dealing with collecting or gathering back the used products from the ultimate users. It can take any forms, with varying costs and benefits to the business. Options include recycling, repairing, re-using or simply

Fig. 2 The logistics network

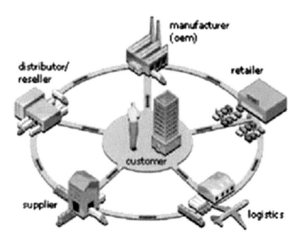

discarding the returned products. This is to be considered as a part of service strategy or part of the business process inherent in any company that makes or sells a product. In the reverse logistics process, a customer returns a product to the company for a various reasons. Although this would usually results in a loss for the business, small business owners can help reduce the costs associated with the reverse logistics process through effective management of available resources.

The process of collecting back the unused or disposable product from the ultimate consumers is *cost cutting and making the disposal environment friendly.* Research has also proved since the customers were not sure of the disposable method they have change their brand of purchasing products. This concept of collecting back the goods can be done with the help of technology. Information Technology has become part and parcel of every business. Therefore the concept of cloud computing can extend a helping hand to the small business who cannot afford a separate set of resources allocated to trace out the reverse logistics process of their products. In such a situation cloud computing can help them to collect the details and dispose the products in a safer way. Or Reuse the products or re-do the products. Cloud computing has given a helping hand to small business houses to make their business effective and successful (Fig. 3).

Defining cloud computing for small business: Cloud computing is a recently evolved computing terminology or metaphor based on utility and consumption of computing resources. Cloud computing involves organize groups of remote servers and software networks that allow centralized data storage and online access to computer services or resources Cloud computing is an emerging area of disseminated computing information that offers many potential benefits to small business houses by permitting access of information technology (IT) services available as a commodity. When the business people get together and represent their requirements

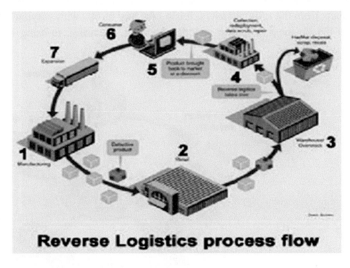

Fig. 3 Reverse logistics process flow

for cloud services, such as applications, software, data storage, and processing capabilities, organizations can improve their competence and capability to respond more quickly and reliably to their customers' needs. At the same time, this system may be associated with risks which are to be considered, including maintaining the security and privacy of systems and information, and assuring the rational expenditure of IT resources. Cloud computing is not a single type of system, but it is a combination of underlying technologies and configuration options.

The strengths and weaknesses of the different cloud technologies, configurations, service models, and practical feasibility methods should be considered by organisations evaluating services to meet their requirements. *By using the concept of cloud computing the manufacturers/business houses can park their big data with safe and reliable resource*. For this the service company would charge a minimum cost called the service cost.

2 Output of This Concept

This concept of implementing cloud computing for reverse logistics is for small business enterprises who cannot afford the cost of implementing the software for their own. It is a suggestion given to the general public how reverse logistics concept of re-do, recycle and re-use concept can be best implemented (Fig. 4).

The concept of cloud computing if associated and communicated properly for reverse logistics would make a great difference for business to save cost and serve their customers better. Develop a strategy: which processes you should retain internally and which processes would benefit from cloud computing. Develop a detailed business case with quantified anticipated benefits, ROI and risk analysis. Agree the standards for success: flexibility, scalability, speed to market and costs. Start with the low-hanging positive results.

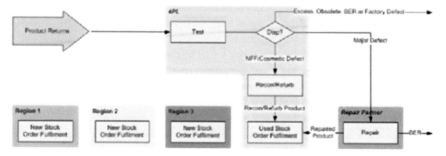

Fig. 4 Reverse logistics model for small logistics partners

3 Advantages of Reverse Logistics for Small Companies

Reverse logistics management has developed into a discipline that produces cost reductions, adds efficiencies and improves the consumer experience. Producers have discovered value within returned assets and the benefits of streamlining repair, return and product reallocation processes. For many organizations, the returns management process has remained a cost center with low visibility that contains products to be restocked, repaired, recycled, repackaged or disposed of appropriately. Conventional logistics service providers had very few alternatives for reversing the channel; however, as the technology has progressed, companies using robust, proficient reverse logistics management have benefited from:

- Reduced administrative, transportation and aftermarket support costs.
- Increased velocity.
- Increased service market share.
- Higher achievement of sustainability goals.
- Greater customer service and higher retention levels.

Once a niche in supply chain management, reverse logistics is emerging as a supply chain corporate strategy supporting everything from environmental sustainability to bottom line profitability. Several forces are combining to drive interest in reverse logistics including: the pressure for profitability, volatility in the fuel markets, community interest in sustainability, changing customer needs, and rising levels of regulation.

Reverse logistics offers a way to help alleviate poor utilization and deadheading issues. Ideally, reverse logistics programs will keep trailers full while also providing continuous moves. Some carriers have milk runs involving simple raw materials (RM) and finish goods (FG) exchanges. Many times this can also include returning specialized pallets and racks, recyclables, cages, and totes.

Mike Dewey is a Cincinatti-based home brewer who decided to get serious after his favorite local brewpub closed about five years ago. Today Dewy's Mt. Carmel brewery is still a small brewing company but it does much higher volume. Dewey sends growlers out on pallets and has to take return on the pallets. A typical load of 16 pallets at $20 each represent $320 in needed cash flow. He uses about 40 pallets per month. More small brewers in the market makes for a more competitive business situation. A dollar saved is as important as a dollar earned. Shipping and logistics costs are important areas in which every small businessman needs to control costs.

Another important advantage of elevating reverse logistics into the strategic planning process is the issue of environmental sustainability. Compliance with environmental regulations is the most often cited reason for undertaking sustainability improvements. Leading edge thought on reverse logistics has now evolved to an understanding that compliance can lead to lower costs and increased productivity.

Reverse logistics can include recycling of packaging and leftover resources found in used containers. As population increases and resources become more scarce, the value of recycling and repurposing resources becomes more apparent. Efficiency in the supply chain eliminates waste and supports green initiatives that can provide differentiation and competitive advantages in crowded markets.

Implementing a strategic reverse logistics program can offer both reduced costs and increased customer value. It is necessary that the value to be obtained through reverse logistics is understood and evaluated in the broadest perspective by senior management as a part of the strategic plan. Implementing a reverse logistics strategy and operational plan needs to start with the realization that there are significant corporate, environmental, and business development opportunities within your value chain. Once you have identified those opportunities and your supply chain group has a good plan, seek an executive sponsor within your organization.

By plotting your shipping and logistical nodes, gain an understanding of your transportation moves and their frequency, you can start looking at synergies and opportunities within your "value chain" which includes your own organization, your partners, your suppliers, and also your neighbors. The key is to realize there are multiple opportunities with reverse logistics especially in returns management (including recalls), obsolescence and damage claims, refurbishing/upgrading of products, MRO of equipment, container movements (racks, crates, totes, etc.), and sustainability/recycling initiatives including the proper and optimal disposal of waste. Many of these opportunities can be found through synergies and location.

Once a reverse logistics initiative commences, you and your organization will need to be flexible in making adjustments to your organization (change management), your supply chain (capacity constraints), and measurement (metrics and financial benefits). Determining what success looks like is based on your own internal views, benchmarks, and participants of your reverse logistics value chain. Likewise, there might be additional financial and community impact/ "green supply chain" opportunities available in secondary markets of products, recyclables and reusable components, and waste. Equally impressive is that these reverse logistics benefits can be shared financially (either through additional revenues, savings, or credits/rebates) or through public relations and brand awareness initiatives associated through sustainability. The time is optimal more than ever to develop a reverse logistics program for your company and community.

4   Benefits of Cloud Computing for Small\Business

- Cloud computing represents a great opportunity to reduce total cost for the users (manufactures or small scale business house). Providing secure computing capacity on a Rental basis as required with no capital costs and with comprehensive technical support provided as part of the service.
- Companies can gain significant advantages by using cloud computing to upgrade their capabilities e.g. by using smart phone, collection and delivery of goods can be in time. Not only this it can also progress of the en route goods be shown on the map in real time.

- By working "in the cloud," companies have shown they can rapidly implement andoperate applications that are secure and inexpensive, while enjoying lower IT maintenance and upgrade costs.
- Delivery process may also be speedup since cloud computing offers a highly collaborative framework with centralized storage and contact points, ease of access and the opportunity to enact simplified, standardized processes e.g. by using KeyTelem remote site measurements can be collected, displayed centrally and notifications sent e.g. for a tanker to go and replenish a storage tank.
- A further advantage is flexibility by making it possible to scale up at par with minimal waste of time and other resources, and even to switch applications entirely without a lot of added cost or complexity. This brings the ability for small business houses to enter into new markets or launch new services quickly.

It would also give the below mentioned benefits:

- Expecting returns as predicted can be done by additional revenue.
- Drive significant cost savings,
- Improve customer satisfaction and deliver a competitive edge
- Reduce wastage
- Implement the concept of make in India and green India.
- Help in improving saving concept of renewable resources for the next generation.

5 Scope for Further Development

Supply Chain Management aims at managing supply and demand, optimizing resources, reducing cost and providing efficient customer services. Keiretsu is a Japanese word that refers to powerful business groups. It depicts how businesses share each other resources. Keiretsu is a relationship among all stake holders of Supply chain. They include suppliers, partners, manufacturers and customers who do business with each other. It is basically a vertically related group with sound manufacturing and a large network of suppliers and sub-contractors. Under the keiretsu arrangement, major suppliers and sub-contractors do business with only one of the producers. Keiretsu shares many of the goals of SCM as the latter is based on the management of relationships both between corporate functions and across companies. SCM offers an opportunity for firms to enjoy many of the benefits of Keiretsu. This is followed by japans. A sincere suggestion is give for the Indian business house and cloud computing service providers to ensure this system can be implemented in India also.

Acknowledgements I sincerely thank my co-author Ms. Arti. I would also like to thank Dr. MGR Educational and Research Institute, University, Maduravoyal who has given me an opportunity to bring this paper in the present format by providing me the necessary time and infrastructure.

Sincere work has been done to highlight the importance of cost reduction through paper. I also would like to thank the convenor and organising team of the conference ISBCC-16, VIT Chennai. My sincere thanks to all my faculty members of Department of Management Studies.

References

1. http://en.wikipedia.org/wiki/Cloud_computing#mediaviewer/File:Cloud_computin_layers.png
2. http://www.trucinginfo.com/article/story/2012/11/5-things-you-should-know-about-cloud-computing.aspx
3. http://gcinfotech.com/cloud-computing-small-business-ct-nyc/
4. http://www.logisticsmgmt.com/article/reverse_logistics_closing_the_loop
5. Dobler, D.W., Strling, S.L.: World class supply. McGraw Hill, New York
6. http://cscmp.org/sites/default/files/user_uploads/educational/downloads/logistics-scsession1 paper3.pdf
7. http://www.reverselogisticstrends.com/rlmagazine/edition04p10.php
8. http://www.keyfort.net/
9. http://rlmagazine.com/edition63p32.php

Improving User Participation in Requirement Elicitation and Analysis by Applying Gamification Using Architect's Use Case Diagram

B. Sathis Kumar and Ilango Krishnamurthi

Abstract Customer involvement and lack of expressive communication mechanism are the major problems in capturing Architecturally Significant Requirements (ASR) in requirement elicitation and analysis stage in the software development. In our earlier work we have introduced Architect Use Case Diagram (AUCD) an enhanced version of use case diagram, which is useful to specify functional and quality attribute requirements in one diagram. Gamification is an emerging technique used in the software industry to improve the user's commitment, inspiration and performance. In this paper we proposed gamification using AUCD, a new methodology for requirement elicitation and analysis. The combination of gamification and AUCD motivates the users to participate in the requirement elicitation and analysis actively, which helps the analyst to capture more architecturally significant requirements. The proposed gamification using AUCD method is applied in different projects for validating the effectiveness. The research result shows that gaming using AUCD improves user participation in requirement elicitation.

Keywords Gamification · Architect's use case diagram (AUCD) · Architecturally significant requirements (ASR)

1 Introduction

The success of software development lies in the clear understanding of client requirement. Requirement elicitation is a process of understanding the requirements from customers and end-users in a software project. It is followed by requirement

---

B.S. Kumar (✉)
School of Computing Science and Engineering, VIT University,
Chennai, Tamil Nadu, India
e-mail: sathiskumar.b@vit.ac.in

I. Krishnamurthi
Sri Krishna College of Engineering and Technology, Coimbatore, Tamil Nadu, India

© Springer International Publishing Switzerland 2016
V. Vijayakumar and V. Neelanarayanan (eds.), *Proceedings of the 3rd International Symposium on Big Data and Cloud Computing Challenges (ISBCC – 16')*,
Smart Innovation, Systems and Technologies 49, DOI 10.1007/978-3-319-30348-2_39

analysis and requirement specification. Requirement analysis is a process of examining the correctness of the elicited requirements and identifying the missing requirements. Requirement specification is a process of documenting the requirement in any prescribed form. This document needs to be agreed upon by the requirement analysts and the customers. This is the foundation for the subsequent architectural design, detailed design and building the software project. "Software Architecture is the fundamental organization of a system embodied in its components, their relationships to each other and to the environment and the principles guiding its design and evolution" [1]. Based on the nature of the requirement and business goal, an architect will design the appropriate architecture. A software architect, designs the software architecture of the system using the functional and nonfunctional requirements. Hence these requirements document should be complete, unambiguous, correct, understandable and verifiable.

A functional requirement describes an expected functionality or feature of the system. A non-functional requirement defines the effectiveness of the function provided by the system. Security, supportability, usability, performance and reliability are important quality attributes of a software system. These quality attributes are also referred as non-functional requirements.

Only a few functional and quality attribute requirements help us to identify and shape the software architecture. If the requirement influences the architectural design decision, then it is referred to as Architecturally Significant Requirements (ASRs) [2]. Quality software product depends on quality software architecture which in turn depends on the quality requirement specification. However, most of the customers fail to express the requirements properly and particularly ASR during requirement elicitation.

Interviews, questionnaires, user observation, workshops, brainstorming, use cases, roleplaying and prototyping are various requirement elicitation methods used in the software industry [3]. Effective requirement elicitation and analysis requires strong communication between system analyst and stake holders. However all these techniques focusing on describing functional requirements using natural language and most of the quality attributes requirements are hidden in the functional requirement statements. To create a better communication between user and developing team, In our earlier work we have introduced Architect Use Case Diagram (AUCD) [4, 5]. This enhanced version of use case diagram is useful to specify functional and quality attribute requirements in one diagram.

Gamification is an emerging technique used by most of the software industry to improve the user's commitment, inspiration and performance. Hence introducing effective communication mechanism with gamification, will improvise the ways of capturing more ASR effectively.

This paper proposes a new methodology of requirement elicitation technique using gamification using AUCD. The combination of gamification and AUCD will motivate the users to participate in the requirement elicitation and hence find more requirements especially architecturally significant requirements.

2 Usage of Gamification and Architect Use Case Diagram in Requirement Engineering

Gamification is a powerful approach to involve and encourage people to attain the organizational goals. The key for success of the gamification is involve people on an emotional level and encouraging them to achieve their aims [6]. Gamification is defined as, "The process of game-thinking and game mechanics to engage users and solve problems" [7]. The main challenge of any organizations is how they effectively involve employees and customers to achieve basic business goals. Gamification is adding gaming mechanics for a complex task with awards which makes that task more attractive.

In our earlier research we have introduced AUCD which is useful to specify both functional and quality attribute requirement in one diagram. In AUCD we have introduced solid dark circle for sub flow connector, dotted circle for alternative flow connector, double circle connector for security flow connector, double square for scalability indicator and triangular for response time indicator which are useful to specify ASR in Use Case Diagram (UCD).

Sub flow connectors are used to state optional functionalities of the use cases. Alternative flow connectors are used to state the possible exceptional and alternative flows. A security flow connectors are used to state the security requirement. Response time indicator is useful to state the performance requirement. Scalability indicator is useful to specify maximum data transfer or maximum number of user access in a particular time. The double square symbol is a scalability indicator which can be added to the model in the needed position.

Requirement visualization attracts the participant to actively involves in requirement elicitation. The Scope of different requirement elicitation techniques are described in Table 1. Comparing different elicitation methods, Architect Use Case Diagram is suitable for requirement elicitation using gaming method because both functional and quality attribute requirements are represented visually in one diagram.

Table 1 Requirement elicitation techniques scope

Elicitation technique	Communication media	Scope
Interviews	Dynamic dialogue	Simple, depends on participants involvement, natural language ambiguity
Observation	Video	Ambiguity in explanation
Workshops	Dynamic dialogue	Depends on the participants selection, natural language ambiguity
Scenarios (use case)	Text, images	Depends on participants involvement, natural language ambiguity
Prototypes	Visual media	Limited ecological evaluation
AUCD	Images	No natural language ambiguity, ASR requirements are represented visually

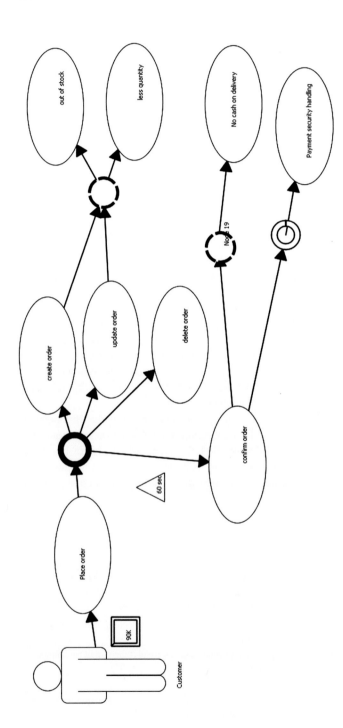

Fig. 1 AUCD for place order use case for online sales portal project

Gamification is useful to motivate the participation and AUCD creates a better communication between stakeholders and the development team.

Lack of motivation and communication are the major issues that prevent effective user participation [2]. Hence gamification using AUCD will remove the difficulties of lack of motivation and active participation of the users.

Figure 1 shows the AUCD for place order use case for online sales portal project. In Fig. 1 create, delete, update and confirm orders are sub flow of place order use case. Out of stock, less quantity and no cash on delivery are alternative flows. Payment handling is a security flow use case. Maximum of 90,000 users will access the portal is specified using scalability indicator and maximum time for confirming the order is 60 s and it is specified using performance indicator. The AUCD notations are very simple to understand by the participants and using these notation most of the ASR are visually specified. This graphical representation will increase the active involvement of the customers for identifying more ASR.

3 Methodology of Requirement Elicitation and Analysis by Applying Gamification Using Architect's Use Case Diagram

The methodology of requirement elicitation and analysis by applying gamification using AUCD are described by using 7 steps as follows.

Step 1: Set the goal of the game.

Identify the objective of the software project is helpful to find the ASR. The cost, quality, security, time and scope are important influence factors for the software goal.

Step 2: Select the Players for the game and form the team.

The following factors are considered for selecting the players and forming the teams.

Select the players from each category.

The advantage of including players from each category is useful to get the requirement information from each category of users. By involving each category of players, the nonfunctional aspects of the system such as functional, reliability, usability, performance, efficiency, compatibility, security, maintainability and portability can be elicited which may not have been expressed explicitly. Acceptance test is easy because users approve the final system according to their workplace needs and requirements.

Table 2 Reward points table

Requirements type	Points
Functional requirement's	2
Sub flows	3
Alternative flows	4
Security flows	5
Performance requirements	5
Scalability requirements	5
Bonus point	5

Group the team based on player's category.

If the team is formed based on the player's category, democracy will be ensured. A mix up of the different category of players affects better communication and chance for influencing their decision by higher category of players. Grouping the team based on the category of players will create competition among the players in the same team. This will also create a healthy completion between the team also.

Step 3: Explain the game rules and the points system to the players.

The game is divided into 6 stages. In stage 1 players need to identify all the possible basic functionalities of the proposed system. In stage 2, stage 3, stage 4, stage 5 and stage 6 players' needs to identify the sub flow, alternative flow, security flow, performance and scalability requirements respectively. All the identified requirements are represented by using AUCD diagram tool by each team in each stage. In each stage, the moderator allots a specific time period to identify the requirements. After the deadline, the moderator collects the AUCD from each team in every stage. Each team needs to present the AUCD for discussion in every stage. Moderator notes down the functionality approved by the group members and reject the functionality which is irrelevant to the project and allots points as described in Table 2 in each stage.

Step 4: Start the game.

Stage 1: Find all the base use case.

(a) Instruct the user to find all the basic functions (base use case) expected from the system in the form of use case.
(b) Give a specific time to work on group and draw the AUCD.
(c) After the deadline the moderator will collect the AUCD from each group.
(d) Members from each group will present the AUCD for discussion.
(e) The moderator notes down the base use case approved by the group members and reject the base use case which is irrelevant to the project.
(f) The moderator allots the points to each group as per the points table.
(g) Based on the collected functionality moderator finalize the base use case diagram and present it to the group members for finding out missed requirements.

(h) If group members identify any missed out requirement then they will get additional 5 bonus point.

(i) Steps g and h are repeated until all the base use case are identified.

Stage 2: Find all the sub flows.

The objective of this stage is to identify all possible sub flow use case of base use case and sub flow use case. A sub flow use case may have any number of sub flow use case.

(a) Instruct the user to find all the sub flows of the base use case.

(b) Give a specific time to work on group and draw the AUCD.

(c) After the deadline, the members from each group will present their AUCD to other members.

(d) The moderator notes down the sub flow use case approved by the group members and reject the sub flow which is irrelevant to the project.

(e) Allot the points to each group as per the point table.

(f) Based on the collected sub flow, the moderator finalizes the sub flow use cases and presents it to the group members for finding the missed out sub flow requirements.

(g) If the group members identify any missed out sub flows and then they will get additional 5 bonus points.

(h) Steps f and g are repeated until all the sub flow use cases of base use case are identified.

(i) Instruct the user to find all the sub flows of the sub flow use case.

(j) Repeat the Steps b to e to find all the sub flows of the sub flow use case.

(k) Steps f and g are repeated until all the sub flow use cases of sub flow use case are identified.

Stage 3: Find all the Alternative flows.

(a) Instruct the user to find all the alternative flow of the base use case and sub flow use case.

(b) Give a specific time to work on group and draw the AUCD.

(c) After the deadline the members from each group will present the AUCD to other members.

(d) The moderator notes down the alternative flow use case approved by the group members and reject the alternative flow which is irrelevant to the project.

(e) The moderator allots the points to each group as per the point table.

(f) Based on the collected alternative flow use case, the moderator finalizes the alternative flow use case for each base use case and sub flow use case. The moderator presents the AUCD to the group members for finding the missed out alternative flow.

(g) If the group members identify any missed out alternative flow, then they will get an additional 5 bonus points.

(h) Steps f and g are repeated until all the alternative flow use cases are identified.

Stage 4: Find all the security flow use case.

(a) Instruct the user to find any security requirement which is required for the base use case/sub flow use case/alternative flow use case.
(b) Give a specific time to work on group and draw the AUCD.
(c) After the deadline members from each group will project the AUCD to other members.
(d) The moderator notes down the security flow use cases approved by the group members and reject the security flows which are irrelevant to the project.
(e) The moderator then allots the points to each group as per the point table.
(f) Based on the collected security flow use case moderator finalizes the security flow use case for each base, sub flow and alternative flow use case. The moderator presents the AUCD to the group members for finding the missed out security flow.
(g) If the group members identify any missed out security flow, then they will get additional 5 bonus points.
(h) Steps f and g are repeated until all the security flow use cases are identified.

Stage 5: Find all the performance requirements.

(a) Instruct the user to find any performance requirement which is required for the base use case/sub flow use case/alternative flow use case.
(b) Give a specific time to work on group and specify the performance requirement in the AUCD.
(c) After the deadline members from each group will present the AUCD to other members.
(d) The moderator notes down the performance requirement approved by the group members.
(e) The moderator then allots the points to each group as per the point table.
(f) Based on the collected performance requirement moderator finalize the performance requirement and presents the AUCD to the group members for finding the missed out performance requirement.
(g) If the group members identify any missed out performance requirement, then they will get additional 5 bonus points.
(h) Steps f and g are repeated until all the performance requirements are identified.

Stage 6: Find all the scalability requirements.

(a) Instruct the user to find the scalability requirement which is required for the base use case/sub flow use case/alternative flow use case.
(b) Give a specific time to work on group and draw the AUCD.
(c) After the deadline, the members from each group will project the AUCD to other members.
(d) The moderator notes down the scalability requirement approved by the group members.
(e) The moderator then allots the points to each group as per the point table.

(f) Based on the collected scalability requirement moderator finalize the scalability requirement and presents the AUCD to the group members for finding
the missed out scalability requirement.
(g) If the group members identify any missed out scalability requirement, then
they will get additional 5 bonus points.
(h) Steps f and g are repeated until all the performance requirements are identified.

Step 6: Consolidating the AUCD.

The moderator and players check whether the AUCD is complete, understandable, unambiguous, correct and traceable.

Step 7: Analyses the game result.

The moderator finally announces the winner of the game and also recommends
the key players to the management for rewards.

4 Discussion

The use of the AUCD tool as a game mechanic was mainly chosen because it will
act as a better communication medium between the stakeholders for discussion
which is needed in the requirement elicitation process. This game motivates the
participants to find the requirement in different stages with different perspectives. In
the first stage, the participants focus on identifying the basic functionalities (base
use case). In the subsequent stages, the participants focus on finding sub flow,
alternative flows, security flows, and performance and scalability requirements.

The proposed gamification using AUCD method introduces the team forming
mechanism of same category stakeholders as one team. This method creates a
healthy competition and motivates the participant to communicate their views about
the requirement without any fear.

The proposed gamification using AUCD method is applied in different projects
in different environments. Web based skill assessment for the software professionals project and web based HR management system for the University project
took place at Goldan infotech (www.goldeninfotec.com) client premise. Web based
Student Information system for University project took place in a university class
room for a course of the II Year M.Tech in Computer Science and Engineering.

In order to validate Gamification using Architect's Use case diagram two different questionnaires were prepared, one for the players (Table 3) and one for the
project manager/software architect (Table 4). The players questions focuses on
whether the game is easy to specify the functional requirement, easy to play,
amusement level of the game, motivates the participant, suitable for requirement
elicitation, extract more requirements and also suitable for specifying quality
attribute requirements. The project manager questions focuses on satisfaction level

Table 3 Gamification using AUCD—players questioner

Player question id	Questions
PQ1	Do you consider that the AUCD is easy to specify your functional requirement?
PQ2	Do you consider gaming using AUCD is simple to play?
PQ3	Rate the enjoyment level of the game
PQ4	Do you consider gaming using AUCD increase the motivation of the participants in requirement elicitation?
PQ5	Do you consider gaming using AUCD is a suitable method for requirement elicitation?
PQ6	Do you consider gaming using AUCD is suitable to find more requirements?
PQ7	Do you consider that the AUCD is easy to specify your quality attribute requirement?

Table 4 Gamification using AUCD—project manager questioner

Project manager question id	Questions
PQ1	Are you satisfied with the identification of quality attributes of the requirements?
PQ2	Are you satisfied the relevance of each requirement presented in the AUCD?
PQ3	The functional and nonfunctional requirements obtained by Gaming using AUCD have helped to define the project scope effectively?
PQ4	Do you consider that AUCD is useful for calculating the use case transactions?

of quality attribute requirement's identification, relevance of each requirement, whether the identified requirements are helpful to define the project scope and useful to calculate use case transactions.

The questionnaire is given to each player after the completion of the game. The questionnaire answer scale is between 0 and 5.0 for low and 5 for high.

The questioner result shows that all the players and project managers showed high level of equal positive response in gaming using AUCD which improves user participation in requirement elicitation.

From the Fig. 2 it is clear that Gamification using AUCD is easy to specify, easy to play, suitable for requirement elicitation, motivates the participants, useful to find more requirements and the amusement rate of the game is also high.

From the Fig. 3 it is clear that Gamification using AUCD is useful to specify the relevance of the requirement, defining functional and nonfunctional requirements clearly, useful to find more nonfunctional requirements and also useful for calculating use case transactions. The questioner result shows that all the players and project managers showed high level of equal positive response in gaming using AUCD which improves user participation in requirement elicitation.

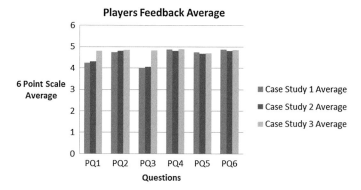

Fig. 2 Players feedback average

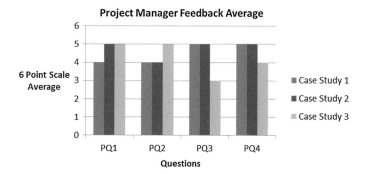

Fig. 3 Project manager feedback average

Analyzing the three case studies results of gamification using AUCD, the stage by stage identification of requirements by stake holders and discussion helps to identify several requirements. Therefore, new requirements were generated with the collaboration of participants using AUCD. Moreover, participants also evaluated other's opinions by arguments.

Participant's feedback shows that gamification using AUCD is exciting and increasing the motivation than existing approaches of requirement elicitation. This means that AUCD was appealing and accepted as a requirement elicitation tool to those participants.

In the field studies we observed that the project managers expressed high satisfaction of the number of quality characteristics identified. They also expressed that Gamification using AUCD method identifies more requirements than the traditional approach. The result also shows that the number of use cases identified using proposed method is 32 % higher than traditional use case model. The number of sub flow, alternative flow, security flow, performance and scalability requirements identified by using proposed method are 33, 54, 66, 62 and 62 % respectively, which are higher than the traditional use case model.

We observed that representing their ideas using AUCD tool and discussing requirements with other stakeholders, getting reward points for the accepted requirements by other stakeholders motivates and improve group discussions. This discussion helps to extract participant's knowledge' and perspectives. Gamification using AUCD increases the group synergy that brings unexpected and new results. These results are generally unforeseen ideas that could not be elicited otherwise.

Acknowledgements This work was supported by the Golden infotech (goldeninfotech.com) software organization. The authors also would like to acknowledge to the Golden infotech for the support.

References

1. http://www.iso-architecture.org/ieee-1471/defining-architecture.html
2. Chen, L., Babar, M.A., Nuseibeh, B.: Characterizing architecturally significant requirements. IEEE Software **30** (April 2013)
3. Suteliffe, A., Sayer, P.: Requiremnts Elicitation: Towards the Unknown Unknowns, Requirements Engineering Conference (RE), 2013 21st IEEE International
4. Aurum, A., Wohlin, C.: Engineering and Managing Software Requirements. Springer, Berlin (2005)
5. Sathis Kumar, B., Krishnamuthi, I.: Capturing architecturally significant requirements using architect's use case diagram. Int. J. Appl. Eng. Res. **10**(6), 15141–15164 (2015)
6. Ribeiro, C., Farinha, C., Pereira, J., da Silva, M.M.: Gamifying requirement elicitation: practical implications and outcomes in improving stakeholders collaboration. Entertain. Comput. **5**, 35–345 (2014)
7. Pedreia, O., Garcia, F., Brisaboa, N., Piattini, M.: Gamification in software engineering—a systematic mapping. Inf. Softw. Technol. **57**, 157–168 (2015)
8. Sathis Kumar, B., Krishnamuthi, I.: CloudASR an android application for capturing architecturally significant requirements visually in cloud computing applications using platform as service. Int. J. Appl. Eng. Res **10**(5), 13505–13514 (2015)
9. Lange, C.F., Chaudron, M.R., Muskens, J.: In Practice: UML software architecture and design description. IEEE SOFTWARE **23**, 40–46 (2006)
10. Booch, G., Rumbaugh, J., Jacobson, I.: The Unified Modeling Language User Guide. Addition Wesely (2001)
11. Somerville, I.: Software Engineering. Pearson Education India (2011)
12. Bittner, K., Spence, I.: Use Case Modeling. Addison Wesley (2002)
13. Lim, L., Damian, D., Finkelstein, A.: Stakesource2.0: Using social networks of stakeholders to identify and prioritise requirements. In: 33rd International Conference on Software Engineering, pp. 1022–1024 (2011)
14. Bano, M., Zowghi, D.: A systematic review on the relationship between user involvement and system success. Inf. Softw. Technol. **58**, 148–169 (2015)
15. Seyff, N., Graf, F., Maiden, N.: End-User Requirements Blogging with iRequire, ICSE'10, ACM, 2–8 May 2010
16. Asarnusch, R., David, M., Jan, W., Astrid, B.: Visual requirement specification in end-user participation. In: Proceedings of the First International Workshop on Multimedia Requirements Engineering: IEEE Computer Society

A Study of the Factors Which Influence the Recruiters for Choosing a Particular Management Institute

Rajiv Divekar and Pallab Bandhopadhyay

Abstract Placements are an important facet of Management education. With placements taking the centre stage and very vital, several institutions are aligning with market forces in order to ensure good placements. Increasingly institutions are relying on placement as a marketing strategy to influence the student influx as well as position themselves in this competitive market. This study was conducted in order to understand the recruitment dynamics; mainly the recruiter's evaluation of the management institute for campus recruitment of management trainee. The study was carried out taking a sample of 42 companies which go to management institute campuses for trainee recruitment. Study led to some interesting findings namely quality of student intake, teaching faculty quality, curriculum quality design and delivery as important factors which influence recruiters. Recruiters also value the parameters which influence the personality of the students, how they perform on different managerial skill sets to perform in a team and work place. Placements is most commonly executed in college's campuses rather than in companies and role of infrastructural support for placements also play an important role. Most importantly it is clear from the study that for companies the skill sets in a candidate are the deciding and influencing factor and not so much the knowledge based outcome.

Keywords Management trainee · Placements · Recruitment

1 Introduction

Indian education system has witnessed a prolific growth is last six decades with business programs being one of the top growth drivers. Today there are 4000 (approximate) Business Schools in India. Masters in Business Administration

R. Divekar (✉) · P. Bandhopadhyay
Symbiosis Institute of Management Studies, Symbiosis International University,
Range Hills Road, Khadki, Pune, India
e-mail: director@sims.edu

P. Bandhopadhyay
e-mail: arti.chandani@sims.edu

© Springer International Publishing Switzerland 2016
V. Vijayakumar and V. Neelanarayanan (eds.), *Proceedings of the 3rd International Symposium on Big Data and Cloud Computing Challenges (ISBCC – 16')*,
Smart Innovation, Systems and Technologies 49, DOI 10.1007/978-3-319-30348-2_40

program has positioned itself as a strong program for career growth and employ-ability. India was amongst the first country to recognise the need for a proper and structured teaching of management. The first Indian B-School, Commercial School of Pacchippa, Madras was established in 1886. India's first management college was founded in 1948 (Institute of Social Science) which was followed by Indian Institute of Social Welfare and Business Management (IISWBM) established in 1953 as a graduate business school in Kolkata. It has the distinction of being the first institute in India to offer an MBA degree. Xavier Labor Relations Institute (XLRI) was started in 1949 by Jesuit community, followed by IIM Calcutta and IIM Ahmedabad in 1960 and 1961 respectively.

The second wave of establishment of Management Institutes was around 1978 when JBIMS, SIBM, MDI, IMT, SP Jain and others were set up. With the Indian economy opening up post 1991, the demand for MBA graduates kept increasing leading to mushrooming of several other management institutes. The growth in number of management institutes can be attributed to strong growth in Indian economy and Indian industrial demand. With so many B Schools churning out approximately 4 lakh graduates each year the recruiters have enough choices in recruiting. However recruiting has its associated cost and time constraints and hence the problem of filtering which institutes to consider for recruitment drive. Most of the recruiters identify campuses where they have had successful experience of recruiting good candidates who meet the parameters sought for/meet their specific need. The variation in recruitment parameters is coupled with market forces. Even though the job profiles, criteria or type of skill sets required in a candidate are dynamic what remains unchanged is the campus for recruitment where the companies go to.

The companies which go for the placement to these business schools are con-tinuously looking to find the students who are not only academically good but are able to change the roles, have adaptability and ability to learn in the different atmospheres. The research has been carried out with the objective to find the parameters on which the companies select the candidates by administering a structured questionnaire.

2 Literature Support

Trank and Rynes (2003) emphasized the importance of managerial skill over the-oretical knowledge for selection of a management trainee. Their concept exposed the weak link between graduate learning and ability of a management trainee to perform. In addition they rewarded managerial skills more in order to perform better and work in a team. Safon [1] in his study found that media rankings had a significant influence over recruiters decision on selection of right B-school. Kowar and Barman [2] studied the global Competency Scales for Management education in India and found that skill based outcomes are important to industry but that many

B-School graduates lack them or have a dismal score. Saha [3] had studied the issues and concerns of Indian management education and observed that most B-Schools only concentrate on imparting theory and do not have/have very little training in skills. His work laid a paramount importance on performance and multi-tasking skills rather than simply knowledge management orientation. Dayal [4] pointed out the relevance of curricula, content, methodology and coverage of the academics and how they can improve quality of education as well as industry perspective of management education. Richardson [5] discussed the strategies which the companies can ideally employ in order to ensure, not only the best possible pool but also the select the candidate which are best.

The presence of such studies brings forth vital suggestions in improving B-School outcomes in terms of student performance and skill. However there is a gap in understanding recruitment dynamics, mainly the factors that affect recruiter's ability to select a B-school. Many B-schools still wonder as to what the recruiters look for when they visit the campus whether it is the curriculum, experiences, skill etc. This study was conducted to address that gap. This knowledge and outcome of this research paper will help the B-Schools understand the view point of the recruiters and accordingly they can education and other process to fit the bill.

3   Research Design

In order to understand the recruitment dynamics the following objectives were formulated:

1. To analyse the thought process of the recruiters to recruit management graduates
2. To list the factors which influence the recruiters to select a particular college/institute

3.1   Hypothesis

Ho1:   There is no significant association between knowledge based outcome and skill based outcome
Ho2:   There is no significant association between recruiters preference with respect to factors for selection of MBA College

In order to fulfil the aforementioned objectives a descriptive research, along with excel, was performed with help of judgemental sampling. Recruiters with experience of 5 years and above in recruiting management trainee directly from management institute campuses were considered for the study. 42 recruiters from organizations in Pune were administered the questionnaire for the study.

4   Results and Analysis

The result of the data collected using the questionnaire has been analysed with the help of MS-excel and it has been presented below where the graphical representation of the few of the responses has been given here (Fig. 1).

The descriptive results based on 42 recruiters indicate that 53 % of them frequently recruit marketing management trainees more than other specialization trainees. 19 % go for financial management trainees and 17 % of them for human resource management trainees. This also reveals that the recruiter does the recruitment much more in the field marketing than any other field (Fig. 2).

Analysis shows that of the companies visiting campuses for recruitment 41 % are Indian MNC and 33 % are foreign MNC. 17 % are Public sector companies and 7 % are Listed Companies. The data reflects great diversity in types of companies that prefer campus recruitment. Though the students as well as the B-Schools may have the dream company which is a foreign MNC but the fact remains that the companies which comes to these B-Schools for recruitments are Indian MNCs majorly. This can be considered as one of the important findings of the research.

The factors preferred for selection of MBA College for recruitment by recruiters was analysed, in Tables 1 and 2. Student Performance, Curriculum Structure and Infrastructural Support (Resource to conduct the process of recruitment) are factors preferred for selection of MBA College for recruitment. Status/Reputation/ Rankings Published through Media, Competitor's recruitment preference, Location convenience of the college from Industry belt and similar organizational

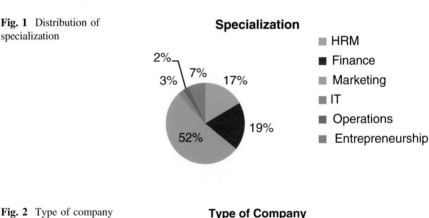

Fig. 1 Distribution of specialization

Fig. 2 Type of company

Table 1 Factors preferred for selection of MBA College

Sr. No.	Parameter	Factors preferred for selection of MBA College for recruitment by recruiters (satisfaction level on particular parameter)				
		1: highly preferred	2: preferred	3: neutral	4: not preferred	5: highly not preferred
1.1	Student performance	13	14	7	3	5
1.2	Status/reputation/ranking publishes through media	1	0	0	16	25
1.3	Curriculum structure	24	7	0	8	3
1.4	Competitors recruitment preference	0	9	9	24	0
1.5	Location convenience of the college from Industry belt	9	3	0	21	7
1.6	Infrastructural support (resource to conduct the process of recruitment)	20	4	0	0	18
1.7	Similar organizational structure and culture	0	4	13	19	6

Table 2 Mean of factor preferred for selection of MBA College

Factor	Mean
Factor_student_performance	2.36
Factor_student_status_reputation_media	4.52
Factor_student_curriculum_stucture	2.02
Factor_student_competitors_recruitment_preference	3.36
Factor_student_location_convenience	3.33
Factor_student_infrastructural_support_from_college	2.81
Factor_student_similar_org_stuctur_culture	3.64

structure and Culture are not so preferred parameters for selection of colleges. As the neutral point is 3, a score above 3 (towards 5) of any parameter means it is tending more to a not preferred parameter. In case a parameter has a score below 3 (between 1 and 3) it implies that recruiters prefer the parameters.

The least preferred factor is Status/Reputation/Rankings Published through Media while the most preferred factor is' curriculum structure followed by students' performance. This reiterates the facts that the academics and curriculum remains the most sought factors as the companies would like to hire those students who have studied and excel the subject and domain knowledge.

The factors influencing MBA Graduate performance as a student for getting recruited were analysed in Tables 3 and 4. Skill based performance is more important for recruiters at the initial stage of recruitment compared to knowledge

Table 3 Factors influencing MBA graduate performance

Sr. No.	Parameter	The factors influencing MBA Graduate performance as a student for getting recruited				
		1: highly influencing	2: influencing	3: neutral	4: not influencing	5: highly not influencing
3.1	Knowledge based out-come after MBA course	3	2	0	10	27
3.2	Skilled based out-come demonstrated or possessed by students	23	10	4	1	4

Table 4 Mean of factors influencing MBA graduate performance

	Mean
Factors_influencing_performance_of_student_knowledge_based	4.33
Factors_influencing_performance_of_student_skill_based	1.88

based performance from the recruiters point of view. Skill based performance preference score is 1.88 out of 5 which is lower than neutral point of 3 which implies that recruiters are agreed to that preference whereas score for knowledge based performance is 4.33 which is nearer to 5 more on the negative preference of the recruiters.

4.1 Hypothesis Testing

Ho1:   There is no significant association between knowledge based outcome and skill based outcome (Tables 5 and 6)

Chi square test was conducted at a significance value of probability (p) less than 0.05. The chi square value obtained was 19.275 for $p > 0.05$. Hence the null hypothesis is accepted. There is no significant association between knowledge based outcome and skill based outcome. The two factors are seen as independent factors by the recruiters.

Ho2:   There is no significant difference between recruiters preference with respect to factors for selection of MBA college

Table 5 Knowledge based performance's skill based performance cross tabulation

Count

		Skill based performance					Total
		Highly influencing	Influencing	Neutral	Not influencing	Highly not influencing	
Knowledge based performance	Highly influencing	3	0	0	0	0	3
	Influencing	2	0	0	0	0	2
	Not influencing	10	0	0	0	0	10
	Highly not influencing	8	10	4	1	4	27
Total		23	10	4	1	4	42

Table 6 Chi-square test

	Value	df	Asymp. Sig. (2-sided)
Pearson chi-square	19.275[a]	12	0.082
Likelihood ratio	25.027	12	0.015
Linear-by-linear association	6.552	1	0.010
N of valid cases	42		

[a]17 cells (85.0 %) have expected count less than 5. The minimum expected count is 0.05

ANOVA test was conducted at a significance value of probability p less than 0.05. The results indicated that the factor "curriculum structure" differed significantly from the other factors for selection of MBA College for $p < 0.05$. Hence the alternate hypothesis is accepted. The "curriculum structure" is seen as the most influential factor.

5 Conclusion

Placement process is a yearly feature and a critical process of MBA curicullum for Management Institutes. Recruiters prefer to go to colleges for their recruitment rather than doing the same by calling the candidates to their organisation. Students Performance, Curriculum quality and campus infrastructure support are the motivational factors for recruiters to come for campus placement to a particular campus. Factors which have more relevance for recruiter's are those which influence the student quality of knowledge and skills to perform like scope of industry interface, Skill based training, additional innovative thinking, quality of curriculum and support provided by institute to the recruiters to execute the selection process.

Also companies on coming to a campus are able to see the infrastructure, meet with the faculty and also study the curriculum being taught. Companies while selecting institutes prefer those where they have had good experience before and where the teaching and student development is oriented not just towards knowledge outcomes but also towards skill based outcomes. Organizations who come for campus placement are generally listed Indian and foreign MNC's with large no. of employees above 500 and 1000 people working.

Bibliography

1. Safon, V.: Factors that influence recruiters choice of B-Schools & their MBA Graduates: evidence & implications for B-Schools. Acad. Manag. Learn. Educ. **6**, 217–233 (2007)
2. Konwar, J., Barman, A.: Validation global competency scale for Indian business & management education. *Revista Romaneasca Pentru Educatie Multidimensional*, 199–214 (2013)

3. Saha, G.: Management Education in India: issues & concerns. J. Inf. Knowl. Res. Bus. Manag. Admin. **2**, 35–40 (2012)
4. Dayal, I.: Developing management education in India. J. Manag. Res. 98–113 (2002)
5. Richardson, M.A.: Recruitment strategies. Retrieved 17 Dec 2015, from managing/effecting. The recruitment process. http://unpan1.un.org/intradoc/groups/public/documents/UN/UNPAN-021814.pdf (n.d.)

Subset Selection in Market-Driven Product Development: An Empirical Study

S.A. Sahaaya Arul Mary and G. Suganya

Abstract Proper subset selection for next release is one of the key factors responsible for the success of the software projects. Selecting a subset of requirements from a pool of requirements in a market driven product development is found to be challenging since it involves unknown customers who may be geographically distributed. The objective of this paper is to study the importance of requirements prioritization through a exploratory study of literature supported by a survey conducted with industry experts. We have also analyzed and identified the different stakeholders that are involved in the process. The use of social media for gathering information from unknown market and the reasons behind the negligence of prioritization in industries is also been discussed.

Keywords Subset selection · Market driven development · Requirements prioritization

1 Introduction

Statistical studies show that software companies likely fail to produce products that satisfy customers on time and budget [1]. Every real world problem will have a list of software requirements that convey the needs and constraints fixed for the development of that software product. These requirements are gathered from different stakeholders through various requirements elicitation methods. From the customer point of view, every requirement is of utmost importance but all of them cannot be implemented at once because of the existing complexities like, dependencies among requirements, availability of resources in the organization etc.,

S.A. Sahaaya Arul Mary (✉)
Jayaram College of Engineering and Technology, Thuraiyur, Tamilnadu, India
e-mail: samjessi@gmail.com

G. Suganya
VIT University, Chennai Campus, Chennai, Tamil Nadu, India
e-mail: suganya.g@vit.ac.in

© Springer International Publishing Switzerland 2016
V. Vijayakumar and V. Neelanarayanan (eds.), *Proceedings of the 3rd International Symposium on Big Data and Cloud Computing Challenges (ISBCC – 16'),*
Smart Innovation, Systems and Technologies 49, DOI 10.1007/978-3-319-30348-2_41

Hence, product managers must select the functionalities of high priority to get included in the final software product. Requirements selection is considered as the most complex task in every software development since many aspects are involved in this subset selection. A badly chosen choice of requirements may create lot of problems during software development: scheduling problems, customer dissatisfaction, and financial losses to the organization.

Software development falls into two broad categories: Be-spoken and Market-driven. While be-spoken products aim at developing products for specific customer or known set of customers, market-driven products are meant for unknown market. Because of the emergence of markets for packaged software, market driven development is gaining increased importance in comparison to traditional bespoken development [2]. In contrast to bespoken product requirements which come from known set of clients, in market driven development environment, requirements come from different stakeholder groups that includes developers, marketing personnel, sales team, support team, testing team and from customers and competitors through interviews, surveys, competitor analysis etc [3, 4].

The following research questions (RQ) have been addressed in this paper.

RQ1. Is prioritization mandatory for market driven development organizations?
RQ2. What are the challenges faced in the process of prioritization?
RQ3. Who are all involved in prioritization?
RQ4. What are the means of collecting information from stakeholders?
RQ5. What may be the reasons for the negligence of using existing prioritization methodologies suggested by various researchers by the software companies?

2 Methodology

The research questions have been addressed by a methodical analysis of existing methodologies. The analysis is also supported by the views of more than 100 software engineers of various designations from different product organizations. The views are collected through a detailed survey. The questionnaire for the survey has been prepared according to the methodology suggested by Daneva et al. [5].

We conducted a systematic literature study on the existing methodologies proposed by different researchers. For this, a search strategy has been prepared in which we have fixed the search terms and the resources available for search. From the obtained results, duplicates are removed and a filter to select the required documents based on relevance has been applied. The study has been extended to the relevant sources of literature stated in selected documents and a detailed analysis has been done on the methodologies described.

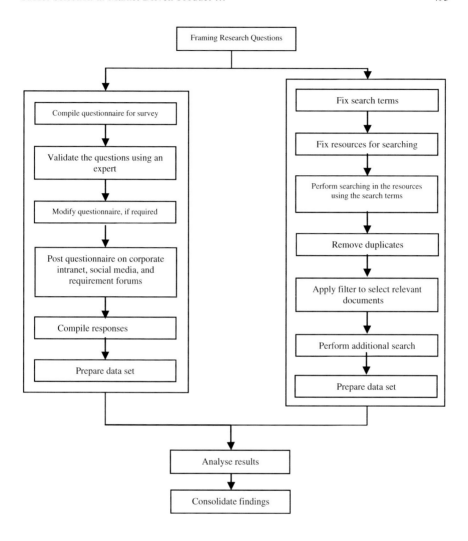

We have conducted a detailed survey to understand the state of mind of software engineers about the prioritization process. The questions have been carefully chosen in such a way to capture the mindset and opinion of all the stakeholders without making them tiresome in the process. The survey has been done by creating a Google form and by circulating it to various product organizations. The use of social media in gathering information has been proved to be successful by Seyff [6] and hence the link has also been posted on social media like Facebook and discussion forums. The questionnaire used for the survey is attached in Appendix.

Achimugu et al. [7] have done a systematic literature on requirement prioritization methods discussed by various researchers. The study includes the research carried out until the year 2013. Out of the 107 research papers discussed in [7], less than 10 researchers talk about requirement prioritization in market driven environment. We extended the research by using the search terms, "Market driven product development", "Mass market", "Packaged products" in the research papers from 2013 to current date. Resources selected for study include both journal and conference publications taken from IEEE Xplore, Science direct, Springer, ACM digital library and Google scholar. We have also extended our search in the blogs posted by various product and project managers in different discussion forums and blogspots.

3 Findings and Discussions (RQ1)

RQ1: Is prioritization mandatory for market driven development organizations?
Analysis: According to a survey conducted by Standish group about the success of IT projects [1], it was found that only 16.2 % of the projects were reported to be successful. The various factors identified for the failure of projects has been listed out in Table 1. 7 out of 10 reasons projected (Incomplete requirements, Lack of resources, Unrealistic expectations, Changing requirements and specifications, Didn't need it any longer, Lack of planning, Technology illiteracy) are due to the lack of significance shown in requirements engineering.

Timely release to market is the aim of most market driven product development organizations. These organizations initially have the motive to establish the product in the open market. Later, they thrive to survive in the market after every release. Requirements prioritization is recognized as a most difficult and important activity in Market-Driven Requirement Engineering [8]. Gomes et al. [8] have also insisted about the importance of prioritization in achieving a quality product [9]. Release planning, hence is an vital activity in software organizations that provide consecutive releases of the software product [10, 11]. This subset selection is very

Table 1 The Standish group report

S.No.	Project impaired factors	% of responses
1.	Incomplete requirements	13.1
2.	Lack of user involvement	12.4
3.	Lack of resources	10.6
4.	Unrealistic expectations	9.9
5.	Lack of executive support	9.3
6.	Changing requirements and specifications	8.7
7.	Lack of planning	8.1
8.	Didn't need it any longer	7.5
9.	Lack of IT management	6.2
10.	Technology illiteracy	4.3
11.	Other	9.9

Source [1]

Fig. 1 Need for prioritization

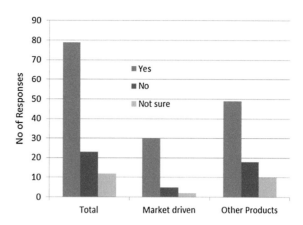

challenging as it is based on undecided predictions of the future, while being vital for the product's survival on the market [10].

Figure 1 depicts the responses for the question, "Do you think prioritization is required for selecting requirements for next release?". The survey clearly states that irrespective of the type of product, about 80 % of the respondents feel that the process is important.

4 Findings and Discussion (RQ2)

RQ2: What are the challenges faced in the process of prioritization?

Analysis:

1. In market driven development, stakeholders usually are geographically distributed [12]. Geographical distribution may mean the locality be within the same region, country, continent or anywhere in the world. The market for the same product may be different with different geographic locations. Hence, the means of collecting information from them and discussion seems to be a challenging task.
2. Stakeholders usually have different concern, responsibilities and objectives. When several stakeholders contribute for decision making, requirements very often conflict. But yet there, concept is indecisive and quality prioritization process is a challenging part [13].
3. Being skillful, the developers consider themselves as the most dominant stakeholders in subset selection [14]. On the other hand, it is widely accepted that the requirements priorities are best determined by customers since they have the authority towards the system to be built [15]. Hence, arriving a consensus between developers and customers is a challenging task.

4. Decisions are always biased by human tendencies to prove that they know everything. Hence, arriving at a optimal solution for a unknown market is hence challenging.
5. In environments where agile methodologies are followed, some requirements may have implicit dependencies and hence has to be delivered in a particular order. Since, agile doesn't support any change in delivered part, prioritization seems to be critical.

5   Findings and Discussion (RQ3)

RQ3: Who are all involved in prioritization?

Analysis: The concept of stakeholders and their importance has been discussed in detail by various researchers in the literature. The meaning of the term "stakeholders" has been discussed by various persons in the literature. One possible definition may be: A stakeholder is a person who is a direct user, manager of users, senior manager, operations staff member, owner who funds the project, end users, indirect user, auditors, portfolio manager, developers working on other projects that interact or integrate with the one under development, or maintenance professionals who will be affected by the development and/or deployment of a software project. According to this definition, stakeholders include all those who directly affects or getting affected by the product and hence designers, developers, testers, UI Designers, Customers, end users etc., will make the representative group. The systematic study has been conducted with the search terms, "stakeholders for next release", "stakeholders for subset selection", "stakeholders for requirement prioritization". A decision from whoever is included in the project is considered to be more useful. Mary and Suganya [16] discussed about the process of selecting stakeholders for prioritization.

Theoretical explanation has to be justified by practical survey and hence, the list of all stakeholders to be included for the process has been taken from the survey results. Analysis state that, utmost everyone feels that they are part of selection. Figure 2 shows the responses for the question, "Are you a part of discussion to select the requirements to be developed for next release?" raised in the questionnaire. It is observed from the study that all stakeholders are involved in the process of prioritization.

6   Findings and Discussion (RQ4)

RQ4: What are the means of collecting information from stakeholders?

Analysis: Gathering information from unknown market is actually challenging since, the customers are not fixed and in most of the times customers are not known

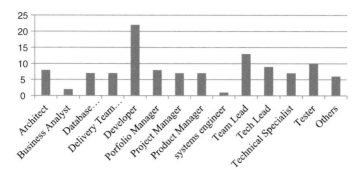

Fig. 2 Stakeholders involved in prioritization

in advance. Researchers have proposed the way of involving and collecting the information from end users in a traditional way [17]. But, in today's Cloud computing era [18], those traditional procedures found to sink themselves. Customers and end users are moving towards mobile computing and cloud computing and hence, to survive in the market, product organizations nowadays employ the way of collecting information through feedbacks provided in their websites, by calling customers etc., Seyff et al. have explored the use of social media like Facebook, Twitter, LinkedIn and Google+ in the use of requirement engineering. They have experimented the use of these social media and proved that they are the easier way of gathering information. Chain of people can be related to each other through these social media. We have felt the practical advantage of using social media while taking the survey, since the link has been forwarded like a chain using the concept of friends in Facebook. Hence, social media can act as a perfect medium for gathering the likes and dislikes of people about a product.

7 Findings and Discussion (RQ5)

RQ5: What may be the reasons for the negligence of using existing prioritization methodologies suggested by various researchers by the software companies?

Analysis: Based on the studies, we observed the following issues in using existing methodologies. The studies include the available literature about the practical use of prioritization approaches with a thorough analysis in the related blog spots.

1. Most organizations feel that the time invested for prioritization is merely a waste of time and money. But Standish survey [1] clearly states that, the success of project depends only on the proper requirements engineering.
2. Every human being has the tendency to prove that their's is the best and this tendency plays a vital role in decision making. Hence, product and project

managers generally think that with due experience, their decision will succeed and hence avoids the process of prioritization.

3. In market driven development, stakeholders usually are geographically distributed [12] and the method of gathering the preference is not clearly stated in the literature.

4. Framework for prioritization which includes the process of gathering information, processing it based on organizations decisive situation at that time without human intervention is lacking and this also leads to negligence of using proper automated methods for prioritization.

8   Conclusion and Future Works

From the survey and a detailed analysis on existing literature, the necessity of a clear framework for prioritization is clearly understood. In future, we will propose such a framework that can be used by software product organizations for decision making. The model will also have a clear cut automation methodology for ranking requirements and hence can create a subset of requirements from a large set of available backlog.

References

1. Standish survey: The Standish group report chaos. http://www.projectsmart.co.uk/docs/chaos-report.pdf (2014)
2. Karlsson, L., et al.: Challenges in market-driven requirements engineering—an industrial interview study. In: Proceedings of the Eight International Workshop on Requirements Engineering: Foundation for Software Quality (REFSQ'02), pp. 101–112 (2003)
3. Carlshamre, P.: Release planning in market-driven software product development: provoking an understanding. Requir. Eng. 7(3), 139–151 (2002)
4. Karlsson, J., Gurd, A.: Introduction to market-driven product management. Telelogic White Paper (2006)
5. Daneva, M., van der Veen, E., Amrit, C., Ghaisas, S., Sikkel, K., Kumar, R., Ajmeri, N., Ramteerthkar, U., Wieringa, R.: Agile requirements prioritization in large-scale outsourced system projects: an empirical study. J. Syst. Softw. 86, 1333–1353 (2013)
6. Seyff, N., Todoran, I., Caluser, K., Singer, L., Glinz, M.: Using popular social network sites to support requirements elicitation, prioritization and negotiation. J. Internet Serv. Appl. 6, 1–16 (2015)
7. Achimugu, P., Selamat, A., Ibrahim, R., Mahrin, N.M.: A systematic literature review of software requirements prioritization research. Inf. Softw. Technol. 568–585 (2014)
8. Gomes, A., Pettersson, A.: Market-Driven Requirements Engineering Process Model—MDREP. Blekinge Institute of Technology, Karlskrona (2007)
9. Potts, C.: Invented requirements and imagined customers: Requirements engineering for off-the-shelf software. In: Proceedings of the Second IEEE International Symposium on Requirements Engineering (RE'95), pp. 128–30 (1995)

10. Carlshamre, P., Regnell, B.: Requirements lifecycle management and release planning in market-driven requirements engineering processes. In: IEEE International Workshop on the Requirements Engineering Process (REP'2000), Greenwich, UK, September 2000
11. Carshamre, P., Sandahl, K., Lindvall, M., Regnell, B., Natt och Dag, J.: An Industrial survey of requirements interdependencies in software release planning. In: 2001 IEEE International Conference on Requirements Engineering (RE'01), pp. 84–91 (2001)
12. Ahmad, A., Shahzad, A., Padmanabhuni, V.K., Mansoor, A., Joseph,S., Arshad, Z.: Requirements Prioritization with Respect to Geographically Distributed Stakeholders, IEEE (2011)
13. Majumdar, S.I., Rahman, Md.S., Rahman, Md.M.: Stakeholder prioritization in requirement engineering process: a case study on school management system. Comput. Sci. Eng. **4**, 17–27 (2014)
14. Bakalova, Z., Daneva, M., Herrmann, A., Wieringa, R.: Agile requirements prioritization: what happens in practice and what is described in literature. In: Requirements Engineering: Foundation for Software Quality, pp. 181–195 (2011)
15. Svensson, R.B., Gorschek, T., Regnell, B., Torkar, R., Shahrokni, A., Feldt, R., et al.: Prioritization of quality requirements: state of practice in eleven companies (2011)
16. Sahaaya Arul Mary, S.A., Suganya, G.: Requirements prioritization—stakeholder identification and participation. Int. J. Appl. Eng. Res. 27230–7234 (2015)
17. Boehm, B., Gruenbacher, P., Briggs, R.: Developing groupware for requirements negotiation: lessons learned. IEEE Softw. **18**, 46–55 (2001)
18. Todoran, I., Seyff, N., Glinz, M.: How cloud providers elicit consumer requirements: an exploratory study of nineteen companies. In: Proceedings IEEE International Requirements Engineering Conference (RE), IEEE, pp. 105–114, (2013)

A Survey on Collaborative Filtering Based Recommendation System

G. Suganeshwari and S.P. Syed Ibrahim

Abstract Recommender system (RS) is a revolutionary technique which has transformed the applications from content based to customer centric. It is the method of finding what the customer wants, it can either be data or an item. The ability to collect and compute information has enabled the emergence of recommendation techniques, and these techniques provides a better understanding of users and clients. The innovation behind recommender frameworks has advanced in the course of recent years into a rich accumulation of tools that induces the researcher and scientist to create precise recommenders. This article provides an outline of recommender systems and explains in detail about the collaborative filtering. It also defines various limitations of traditional recommendation methods and discusses the hybrid extensions by merging spatial properties of the user (item-based collaborative filtering) with users personalized preferences (user-based collaborative filtering). This hybrid system is applicable to a broader range of applications. It helps the user to find the items of their interest quickly and more precisely.

Keywords Bigdata analytics · Recommender systems · Content based · User based collaborative filtering · Item based collaborative filtering · Hybrid RS · Similarity measures · KARS · LARS · Cold-start

G. Suganeshwari (✉) · S.P. Syed Ibrahim
School of Computing Sciences and Engineering, VIT University,
 Chennai Campus, Chennai, Tamilnadu, India
e-mail: g.suganeshwari2015@vit.ac.in

S.P. Syed Ibrahim
e-mail: syedibrahim.sp@vit.ac.in

© Springer International Publishing Switzerland 2016
V. Vijayakumar and V. Neelanarayanan (eds.), *Proceedings of the 3rd International Symposium on Big Data and Cloud Computing Challenges (ISBCC – 16')*,
Smart Innovation, Systems and Technologies 49, DOI 10.1007/978-3-319-30348-2_42

1 Introduction

Large volume of data is generated every day from IOT, RFID and other sensors. In order to handle such enormous data and produce recommendation, bigdata is used. Bigdata refers to huge volume of data that cannot be handled by traditional database methods. The information can be Explicit (Reviews, Tags, and ratings) or Implicit (Blogs, Wikis, and Communities) [1, 2]. In order to increase the scalability, bigdata is employed with recommendations. The recommender framework takes the input data as User Profile (Age, gender, topographical area, total assets), Item (Information about the item), User Interaction (browsing, bookmarking, labeling, ratings, saving, messaging,) and Context (Where the thing is to be put away) and yields the predicted ratings as output. Collaborative Filtering (CF) is most widely used recommendation technique along with content based and knowledge based methods.

Amazon and other e-commerce sites tracks the behavior of every customer [3], and when signed into the site, it utilizes the logged information to recommend items that the client may like. Amazon can recommend films that the user may like, regardless of the possibility that the client has just purchased books from it recently. Some online show ticket organizations will take a glimpse at the history of shows that the client has seen before and alarm him/her to the upcoming shows that may be of interest to the clients. Reddit.com lets client vote on links to other sites and afterward utilize the votes to propose different links that may be interesting. There are numerous applications that uses Item based CF, for example recommending items for online shopping [1, 3], books [4–6], finding music [7–9], recommending exciting sites, movies [10–13], e-business [14, 15], applications in business sectors [16] and web seek [17].

The evolution of RS has demonstrated the rise of hybrid methods, which blend diverse procedures so as to get the benefits of each of them. A study concentrated on the hybrid RS has been introduced in [18]. In this paper distinctive approaches to increase the accuracy of the recommender system are discussed. The emphasis of this paper is to build a hybrid RS by merging the spatial properties of users and items [19] along with the reviews given by the user [20]. The precision of RS is likely to increase by recommending the items that are specific to a region or location. This will also yield a personalized recommendation by exploiting all the reviews given by the client. This hybrid method is expected to recommend the items that are of more interest to the users than the traditional CF techniques. Section 2 describes the various methodologies in recommender framework along with different weaknesses of the traditional recommender strategies. Section 3 describes proposed work and the extensions of collaborative filtering. In conclusion the emphasis is on the hybrid method of combining the spatial properties and personal preference of user.

2 Overview of Recommender System

The root of the recommender systems is back from the extensive work in Collective Intelligence. The recommender system emerged in 1990s, during which recommendations were provided based on the rating structure. In recent years there are myriad ways to get details about user's liking for an item. The details can be retrieved by voting, tagging, reviewing, or the number of likes the user provides. It can also be a review written in some blog, message about the item or it can be a reply to an online community. Regardless of how preferences are communicated or represented they must be expressed as numerical values as described in [21].

Recommender frameworks are normally classified as per their way to deal with rating estimation, an overview of different sorts of recommender frameworks, are classified: as Content based, Collaborative Filtering and Hybrid approaches [22]. In online shopping the amount of items and users can increase tremendously. Maintaining these users and items is tough because while representing them in matrix they produce sparse data. Users may be interested only in few items. So the relationship between users and the items is always sparse. Sites like amazon produces precise recommendations with massive data with the help of item based CF. For such large scale systems new recommender systems are needed that can handle huge volume of data. Item based CF in [19, 22, 23] are used to address these issues and produces high quality recommendations.

The CF has been a significant method; it works on the assumption that if users has similar taste regarding choosing or liking of an item in the past then are likely to have similar interest in future too. [24] In his paper Automatic Text Processing put forth the ideas of content based model. Whereas CF was coined by David Goldberg at Xerox PARC in his paper "Using collaborative filtering to weave an information tapestry" [25]. His work allowed users to comment on the document as either "interesting" or "uninteresting". Depending on the comments, the documents were suggested to other users who are similar. There are currently many sites that utilizes CF algorithms for suggesting books, dating, shopping, articles, movies, music and jokes [22, 26]. Clearly states all the recommendation technique along with their pros and cons. The other great techniques that created revolution in recommender system are Group Lens—the first technique based on the ratings of the user [27], MovieLens—the first Movie recommender system. Fab—the first Hybrid RS framed by Marko [28]. Amazon got the patent rights by implementing item based CF [3]. Pandora was the first RS in music world (Table 1).

CF filtering is further classified into Memory based and Model based CF. In memory based CF the input should be in the form of matrix. A movie recommendation application system can be represented as user-movie rating matrix. The set of users are represented as columns and items (movies) as rows. Depending upon the ratings given by the user, it can be shown in the scale of 1 to 5. If the user has not rated a movie it is shown as "?". The goal or purpose of the recommendation engine is to produce appropriate movie recommendation for the user which the user has not seen. The user-movie rating matrix is represented in Table 2. These

Table 1 Recommendation algorithms

Type of algorithm	Description
Content based	Recommends the items based on similarity of the user or item
Collaborative filtering	Recommends items based on the purchase history, user profile and interest
Hybrid method	It merges both the techniques

Table 2 User-item matrix for movie recommendation

Movies	Sam	Iren	Lara	Harry
Movie1	5	5	0	0
Movie2	5	?	?	0
Movie3	?	4	0	?
Movie4	0	0	5	4
Movie5	0	0	5	?

matrix values serve as input for the recommender systems. The output of the system will be list of predicted values for the movies which the user has not yet watched. Depending on these predicted values recommendation model will be built and top-k items in the list will be recommended to the user. Model based CF is further classified as Matrix factorization and Association rule mining.

2.1   Content Based Filtering

Content is the foundation for building recommender applications. These recommendations are based on item description rather than similarity of other users. It uses machine learning algorithms to induce a model of user preferences based on feature factors. There are numerous contents available, such as articles, photos, video, blogs, wikis, classification terms, polls and lists. Metadata can be extracted by analyzing text. The process consists of creating tokens, normalizing the tokens, validating them, stemming them, and detecting phrases.

Depending on how metadata is extracted content is classified into simple (Articles, photos, video, blogs, polls, products) and composite (Questions with their associated set of answers, Boards with associated group of messages Categories). This form of filtering is useful to derive intelligence from unstructured text mostly from articles and documents. The words that appear in the text is called as term. When they are put together they form phrases. E.g., the title "Using Collaborative filtering for recommendation" consists of 5 different words and the useful phrases are Collaborative Filtering, Recommendations and collaborative Filtering for Recommendations. The vector space Model is given in Fig. 1.

The number of occurrences of a word in a document is given by Term frequency (TF). The count of word "Recommendation" in the current paper will be more, since it is about the recommendation system. The idea is to categorize the less

Fig. 1 The vector space
model

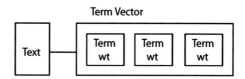

common words. The less common words in this article are Similarity, Factorization
and Intelligence. In order to boost these less common words Inverse document
frequency is used. If n represents the total number of documents of interest and if n_i
represents the total count of the term in the document, then IDF for a term is
computed as follows:

$$idfi = \log\left(\frac{n}{n_i}\right) \qquad (1)$$

Note that if a term appears in all documents, then its IDF is log(1) which is 0.
Content-based filtering [29–31] makes recommendations based on user's history.
(e.g. in a web-based e-commerce RS, if the user purchased some romantic films in
the past, the RS will probably recommend a recent romantic film that he has not yet
watched). Text, images and sound can be analyzed and recommendations can be
made. From the analysis items are recommended that are similar to the items that
the user has bought, visited, browsed, liked and rated positively in the past. Figure 2
represents the inputs of content based system.

Different predictive models can be used to extract information from the content.
They are Regression: The output variable is predicted from the input variables and a
predictive model is build (e.g. Regression), Classification: Predicts the output value
for the distinct output variable (e.g., Decision tree, Naive Bayes) Clustering:

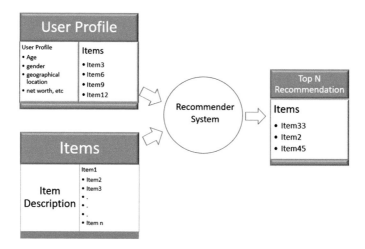

Fig. 2 Content based filtering

Table 3 Feature factor for user-movie matrix

Movies	Romance	Action
Movie1	0.9	0
Movie2	1	0.01
Movie3	0.99	0
Movie4	0.1	1.0
Movie5	0	0.9

Creates clusters in data to find patterns (e.g., k-means, hierarchical clustering, and density-based algorithms), Attribute importance: Determines the importance of attributes with respect to predicting the output attribute (e.g., Minimum description length, decision tree pruning) and Association rules: Finds interesting relationships in data by looking at co-occurring items (e.g., Association rules, Apriori).

For the example of user-movie matrix represented in Table 2, number of feature factors can be considered such as actors, director, story, song sequence and many more which can be used to recommend the movies based on the user's interest. Let's assume two feature factors, "Romance" and "Action" to categorize the movies in our example. The feature factors and their corresponding values are represented in Table 3. Then by comparing the user profile interest and feature factor, movies can be recommended to the user. The user Sam shows more liking towards romantic movies than the action movies. Sam has rated Movie1, Movie2 as 5 and Movie4, Movie5 as 0. By comparing the feature factor values, Movie3 can be recommended to Sam as it is tagged as romantic movie with feature factor value of 0.99. Similarly, Harry shows much interest in action movies as he has rated fair values for action movies and hence he will be recommended movie4 which has a feature factor of 1.0.

(1) *Limitations*:

There are many number of approaches to learn the model of Content based recommendations, but still content based recommendations do not guarantee accurate predictions if the content doesn't contain enough information [32]. The main limitations of content based filtering are listed below:

1. Limited Content Analysis [33]:
   It's tough to analyze sufficient set of features that would represent the complete content. It's very difficult to analyze whether it is a good or bad document if they contain same terms and phrases. It doesn't guarantee the quality of the document or item. For multimedia data, audio streams, video streams and graphical images it is tough to extract the feature factors.
2. Over-Specialization:
   Same event can be referred with different title. e.g., different news article can have same content with different titles. Billus and Pazzani in [34] suggested to filter out items that are too similar along with the items that are too different from each other.

3. Cold start:

In cold start items cannot be recommended due to lack of information about the user and the item. There are 2 types of cold start

(i) New User: When a new user enters the system there will not be any previous information such as browsing history, interested items and so on. In such case it is tough to recommend items to the users.

(ii) New Item: No rating will be there for new item entering the system and likely they will not be entered.

To address these issues collaborative filtering was introduced.

2.2   Collaborative Filtering

There is a quantifiable amount of progress in CF techniques as shown in [22, 35, 36]. CF model is classified into memory based and model based techniques. Memory based technique is further classified into user based and item based methods. In memory based methods similarities of user and item are calculated depending on whether it is item or user based techniques. They depend on their neighbors. So they are also called as "k Nearest Neighborhood method (KNN)". KNN filtering recommends items to user, based on the similarity measures [37–41]. It is the most popular technique and it performs the following three tasks to produce recommendations [42–44].

(1) Find similar users (neighbors) for the user a,
(2) Apply an aggregation approach,
(3) From step 2 select the top N recommendations (Fig. 3).

In model based, a model is constructed with matrix factorization or association rule mining. References [42, 45] shows different Model-based methods. The most popular model based methods are matrix factorization [46], fuzzy systems [47], Bayesian classifiers [18] and neural networks [48] etc. The RS database is very huge and consists of sparse data. In order to deal with this sparse database dimensionality reduction technique are employed [23]. These techniques are based on Matrix Factorization [49, 50]. Even though the prediction accuracy is high for

Fig. 3 Cosine similarity

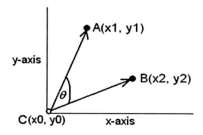

Singular value decomposition the computation cost is expensive. Memory based [12, 41, 51, 52] methods utilizes the user ratings as input in the form of matrix representations only. They use similarity metrics to calculate the similarity score between two users or items. Similarity between users or items can be calculated by many techniques. Some of them are described below.

(1) Similarity Measure

In user based CF similarity between users is calculated and in Item based CF similarity between items are calculated. There are many functions to calculate similarities, such as the Pearson correlation, cosine based function, *Jaccard coefficient* and *Manhattan distance*. Any of these functions can be used as long as they have the same input format within the same range and return a ratio which shows a high degree of similarity for higher values. These approaches are given in detail in [22, 26].

(a) Cosine Based Similarity:

Is a measure of similarity between two vectors of an inner product space that measures the cosine of the angle between them. The cosine of $0°$ is 1, and it is less for any other angle [53].

$$\cos(\theta) = \frac{x \cdot y}{\|x\|\|y\|} = \frac{x1 * x2 + y1 * y2}{\sqrt{x1^2 + y1^2}\sqrt{x2^2 + y2^2}} \tag{2}$$

(b) Pearson Correlation

The similarity between any two vectors can also be calculated using Pearson correlation. The output ranges from -1 to 1. Where 1 and -1 represents that they are strongly correlated. If 1, then they are positively related. If -1 then they are negatively related. In negative correlation the value of one vector decreases as other vector increases. Towards 0 the degree of correlation is null. It is mainly used in applications when two users give different ratings but their difference are equal. For example, if critic1 rates (10, 8, 6) for Movies1, Movie2 and Movie3 and if critic 2 rates same Movies as (8, 6, 4). Then they are strongly correlated. Correlation is calculated only by the ratings of the items which are valued by both the vectors.

Let U be the set of users that have rated both item i and j.

$$r = \frac{\sum_{i=1}^{n}(X_i - \bar{X})(Y_i - \bar{Y})}{\sqrt{\sum_{i=1}^{n}(X_i - \bar{X})^2}\sqrt{\sum_{i=1}^{n}(Y_i - \bar{Y})^2}} \tag{3}$$

From the user-movie matrix representation of Table 2 a graph is drawn to show the similarity between users. Figure 4a shows the similarity between Sam and Iren and Fig. 4b shows the similarity between Sam and Lara. Sam and Iren have rated Movies 1, 3, 5 and they have rated them alike. It shows

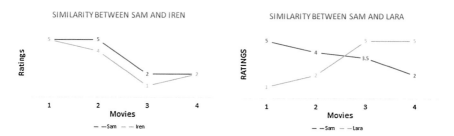

Fig. 4 **a** Correlation relation between Sam and Iren. **b** Correlation relation between Sam and Lara

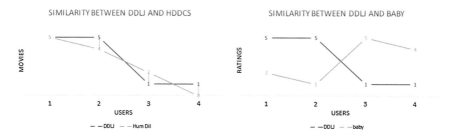

Fig. 5 **a** Correlation relation between Movie1 and Movie2. **b** Correlation relation between Movie1 and Movie4

the positive correlation with value 0.9487 and from the graph it can be inferred that these two users are similar. Figure 4b shows the relationship between Sam and Lara. They both have rated Movies 1, 4, 5, and their ratings are totally different. From second graph it is inferred that both these users are dissimilar.

In similar fashion similarity between items can also be found. The correlation value of Movie1 and Movie2 is 0.9113 and is represented in Fig. 5a, whereas for Movie1 and Movie4 is −0.9487 and are shown in Fig. 5b. This shows that Movie1 and Movie3 are very similar movies and Movie1 and Movie4 are dissimilar.

There are two main approaches in building memory based CF recommendation systems, based on whether the system searches for related items or related users. The prior is called as item based collaborative filtering and later is called as user based collaborative filtering.

2 User based Collaborative Filtering

The user is recommended items based on similarity with their neighbors. Neighbors are the similar users who share similar interest with the active user for whom the recommendations are to be produced. If U_a is an active user, then the neighbor of U_a is U_i, if they share similar taste. The prediction of the active user depends on the ratings given by their neighbors for an item [54] and that is displayed in Fig. 6. KARS [20] is user based CF technique that produces the recommendations based on the preferences of the user. KARS

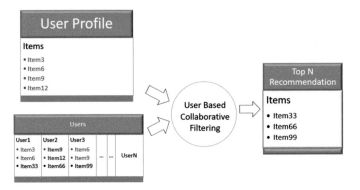

Fig. 6 User based CF

analyzes the reviews of previous users and stores it. The keywords which are extracted from these reviews will serve as a factor to recommend a new user.

3 Item Based Collaborative Filtering

User based techniques creates an over-load on the system, it has to find similar users and then compute items that are similar to users. User based CF has to evaluate large itemsets, whereas users would have purchased only less number of items. It is less accurate when compared to the item based filtering. In item based filtering items similar to other items are discovered by using any of the similarity measures. When a user shows an interest by browsing or searching for a particular item, items similar to the searched items are popped as recommended items [55] and this is shown in Fig. 6. LARS [19] is an item based collaborative filtering method which produces recommendation based on the spatial properties of the user and item. This reduces the number of items to be compared when compared to the traditional item based collaborative filtering. The model is same as that of item based collaborative filtering but only for the items that belong to some spatial limits are defined (Fig. 7).

The main goal of item-based analysis is to find similar items. If user purchase an item A, then the user who bought item A is found. Then other item say item K which the user has bought is found and both are recorded as purchased. Then similarity is calculated for all the items and top-k items are recommended depending on the similarity score. A detail description of Collaborative filtering along with their pros and cons are explained in the Table 4.

Depending upon the application either user based or item based collaborative filtering can be used. If item list doesn't change much (amazon)—item-to-item CF can be used. If item list changes frequently (news)—user based CF can be used. If recommended item is a user (social networking) user based CF should be used. Item-based algorithms are computationally faster to implement than user-based algorithms and provide better results, because similarity of users needn't be calculated.

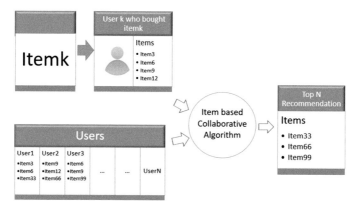

Fig. 7 Item based CF

Table 4 CF approaches

CF technique	Representative algorithm	Advantages	Limitations
Memory-based collaborative filtering (neighborhood based)	• User-based CF • Item-based CF	• Easy implementation • New data can be added easily and incrementally • Need not consider the content of items being recommended • Scales well with correlated items	• Are dependent on human ratings • Cold start problem for new user and item • Sparsity problem of rating matrix • Limited scalability for large datasets
Model-based collaborative filtering	• Slope-one CF • Dimensionality reduction (matrix factorization) • Eg. SVD	• Better addresses the sparsity and scalability problem • Improve prediction performance	• Expensive model building • Trade-off between the prediction performance and scalability • Loss of information in dimensionality reduction technique (SVD)
Hybrid collaborative filtering	• Combination of memory-based and model-based CF	• Overcome limitations of CF such as sparsity • Improve prediction performance	• Increased complexity and expense for implementation

3 Proposed Work

RS uses more than one recommending techniques from different classes of recommender classifications. The advantage of merging two techniques always has more advantage and takes the benefits from both the classification types. In [56] the collaborative filtering is portrayed by merging demographic filtering with content based filtering. The hybrid technique produced exploits the advantage of both the techniques.

Fab is a recommendation framework intended to help users to isolate the most vital event from the vast amount of data accessible from the internet [28]. Though there are numerous methods to predict items for the users, still there is a large space of research needed to provide accurate recommendation for the user. The Netflix price context which was held in 2006 motivated many scientist and researchers to work in recommender systems. Even the prize winning team was not able to predict the accurate ratings for few movies such as "Napolean". In order to get accurate predictions several hybrid approaches evolved to eliminate the drawbacks of each techniques [2, 18, 41, 57, 58, 59]. The main source of hybrid technique can be formed by merging the content and collaborative filtering. This can be done in several ways. The main steps to be considered are

1. Combining their predictions after implementing them separately.
2. Merging some collaborative characteristics into a content-based approach, and
3. Merging some content-based characteristics into a collaborative approach,
4. Integrating both content-based and collaborative methods to form a unified model.

Data from diverse sources such as user contents, social relations, explicit ratings, knowledge-based information and locations based systems increases every day. This forces the current technique to produce a high recommendation with quick response time for the users. In this paper we try to merge LARS and KARS. LARS [19] considers spatial properties of both the items and the users. The future of recommendation system will be based on the location of the user and the item. The user is recommended the items based on his current location. A partial pyramid data structure is maintained and highest recommendation is suggested based on locality. It combines the spatial properties and the item based CF. Whereas KARS captures the user preference by a keyword aware approach [20]. Each review by the user will be transformed into corresponding keyword set according to keyword candidate list and domain thesaurus. In this paper Keyword and Spatial properties of user are merged. Both belong to the category of Collaborative recommendation. Location based RS is Item based CF and Keyword based RS is User based CF. Both the RS technique have their own advantage of producing personalized recommendation with accuracy. In papers [58, 59, 60, 61], the hybrid approaches are compared with traditional approaches and the result has been proved that hybrid approaches are more accurate than the original techniques.

4 Conclusion

Recommender frameworks gained noteworthy ground over a decade ago when various recommendation techniques were proposed and a few "modern quality" frameworks were created. Though there is great amount of advances in the content based and different collaborative filtering methods, the increase in the data forces the researchers and scientist to upgrade the system and make it more utilizable. To reduce the complexity of the user based CF Item based CF is used. Though it produces recommendations accurately when compared to user based, the complexity still remains high. LARS which is the item based collaborative filtering reduces this complexity by reducing the number of users and items by considering their spatial properties. KARS which is user based collaborative filtering handles the unstructured data (reviews) given by the user. Merging these two collaborative filtering would leave behind the traditional recommendation system in all measure of precision.

References

1. Lee, S.K., Cho, Y.H., Kim, S.H.: Collaborative filtering with ordinal scale based implicit ratings for mobile music recommendations. Inf. Sci. **180**(11), 2142–2155 (2010)
2. Choi, K., Yoo, D., Kim, G., Suh, Y.: A hybrid online-product recommendation system: combining implicit rating-based collaborative filtering and sequential pattern analysis. Electron. Commer. Res. Appl. doi:10.1016/j.elerap.2012.02.004 (in press)
3. Linden, G., Smith, B., York, J.: Amazon.com recommendations: item to item collaborative filtering. IEEE Internet Computing (2003)
4. Serrano-Guerrero, J., Herrera-Viedma, E., Olivas, J.A., Cerezo, A., Romero, F.P.: A google wave-based fuzzy recommender system to disseminate information in University Digital Libraries 2.0. Inf. Sci. **181**(9), 1503–1516 (2011)
5. Porcel, C., Moreno, J.M., Herrera-Viedma, E.: A multi-disciplinary recommender system to advice research resources in university digital libraries. Expert Syst. Appl. **36**(10), 12520–12528 (2009)
6. wikipedia.org Recommender system
7. Upendra Shardanand: Social information filtering for music recommendation. MIT EECS M. Engineering thesis, also TR-94-04, Learning and Common Sense Group, MIT Media Laboratory (1994)
8. Nanolopoulus, A., Rafailidis, D., Symeonidis, P., Manolopoulus, Y.: Music box: personalized music recommendation based on cubic analysis of social tags. IEEE Trans. Audio Speech Lang. Process. **18**(2), 407–412 (2010)
9. Tan, S., Bu, J., Chen, C.H., He, X.: Using rich social media information for music recommendation via hypergraph model. ACM Trans. Multimedia Comput. Commun. Appl. **7**(1) Article 7 (2011)
10. Carrer-Neto, W., Hernandez-Alcaraz, M.L., Valencia-García, R., Garca-Sanchez, F.: Social knowledge-based recommender system, Application to the movies domain. Expert Syst. Appl. **39**(12), 10990–11000 (2012)
11. Winoto, P., Tang, T.Y.: The role of user mood in movie recommendations. Expert Syst. Appl. **37**(8), 6086–6092 (2010)

12. Symeonidis, P., Nanopoulus, A., Manolopoulus, Y.: MovieExplain: a recommender system with explanations. In: Proceedings of the 2009 ACM Conference on Recommender Systems, pp. 317–320 (2009)
13. Netflix News and Info—Local Favorites [Online]. Available: http://tinyurl.com/4qt8ujo
14. Huang, Z., Zeng, D., Chen, H.: A comparison of collaborative filtering recommendation algorithms for e-commerce. IEEE Intell. Syst. 22(5), 68–78 (2007)
15. Castro-Sanchez, J.J., Miguel, R., Vallejo, D., Lopez-Lopez, L.M.: A highly adaptive recommender system based on fuzzy logic for B2C e-commerce portals. Expert Syst. Appl. 38 (3), 2441–2454 (2011)
16. Costa-Montenegro, E., Barragans-Mart nez, A.B., Rey-Lopez, M.: Which App? A recommender system of applications in markets: implementation of the service for monitoring users interaction. Expert Syst. Appl. 39(10), 9367–9375 (2012)
17. Mcnally, K., Omahony, M.P., Coyle, M., Briggs, P., Smyth, B.: A case study of collaboration and reputation in social web search. ACM Trans. Intel. Syst. Technol. 3(1) Article 4 (2011)
18. Burke, R.: Hybrid recommender systems: survey and experiments. User Model. User-Adap. Inter. 12(4), 331–370 (2002)
19. Sarwat, M., Levandoski, J.J., Eldawy, A., Mokbel, M.F.: LARS*: an efficient and scalable location-aware recommender system. IEEE Trans. Knowl. Data Eng. 26(6) (2014)
20. Meng, S., Dou, W., Zhang, X., Chen, J.: "KASR": a keyword-aware service recommendation method on mapreduce for big data applications (2014)
21. SATNAM ALAG: Collective intelligence in action 2009 by Manning Publications Co
22. Bobadilla, J., Ortega, F., Hernando, A., Arroyo, A.: A balanced memory based collaborative filtering similarity measure. Int. J. Intell. Syst. 27(10), 939946 (2013)
23. Choi, K., Suh, Y.: A new similarity function for selecting neighbors for each target item in collaborative filtering. Knowl. Based Syst. 37, 146153 (2013)
24. Salton, G.: Automatic text processing. Addison-Wesley, Reading (1989)
25. Goldberg, D., Nichols, D., Oki, B.M., Terry, D.: Using collaborative filtering to weave an information tapestry. Comm. ACM 35(12), 61–70 (1992)
26. Bobadilla, J., Ortega, F., Hernando, A.: A collaborative filtering similarity measure based on singularities. Inf. Process. Manage. 48(2), 204217 (2012)
27. Resnick, P., Iakovou, N., Sushak, M., Bergstrom, P., Riedl, J.: GroupLens: an open architecture for collaborative filtering of Netnews. In: Proceedings of 1994 Computer Supported Cooperative Work Conference (1994)
28. Balabanovic, M., Shoham, Y.: Fab: content-based, collaborative recommendation. Comm. ACM 40(3), 66–72 (1997)
29. Lang, K.: NewsWeeder: learning to filter netnews. In: Proceedings of 12th International Conference on Machine Learning, pp. 331–339 (1995)
30. Antonopoulus, N., Salter, J.: Cinema screen recommender agent: combining collaborative and content-based filtering. IEEE Intell. Syst. 21 35–41 (2006)
31. Meteren, R., Someren, M.: Using content-based filtering for recommendation. In: Proceedings of ECML 2000 Workshop: Machine Learning in Information Age, pp. 47–56 (2000)
32. Bobadilla, F., Ortega, A., Hernando, A.: Gutierrez. Knowl.-Based Syst. 46, 109–132 (2013)
33. Explicit and implicit rating Shardanand and Meas (1995)
34. Pazzani, M.: A framework for collaborative, content-based, and demographic filtering. Artif. Intell. Rev. Spec. Issue Data Min. Internet 13(5–6), 393–408 (1999)
35. Garcia, R., Amatriain, X.: Weighted content based methods for recommending connections in online social networks. In: Proceedings of the 2010 ACM Conference on Recommender Systems, pp. 68–71 (2010)
36. Bobadilla, J., Hernando, A., Ortega, F., Gutierrez, A.: Collaborative filtering based on significances. Inf. Sci. 185(1), 1–17 (2012)
37. Gunawardana, A., Shani, G.: A survey of accuracy evaluation metrics of recommender tasks. J. Mach. Learn. Res. 10, 2935–2962 (2009)
38. Goldberg, K., Roeder, T., Gupta, D., Perkins, C.: Eigentaste: a constant time collaborative filtering algorithm. Inf. Retrieval 4(2), 133–151 (2001)

39. George, T., Meregu, S.: A scalable collaborative filtering framework based on co-clustering. In: IEEE International Conference on Data Mining (ICDM), pp. 625–628 (2005)
40. Candillier, L., Meyer, F., Boulle, M.: Comparing state-of-the-art collaborative filtering systems. Lect. Notes Comput. Sci. **4571**, 548–562 (2007)
41. Gemmell, J., Schimoler, T., Mobasher, B., Burke, R.: Resource recommendation for social tagging: a multi-channel hybrid approach. In: Proceedings of the 2010 ACM Conference on Recommender Systems, pp. 60–67 (2010)
42. Adomavicius, G., Tuzhilin, A.: Toward the next generation of recommender systems: a survey of the state-of-the-art and possible extensions. IEEE Trans. Knowl. Data Eng. **17**(6), 734–749 (2005)
43. Schafer, J.B., Frankowski, D., Herlocker, J., Sen, S.: Collaborative filtering recommender systems (Chap. 9). In: Brusilovsky, P., Kobsa, A., Nejdl, W. (eds.) The Adaptive Web, pp. 291–324 (2007)
44. Bobadilla, J., Hernando, A., Ortega, F., Bernal, J.: A framework for collaborative filtering recommender systems. Expert Syst. Appl. **38**(12), 14609–14623 (2011)
45. Su, X., Khoshgoftaar, T.M.: A survey of collaborative filtering techniques. Adv. Artif. Intell. **2009**, 1–19 (2009)
46. Luo, X., Xia, Y., Zhu, Q.: Incremental collaborative filtering recommender based on regularized matrix factorization. Knowl.-Based Syst. **27**, 271–280 (2012)
47. Yager, R.R.: Fuzzy logic methods in recommender systems. Fuzzy Sets Syst. **136**(2), 133–149 (2003)
48. Ingoo, H., Kyong, J.O., Tae, H.R.: The collaborative filtering recommendation based on SOM cluster-indexing CBR. Expert Syst. Appl. **25**, 413–423 (2003)
49. Vozalis, M.G., Margaritis, K.G.: Using SVD and demographic data for the enhancement of generalized collaborative filtering. Inf. Sci. **177**, 3017–3037 (2007)
50. Cacheda, F., Carneiro, V., Fernandez, D., Formoso, V.: Comparison of collaborative filtering algorithms: limitations of current techniques and proposals for scalable, high-performance recommender systems. ACM Trans. Web **5**(1) Article 2 (2011)
51. Park, D.H., Kim, H.K., Choi, I.Y., Kim, J.K.: A literature review and classification of recommender Systems research. Expert Syst. Appl. **39**, 10059–10072 (2012)
52. Kong, F., Sun, X., Ye, S.: A comparison of several algorithms for collaborative filtering in startup stage. IEEE Transactions on Networks, Sensing and Control, pp. 25–28 (2005)
53. wikipedia.org-wiki-Cosine-similarity
54. Schafer, J.B., Frankowski, D., Herlocker, J., Sen, S.: Collaborative filtering recommender systems (Chap. 9). In: Brusilovsky, P., Kobsa, A., Nejdl, W. (eds.) The Adaptive Web pp. 291–324 (2007)
55. Sarwar, B., Karypis, G., Konstan, J.A., Riedl, J.: Item-based collaborative filtering recommendation algorithms. In: Proceedings of the 10th International Conference on World Wide Web. ACM Press, Hong Kong pp. 285–295 (2001)
56. Fengkun, L., Hong, J.L.: Use of social network information to enhance collaborative filtering performance. Expert Syst. Appl. **37**(7), 47724778 (2010)
57. Burke, R.: Hybrid recommender systems: survey and experiments. User Model. User-Adap. Inter. **12**(4), 331–370 (2002)
58. Al-Shamri, M.Y.H., Bharadwaj, K.K.: Fuzzy-genetic approach to recommender systems based on a novel hybrid user model. Expert Syst. Appl. **35**(3), 1386–1399 (2008)
59. Gao, L.Q., Li, C.: Hybrid personalized recommended model based on genetic algorithm. In: International Conference on Wireless Communication, Networks and Mobile Computing, pp. 9215–9218 (2008)
60. Oku, K., Kotera, R., Sumiya, K.: Geographical recommender system based on interaction between map operation and category selection. In: Workshop on Information Heterogeneity and Fusion in Recommender Systems, pp. 71–74 (2010)
61. Matyas, C., Schlieder, C.: A spatial user similarity measure for geographic recommender systems. In: Proceedings of the 3rd International Conference on GeoSpatial Semantics, pp. 122–139 (2009)

62. Park, D.H., Kim, H.K., Choi, I.Y., Kim, J.K.: A literature review and classification of recommender systems research. Expert Syst. Appl. **39**, 10059–10072 (2012)
63. Porcel, C., Herrera-Viedma, E.: Dealing with incomplete information in a fuzzy linguistic recommender system to disseminate information in university digital libraries. Knowl.-Based Syst. **23**(1), 32–39 (2010)
64. Cho, S.B., Hong, J.H., Park, M.H.: Location-based recommendation system using Bayesian users preference model in mobile devices. Lect. Notes Comput. Sci. **4611**, 1130–1139 (2007)
65. Zhong, J., Li, X.: Unified collaborative filtering model based on combination of latent features. Expert Syst. Appl. **37**, 5666–5672 (2010)

Author Index